How to Estimate Building Losses and Construction Costs

Third Edition

How to Estimate Building Losses
and Construction Costs
Third Edition

Paul I. Thomas

Former   Vice President and General Adjuster of the Kemper
Insurance Companies, Professional Engineer and
Consultant on Property Losses

PRENTICE-HALL, INC.
Englewood Cliffs, N.J.

PRENTICE-HALL INTERNATIONAL, INC., *London*
PRENTICE-HALL OF AUSTRALIA, PTY. LTD., *Sydney*
PRENTICE-HALL OF CANADA, LTD., *Toronto*
PRENTICE-HALL OF INDIA PRIVATE LTD., *New Delhi*
PRENTICE-HALL OF JAPAN, INC., *Tokyo*

Library of Congress Cataloging in Publication Data

Thomas, Paul I

   How to estimate building losses and construction
costs.

   Bibliography: p.
   1. Building--Estimates. 2. Building--Repair
and reconstruction. I. Title.
Th435.T45   1976   692'.5   75-28490
ISBN 0-13-405894-1

Printed in the United States of America

To My Wife
MARIE

ACKNOWLEDGMENTS

Andersen Corporation
Bayport, Minn.

Goldblatt Tool Company
Kansas City, Kansas

Gunn and Thompson Construction Company
Miami, Florida

Independent Nail, Inc.
Bridgewater, Mass.

T.A. Manor, Director of Education
General Adjustment Bureau
National Education Center
Englewood, Col.

John H. McKenrick, and Staff
Associate Director
Vail Technical Institute
Blairsville, Pa.

National Flood Insurers Association
New York, N.Y.

National Oak Flooring Manufacturers Association
Memphis, Tenn.

George A. Pasqualucci, A.S.A.
Louis Pasqualucci & Son, Inc.
General Contractors
Quincy, Mass.

Patent Scaffolding Company
Fort Lee, N.J.

Portland Cement Association
Skokie, Ill.

Rockwell International
Power Tool Division
Pittsburgh, Pa.

Senco Products, Inc.
Cincinnati, Ohio

Sumner Rider and Associates
New York, N.Y.

Underwriters Laboratories, Inc.
Chicago, Ill.

U.S. Department of Labor
Bureau of Statistics
Washington, D.C.

G. Raymond Eisenhauer, Jr.
N. Frank Vought, Inc.
Builders and Appraisers
New York, N.Y.

Frank R. Walker Company
Chicago, Ill.

Dewey C. White
White Repair and Contracting Company
Atlanta, Ga.

Preface to the Third Edition

This Third Edition has been prepared with the benefit of many suggestions sent to me by readers. A major source of comments and advice has been persons who employ the book as a teaching text in schools, seminars, company educational courses and offices of adjusters.

The book has been planned and organized to prepare the reader, in the shortest time possible, to be able to make his or her own estimates of building damage with full confidence. It is true that those who have benefited most and who will continue to benefit most from the book are people whose business interests are related directly or indirectly to the property insurance field. However, it has also been a welcome addition as reference material for builders and contractors who either specialize in, or from time to time are involved in, estimating building damages for insured or insurer or who are actively engaged in the restoration of such properties.

The opening chapters lay the ground rules for estimating building losses by introducing the reader to the 22 major causes of damage to structures, most if not all of them being insurable in one form or another.

This discussion of causes, which also covers effects, is followed by fundamental considerations in making estimates, considerations essential to error-free results. There are literally hundreds of successful building estimators who have never held a hammer, saw or trowel in hand, yet have achieved excellence in their work because they early learned the fundamentals and have consistently applied sound principles that do not change with time.

This book, which started out 15 years ago with 14 chapters, became 18 chapters by 1971 in the Second Edition. Four more chapters have been added in this Third Edition through the urging of readers. Because so many losses originate in kitchens as a result of grease fires, a chapter on estimating kitchen cabinets and built-ins has been added. By the same reasoning, a chapter on electrical wiring has been added as there are few losses of consequence that do not involve to some degree electric wiring or fixtures. For some time we have been urged also to include a discussion on thermal insulation and vapor barriers inasmuch as most residential properties built today are air conditioned or heated or both. Such a chapter has been added. Since the majority of those who use this book are ultimately involved, directly or indirectly, in the adjustment of a loss under an insurance contract, a chapter on the procedure for adjusting building losses has been included.

In addition to the new chapters, the number of labor and material reference tables has been increased to 70 and previously existing tables have been greatly expanded. Contractors who prepared samples of estimates have updated them. Obsolete material or references have been culled and new materials and methods of construction have been introduced. A great many new illustrations are included in this Third Edition.

Following the chapters covering the fundamentals; the examination, inspection and note taking in the field; the development, pricing and completing of a detailed estimate, the book gets into specific construction trades ranging from concrete and masonry, all phases or rough carpentry and finished carpentry, roofing, painting and decorating, plumbing, electrical work and other building elements necessary for its completion.

A Glossary of Building Terms has been included in this 1975 Third Edition, also at the request of a number of instructors using the text.

Because of the number of labor and material reference tables, a guide to the table number, subject and page has been added. This minimizes the time to locate specific labor and material data.

This text represents a first in that it brings together the art of estimating construction costs as it relates to the repair and replacement of buildings damaged by fire, hurricanes, tornados, floods, earthquake, landslides, mudslides, collapse, freeze-ups and a score of other causes of impairment to a building or structure. In so doing it was a collateral necessity to incorporate the susceptibility to damage of all the materials of construction. Each chapter on a specific construction feature concludes with the damageability of the *particular materials,* and outlines the methods of repairing or restoring.

A table for the metric system of linear measurement has been included in the Table of Weights and Measures. Conversion factors have been included to convert metric measurement to U.S. equivalents and vice versa.

Because those who use this text from time to time are called upon to analyze and check an estimate of damage prepared by someone else, special attention has been given to ways and means of making such an analysis. One chapter alone is devoted to analyzing unit costs and lump-sum estimates. In the final chapter 23 sources of error in estimating are discussed in detail, not only to enable the estimator to avoid making these errors but also to enable him to ferret out such errors in estimates prepared by others.

40 WAYS THIS BOOK WILL HELP YOU

1. What can fire, heat and smoke do to building materials...What damage can result from *cold* smoke as distinguished from *hot* smoke...What damage can water from fire hoses do to interior paint and wallpaper...to plastered walls and ceilings...What causes mudslides on the West Coast...What can be done to minimize damage in the early stages? Lightning can cause substantial damage to a building without setting it on fire. How? See Chapter 1.
2. There are at least 16 basic factors to be taken into consideration when estimating the hours of labor. See Chapter 2.
3. There are more than 30 overhead items that are chargeable to a specific job operation. There are 15 overhead items that are not chargeable to a specific operation. See Chapter 2.
4. How has OSHA increased the cost of construction of homes? See Chapter 2.
5. In estimating the cost of demolition and removal of debris after a loss there are minimum of 30 questions to be asked. See Chapter 2.
6. When inspecting building losses to take field notes, you do not always start in the same place. Why? See Chapter 3.
7. Protecting property from further loss after a building has been damaged requires 20 important considerations. See Chapter 3.
8. Frequently when inspecting a damaged building, previously existing damage may be found. There are 10 common types of deterioration possibly unrelated to the casualty and 15 contributing causes. See Chapter 3.
9. See how four outstanding contractors set up their estimates of a building loss in Chapter 14.
10. How is estimating in catastrophes different? What conditions can you expect? How do you separate your estimate of damage when there are mixed causes...one covered by insurance, the other not covered such as wind and wave-wash? What is the National Flood Insurance Association (NFIA)? See Chapter 5.
11. If your high school arithmetic is a bit rusty, particularly as it relates to estimating, see Chapter 6.
12. If you want to determine the number of cubic yards of concrete in a footing, foundation, column, slab or wall simply by multiplying the cross section by a common factor, see Chapter 7.
13. Do you want to know how to repair concrete floors and walls damaged by fire; how to repair cracks in concrete or what to do with spalled concrete? See Chapter 7.

14. You can quickly select from a table the cubic yards of mortar needed to lay 1,000 brick or concrete block—or 100 units. See Chapter 8.

15. There are many types and shapes of concrete block units. There are also many different shapes and sizes of brick. See them in Chapter 8.

16. You can rapidly estimate the board feet of all framing in a wall, floor or roof simply by multiplying the area by a common factor. See Chapter 9.

17. The book shows you two ways to estimate the number of board feet in a 2" x 4", 2" x 8" or even a 3" x 12". One is a table, the other gives you a factor to multiply the length by. See Chapter 9.

18. See the latest lumber industry abbreviations in Chapter 9. See men using automatic hammers in Chapter 9.

19. What is "milling waste"? How does it affect the quantity of boards you should order to cover a measured area? See Chapter 9.

20. You do not have to be a mathematician to estimate the number of studs in a wall, joists in a floor, or rafters in a roof whether on 12," 16" or 20" centers. Chapter 9 shows the easy method.

21. If wood cross-bridging is to be cut on the job all you have to do is multiply the length of the floor or room by a number to get the linear feet of material needed. See Chapter 9.

22. If you have residential roof trusses with spans of 20 to 30 feet you can rapidly estimate the material from a table in Chapter 9.

23. How do you lay strip oak flooring over a concrete slab? See Chapter 10.

24. This book lists 35 ways to repair damaged carpentry. See Chapter 10.

25. How much tape and compound do you need for drywall construction? See Chapter 11. See men on stilts applying drywall in Chapter 11.

26. Why struggle to estimate the roof area of a dormer, an intersecting gable roof or a hip roof? There's no need to because a simple procedure is in Chapter 13.

27. Handsplit shakes are no mystery. See how they are made, estimated and applied in Chapter 13.

28. How do you patch and repair different kinds of roofs damaged by windstorm or hail? See Chapter 13.

29. You can't estimate the amount of paint unless you know the covering capacity of the paint for the first, second or third coats. See Chapter 14.

30. You do not need to go through the laborious task of calculating the perimeter of a room times its height, plus its length times width to get the wall and ceiling area of a room. Simply pick the area from a convenient table. You also can pick out the number of rolls of wallpaper from a table of room sizes. See Chapter 14.

31. Ever try to estimate the amount of paint needed for a stairway, a picket or chain link fence, a lattice or cornice? See Chapter 14 for a quick method used by professional painters.

32. There are five different ways to estimate the cost of interior painting.
33. Kitchen cabinets and built-ins are either panel construction or frame construction. Want to know how to estimate their cost? See Chapter 17.
34. There are six methods of electrical wiring; determining the number of branch circuits is not difficult. See Chapter 18.
35. What are the relative values of thermal insulation; what effect does *density* of loose fill insulation have on the amount needed? How much loose fill insulation can a man pour in an hour; how many square feet of batts can be installed? See Chapter 19.36
   When a unit cost of framing 2" x 10" joists is said to be 50 cents a board foot, how would you analyse it to see whether it is correct? See Chapter 21.
37. Does the "Removal of Debris" clause in an insurance policy cover the removal of *all* debris? See Chapter 22.
38. What is meant by "no increased cost of repair by reason of local ordinances" in an insurance policy? See Chapter 22.
39. There are seven methods of adjustment procedure for a building loss. See Chapter 22.
40. There 23 sources of error in preparing building loss estimates. See Chapter 22.

   The foregoing list represents only a sampling of the things of interest and help in the field of estimating in this new Third Edition. The author sincerely hopes you will find it a genuine asset for learning and teaching others and as a reference.

Paul I. Thomas

Contents

10. Finish Carpentry (continued)

11. Interior Wall Finish **279**

12. Exterior Wall Finish **289**

13. Roofing **307**

14. Painting, Paperhanging and Glazing **357**

14. Painting, Paperhanging and Glazing (continued)

Be Applied 382 ● Paperhanging 382 ● Sanitas and Walltex 385 ● Methods of Estimating Cost 386 ● Damageability 389 ● Glazing 389 ● Estimating Materials 391 ● Estimating Labor 391 ● Methods of Estimating 393

Lathing 395 ● Measurements of Areas 395 ● Wood Lath 396 ● Gypsum Lath 397 ● Metal Lath 398 ● Grounds and Screeds 399 ● Corner Bead 399 ● Plastering 400 ● Interior Plastering 400 ● Materials 402 ● Proportions 403 ● Estimating Quantities 404 ● Estimating Labor 406 ● Estimating Costs 407 ● Job-Mixed Gypsum Plaster Using Sand Aggregate 410 ● Machine Plastering 411 ● Exterior Plastering 412 ● Damageability of Gypsum Plaster 416 ● Methods of Repairing Gypsum Plaster 418 ● Damageability of Stucco Plaster and Repair Methods 419

Flashing and Counter-Flashing 421 ● Estimating Costs 422 ● Gutters and Downspouts 422 ● Metal Skylights 425 ● Metal Ventilators 426 ● Ridge Rolls 426

Materials of Construction 428 ● Methods of Construction 429 ● Countertops 432 ● Cost of Cabinet Including Countertop 433 ● Basic Construction Types 434 ● Estimating a Box Type Cabinet 436 ● Estimating a Frame Constructed Cabinet 437 ● Estimating Tall Cabinets 438 ● Points to Remember When Estimating Cabinets 438

Electrical Terms 441 ● Electrical Symbols 443 ● Sizes of Wire 444 ● Colors of Wire 447 ● Methods of Wiring 447 ● Number of Branch Circuits 449 ● Service Entrance 452 ● Estimating Electrical Wiring 454 ● Specifications 454 ● Materials 456 ● Labor 456 ● Pricing the Estimate 458

Insulating Materials 461 ● Types of Insulation 462 ● Reflective Insulation 466 ● Vapor Barriers 467

Plumbing 470 ● Water Supply Materials 470 ● Drainage Piping Materials 470 ● Plumbing Fixtures 473 ● Damageability of Plumbing 474 ● Heating 475 ● Warm-Air Heating 475 ● Steam and Hot Water Heating Systems 476 ● Electric Heating 477 ● Damageability of Heating Systems 477 ● Structural Steel 478 ● Damageability of Structural Steel 478 ● Ceramic Tile 479 ●

20. Mechanical and Miscellaneous Trades (continued)

Damageability of Ceramic Tile 480 ● Metal and Plastic Wall Tile 480 ● Damageability of Metal and Plastic Tile 481 ● Floor Coverings 481 ● Linoleum 481 ● Floor Tile 482 ● Damageability of Floor Coverings 482

Analyzing Unit Costs 485 ● Analyzing Unit Costs by Simple Formula 489 ● Analyzing Unit Costs of Concrete Forms 490 ● Analyzing Unit Costs of Concrete 490 ● Analyzing Unit Costs of Concrete Block 491 ● Analyzing Unit Costs of Brick Masonry 492 ● Analyzing Unit Costs of Hollow Tile Block 493 ● Analyzing Unit Costs of Framing 493 ● Analyzing Unit Costs of Carpentry Surface Materials 494 ● Analyzing Unit Costs of Carpentry Trim Items 495 ● Analyzing Unit Costs For Wall Boards and Plywood Paneling 496 ● Analyzing Unit Costs of Windows, Doors, Cabinets 497 ● Analyzing Unit Costs For Roll Roofing and Shingles 497 ● Analyzing Unit Costs For Painting 498 ● Analyzing Lump Sum Carpentry Labor Figures 499

Insurance Policies 503 ● Exclusions 505 ● Extensions 506 ● New York Standard Fire Policy 506 ● Check Coverage 512 ● Investigation and Survey 513 ● Method of Adjustment 513 ● Preparation for Discussing the Claim 514 ● Purpose of Analyzing Estimates 515 ● Sources of Error in Estimates 516 ● The Mechanics of Checking Estimates 532 ● Value for Insurance Requirements 534 ● Proof of Loss and Final Papers 535

Chapter 1.

Causes of Damage to Buildings

A person who wishes to learn how to estimate the cost of restoring damaged buildings, or other structures, should familiarize himself with the common causes of damage. He should understand how these causes, or perils, affect different materials of construction.

Policies of insurance covering property damage may be divided into two broad types. The first group insures against named perils, while the second group insures against all risks of physical loss, with specifically named exclusions. The most frequently encountered causes of damage under both groups are: (1) Actual burning during a fire; (2) Heat; (3) Smoke; (4) Water; (5) Activity of Firemen; (6) Chemicals; (7) Falling debris; (8) Explosion; (9) Lightning; (10) Windstorm; (11) Hailstones; (12) Aircraft and Vehicle; (13) Sprinkler leakage; (14) Flood, Backing of sewers, Seepage, etc.; (15) Sonic boom; (16) Earthquake; (17) Miscellaneous causes.

ACTUAL BURNING DURING FIRE

Actual ignition and burning of building materials as distinguished from exposure to heat is confined to the combustible materials. These are mainly wood and wood products, paints, asphaltic types of roofing and siding, electrical insulation and some types of composition wall board and insulation. Masonry, plaster, steel and other metals do not support combustion, but are subject to damage or complete destruction by heat from the burning of the building or its contents. The burning of paint, roofing materials, and insulation usually will require complete replacement of the part or area involved.

When framing lumber or large timbers have been subjected to burning, it is not always necessary to replace them. Much will depend on whether they have been structurally damaged, and on the importance of appearance. There are many instances when slightly charred joists, rafters, or studding, have been left in place. They may be scraped, white washed or painted to seal any odor, and then covered with sheetrock or some other type of wall board. Heavy girders in a dwelling or large timbers in mill-type buildings may be scraped and painted or boxed in, where inspection reveals no important structural damage. Surfaces of millwork, however, such as stairways, cabinets, built-in features, trim, or finished floors which have any degree of charring, in almost every case must be removed and replaced.

HEAT FROM FIRE

The extent of damage to building materials from the heat that is generated by the combustible parts of the building, but more significantly by the burning of the contents, depends on the temperatures reached and on the length of time the materials are subjected to these temperatures. Brick, stonework, structural steel, sheet metal and iron expand in the presence of heat. Lateral expansion of steel girders during a fire has many times caused them to push out exterior masonry walls. Light, unprotected steel trusses, columns or beams will soften and lose their tensile strength. They may sag or collapse even during moderate fires. Most types of masonry will withstand high temperatures for a long period before sustaining structural damage. Spalling of masonry is a condition brought about by long exposure to high temperatures or alternate heating then cooling by water whereby the surfaces exposed "pop-off" in chunks or scales. The thicker the masonry the less likelihood there is of major structural damage from heat.

Heat causes glass to crack at relatively low temperatures, and at about 1600°F. it softens. Plaster tolerates fairly high temperatures before flaking and chalking on the surface, with a change in its chemical composition.

Paint softens and blisters in the presence of heat. Most woods shrink slightly as they lose moisture, and boards or timbers of wider dimension frequently check. Wood ignites at approximately 400°F. after 20 minutes exposure.

While it is impractical to record either the temperatures that are developed or their duration in fires of accidental origin, it is sometimes possible to estimate the degree of heat by observing the physical condition of glass, metals, and ceramics in the area after the fire. The following table lists some of the melting or fusing temperatures of materials in buildings.

APPROXIMATE FUSING TEMPERATURES

Material	F temperature
Tin	449°
Lead	621°
Zinc	786°
Aluminum	1217°
Copper	1981°
Iron (Cast)	2250°
Sand	2588°
Porcelain	2588°
Steel (structural)	2606°
Chromium	2750°

Heat rises during a fire, not only within a room, but also up stairways, elevator shafts, and other areas or vertical openings where its passage is offered the least resistance. For that reason concrete floors on the top of which fire is burning usually show very little and sometimes no damage in fires that may have caused serious loss to the walls and roof. Ceilings in rooms may be seriously injured by heat, and joists may be heavily charred, while the lower regions often escape with hardly more than discoloration of paint, or a blistering of the finish on wooden floors. "Look high for heat damage" is an expression which should be remembered.

SMOKE FROM FIRE

Smoke is the result of incomplete combustion. Heavy smoke develops when burning materials are shut off from adequate oxygen preventing complete combustion.

There are two kinds of smoke that can occur during a fire. One is a *hot smoke,* the other is a cool or *cold smoke.* Hot smoke is that which is relatively close to the fire and carries the heat of the fire with it. Hot smoke causes the greater damage because it penetrates porous surfaces and, in contact with cool surfaces, it condenses, leaving heavy stains of tars and resins. Building decorations are readily damaged by such stains and require washing or redecorating. Wallpapers touched by hot smoke generally need replacing. Window glass becomes coated and must be washed. Ceramic tile normally has hairline cracks that are almost invisible and, under proper conditions, the hot smoke penetrates through the cracks in the glazed surface to the porous backing. This stain is impossible to remove and the tile must be replaced, or an allowance should be made for the damage. Enameled and porcelain plumbing fixtures will suffer less than ceramic tile. Test cleaning is the best way to determine whether there has been any permanent injury.

Stonework, brick, or cement surfaces are difficult and sometimes impossible to clean, once they have been exposed to hot smoke. Experts are frequently called in to test clean before an estimate of damage is made. Usually, if cleaning or sandblasting is impossible, cement washing, coating, or painting may be satisfactory to all interests.

Cold smoke is smoke that is carried some distance from the fire and has cooled and condensed in the air before it reaches the area or surfaces on which it deposits. Cold smoke is often in the form of soot. Vacuum cleaning or washing will many times restore a surface to its former condition. Cold smoke does not leave the characteristic pungent smoke odor that results from hot smoke.

Smoke odor in a building can be removed in nearly every case by specialists offering smoke odor removal service. They maintain service offices throughout the country. This treatment is particularly effective in removing smoke odor from inaccessible areas, under floors, or behind plastered walls and ceilings. Frequently exposed framing, sheathing, and like areas that are charred can be completely deodorized through the use of special liquids.

WATER

Water from automatic sprinkler systems and fire hoses, or from rain or melted snow entering through windows, doors, or openings in roofs and walls after a fire, hail, or windstorm, causes varying degrees of damage to finished millwork, flooring, plaster walls and ceilings, sheet rock, composition wallboards, and interior paint and wallpaper.

Electric cable and conduit in walls or floors may suffer damage from wetting. The extent of injury will depend on how well-sealed or impervious the cable shield is to moisture and water. It is often cheaper to replace them than to repair outlet receptacles, switches or electric fixtures that have been seriously water soaked. Electric motors that have been wet, even those submerged for a short time, can under some conditions be baked out.

Kiln-dried wood or well-seasoned woods absorb water up to about 25 per cent by weight, during which time swelling occurs. After this "fibre-saturation" point is reached, the swelling ceases. Plaster softens in the presence of water, and wallboards frequently expand and warp, some of them permanently. The glue on wallpaper softens and eventually the wallpaper may peel off. Water in back of paint, particularly enamels and semi-glazed surface paints, forms small blisters full of water which ruin the decorations. Dirty water, which might be pumped from rivers or creeks, and dirt-laden water percolating through ceilings to floors below discolors decorations, carpeting and finished floors.

The chemical or physical properties of water used during a fire are seldom factors that increase damageability, except possibly salt water which has a corrosive effect on metals, if left in contact for any length of time.

The weight of water (64.5 pounds per cubic foot) which has been absorbed by plaster and some types of contents such as paper or fabrics, may cause the collapse of wood beams, girders or sometimes entire floor structures.

Water expands various types of materials including newsprint, sisal, and cottonseed and, if it is closely packed in a building it may swell and exert enough side pressure to push out masonry walls.

The freezing of water in the winter during or shortly after a fire causes further damage by expanding in plaster, water pipes, toilet bowls, traps, and hot water or steam heating systems.

ACTIVITY OF FIREMEN

A certain amount of damage is caused by firemen in gaining entrance to a building or to the rooms in it. Doors, door jambs and casings are sometimes broken or chopped, locks are forced, and glass in sash doors is broken. A common practice in good fire fighting is to ventilate the premises. For that reason, firemen break windows or skylights and occasionally chop through a roof to let out smoke and heat. Miscellaneous chopping and cutting may be done by firemen in tracing the course of a fire under floors, or in partitions and walls.

In fire-resistive multiple story buildings, toilet fixtures are sometimes broken or removed so that the accumulation of water on the floor can be squeegeed to the outlet. In buildings that have wood floors, holes may be bored or chopped to drain off water to lower floors.

CHEMICAL EXTINGUISHERS

Most towns and cities equip their fire trucks with chemical extinguishers for the average small fire. These extinguishers usually contain calcium chloride, foamite, carbon tetrachloride or carbon dioxide. While all but the latter of these chemicals may stain or discolor decorated surfaces, they are otherwise not harmful to building materials.

FALLING DEBRIS

During the more serious fires, there is always danger of weakened floor structures collapsing because of overloading with wet plaster and contents and the debris from floors above. Chimneys and brick walls may fall inward, crushing portions of the building that are yet unburned or otherwise uninjured. Charred wood girders, beams or rafters weaken and allow parts of the structure and debris to collapse.

Buildings adjacent to a burning building sometimes are damaged by walls collapsing outward, falling brick, or other debris.

EXPLOSIONS

In buildings of dwelling occupancy, the most common explosions are in furnance boxes of heating plants, caused by delayed ignition of accumulated fuel gases. The majority of these are low-pressure explosions and the damage is usually confined to the oil burner, fire box brick lining, insulated covering, and the smoke stack connecting the boiler to the chimney. Smoke, soot or fumes may follow such an explosion and be carried to other portions of the premises through open doors and hallways; when drawn into heating ducts they may reach into every part of the building through heat registers. The presence of oil in the smoke or fumes can cause substantial loss to decorations. If fire follows an oilburner explosion, more serious damage can be anticipated by burning of controls and by sooting. There have been occasions of violent explosions in the fire boxes of furnances, but they are more prevalent in the large industrial furnances and also in furnaces that burn gas rather than oil or coal.

Explosions in manufacturing and other industrial plants originate from many different causes, the most common being ignition of the vapors of flammable liquids. There is also the hazard of dust explosions in plants that process starch, sugar, grains, cork, spices and numerous other products. Chemical explosions occur in plants experimenting with new products, and occasionally happen in munition plants, fireworks manufacturers, petroleum processors, fertilizer processors or storage plants. Such explosions can demolish buildings, or completely or partially damage them, depending on the force and the distance of the explosion from the risk. Masonry or plaster walls often crack. Windows and plate glass may also crack or shatter from ground shock or concussion.

The damage caused by blasting for roads, foundations, tunnels, and so forth, follows a pattern similar to that mentioned above.

Examination of damage done by explosive forces requires great care. The estimating should be confined to explosion damage only; and the wear and tear or other conditions that may be chargeable to maintenance must not be included. (An excellent reference on this subject is *Blasting Claims—A Guide for Adjusters,* prepared by the American Insurance Association.)

A typical low-pressure explosion, not uncommon during fires, occurs when a heavy concentration of gases of incomplete combustion is ignited. These gases become "pocketed" in blind or concealed spaces, or at one end of a floor with the fire burning in the other end. If the premises are not vented to allow the gases to escape, they can be ignited by flames when the fire reaches the area. Windows, particularly display windows in store buildings, are frequently blown outward by such explosions. Because these explosions are low-pressured, they seldom cause serious structural damage.

LIGHTNING

Damage to buildings and structures by lightning occurs in all parts of the continent, but more frequently in the southeastern section of the United States, where the incidence of thunder storms is higher. Lightning is a high-voltage electrical discharge between the clouds and the earth. The voltage may run into millions, and the amperes into tens of thousands.

Once the discharge takes place, the effect on the object in its path on earth may be the result of a direct hit by a *lightning bolt,* or the result of an *atmospheric surge* of electricity induced in the area, usually by a bolt of lightning which has struck nearby. The latter is sometimes called a *ground surge.*

Damage caused by a direct hit can be trifling (such as knocking a few bricks from the top of a chimney), or devastating. The knowledge that the objective of lightning is the ground, will aid, perhaps, in understanding the unpredictable and sometimes perplexing path that lightning follows from the point of contact to the place of exit in a building. Where resistance to its path is encountered, instantaneous heat is created and, if the bolt of lightning is a powerful one, the result is explosive. Inspection of a building struck by lightning may reveal a few shattered clapboards, or a corner of a roof or a few bricks knocked off, giving the impression of minor, localized damage. A more thorough survey, however, often discloses a greater injury to the interior than anticipated. Electric receptacles may be scorched or blown out of the wall, cracks may appear in some rooms but not in others, and a joist or rafter may be split. When lightning strikes a structure, it frequently fans out rather than pursuing a direct course, making its path more difficult to trace.

Lightning that strikes behind a concrete block retaining wall can lay it flat on the ground. Chimneys struck by lightning require careful inspection from top to bottom, not only to include all damage caused by lightning, but also to detect and distinguish old cracks and damage caused by age and settlement.

Losses to pole transformers, appliances, motors and other equipment can be the result of atmospheric or ground surges, and not necessarily direct hits. In most instances examination will show either fusing of a part or parts or else explosive effects due to violent interior pressure. Because electrical appliances and apparatus are subject to inherent defects and wear, which bring about shorts and breakdowns, the inspection of damage attributed to atmospheric surges of lightning calls for a high degree of care. Expert advice should be sought when there is serious doubt that lightning is involved.

WINDSTORM

It has become a fairly general practice among insurance carriers to consider windstorm as any wind capable of causing damage without regard to a stipulated velocity. Gusts of wind in local areas, between tall buildings, or down alleys on relatively calm days may cause damage by slamming doors open or closed, tearing awnings and even breaking windows. Although this type of damage is contemplated under the policy, the windstorm provisions of a policy contain a number of exclusions not common to the hazard of fire or lightning.

Special Considerations

The person charged with preparing specifications for repairing a structure damaged during any type of windstorm either should be familiar with or instructed in the policy exclusion of certain kinds of losses and specific types of property.

Where windstorm is a *named peril* in the insurance policy the following exclusion is typical though the language may vary from one policy to another:

(a) loss caused directly or indirectly by frost or cold weather or ice (other than hail), snowstorm or sleet, all whether driven by wind or not;

(b) loss to the interior of the building(s), or the property covered therein caused by rain, snow, sand, or dust, all whether driven by wind or not, unless the building(s) covered or containing the property covered shall first sustain an actual damage to roof or walls by the direct force of wind or hail and then this company shall be liable for loss to the interior of the building(s) or the property covered therein as may be caused by rain, snow, sand, or dust entering the building(s) through openings in the roof or walls made by direct action of wind or hail.

Also, unless there is a specific endorsement, all policies exclude damage caused by flooding, surface waters, overflow of streams, wavewash, and tidal waters.

In situations where part of the damage has been caused by a covered peril and part by one not covered (or excluded) the estimate should be prepared to indicate the estimator's opinion of the damage *covered, not covered,* and that which is *questionable.*

In addition to excluding specific hazards, most forms exclude coverage on specified property unless liability is specifically assumed by endorsement. Property more commonly excluded consists of windmills, windpumps, and crop silos. Many forms also exclude such items as awnings and television antennas.

The windstorm provisions are not identical in all states, nor in all forms. The exact wording should always be checked before preparing an estimate of damage.

Tornadoes

The word tornado comes from the Latin word *tornare* meaning to turn. It is cyclonic, and is probably the most violent type of windstorm in nature. Approximately 150 to 200 tornadoes occur annually in the United States, the majority being in the midwestern area. The tornado moves generally from southwest to northeast with forward speeds ranging from a few miles per hour to as much as 60 miles per hour.

The path of a tornado is narrow, usually not much more than a quarter to one-third of a mile wide. The average length of the path of tornadoes is eight to ten miles. The destructive force in a tornado is the high velocity of the cyclonic winds, which often reach speeds up to 450 miles per hour. This counter-clockwise swirling action produces in the center a condition approaching a vacuum. The effect of this vacuum-like center is explosive when passing over any structure. Buildings are reduced to rubble and in many instances they are sucked up with other objects and carried for many miles. Probably no catastrophe so quickly and completely changes the appearance of a community. Trees are stripped of all foliage and many times complete denuded of their bark. Persons who have witnessed tornadoes close at hand describe its sound as "like a million bees" while in the air, and "like a thousand express trains" when it touches the ground. The characteristics of a tornado are its narrow and relatively short path and its destructive violence, which is almost atomic.

Lightning generally accompanies a tornado. Hail and rain also either precede, accompany, or follow this type of storm. Hailstones as large as baseballs have been reported. On one occasion disc-like hail 6 to 10 inches in diameter and 2 to 3 inches thick fell during a tornado.

About 70 per cent of the tornadoes occur in the four months, March through June. Most of the others take place in the late summer; a few in other months. The states which have suffered most in property damage by tornadoes are, in order of the extent of damage: Oklahoma, Illinois, Missouri, Texas, Georgia, Ohio, Indiana, Kansas, and Iowa.

Hurricanes

A hurricane, like a tornado, is also cyclonic with winds traveling in a counter-clockwise motion. However, the intensity is second to a tornado. Officially, hurricane winds are those with velocities in excess of 75 miles per hour. The path of destruction of a hurricane may cover several hundred miles. Our hurricanes generally start in the North Atlantic around the Cape Verde Islands. They travel westerly toward the West Indies, then northwesterly and as a rule curve to the northeast. They may travel inland toward the Gulf States or up the Atlantic Coast, either striking land or veering seaward.

The wind velocities of hurricanes may reach 150 to 200 miles per hour, but generally they do not exceed 100 to 125 miles per hour at the highest point. The width of the path of a hurricane may be 50 or 100 miles. The length of its path may be as much as several hundred or a couple of thousand miles. The forward speed of hurricanes varies from a few miles per hour to 60 miles per hour; as it passes over land, the forward and cyclonic velocities are retarded.

Hail seldom accompanies a hurricane, but extremely heavy rains are not uncommon. Along the coastal shores, high tides generally occur, inundating the low lands and causing serious damage.

The "eye" of a hurricane may be a few miles wide, or it may be as much as 40 or 50 miles wide. The suction effect, so common to tornadoes usually is present in a hurricane to a lesser degree.

Hurricanes occur most frequently in the months of June, July, August and September. There have been exceptions when hurricanes have occurred in October and November. The height of the season is considered to be the month of August and the first half of September. The states visited by hurricanes more than any others are those that lie along the eastern seaboard, and the gulf states.

Where wind velocities reach over 100 to 120 miles per hour for any length of time, the roof structures on many buildings start to lift, show windows are sucked out or blown in, and total destruction—even disappearance of small lighter buildings—may take place. Wind velocities between 70 and 100 miles per hour usually tear off part or all of asphalt type shingles, and can cause extensive damage to wood, asbestos, aluminum, slate and roll roofing. Fences are overturned, depending on their construction and the condition of wooden posts in the ground. Trees, or their heavier limbs, when blown on top of or against buildings cause varying degrees of damage. Car ports or other structures lightly built, may be blown over or collapse. Flying shingles and portions of roofs or other debris cause damage to adjacent and nearby buildings.

Because windstorms are often preceded, accompanied or followed by rain, damage caused by rain entering openings made in the roof or walls (including broken windows) by the action of the wind should be included in an estimate of damage. The extent of such damage depends on the amount of water that gets into the premises, and the portions of the building with which it comes into contact. In most cases the damage by water is confined to ceilings, wallboard or plaster, insulation in attics, and decorations in those rooms affected. Occasionally, water on finished floors causes them to "cup," but in most cases they can be resanded and refinished. Wooden doors and windows may absorb water through bare or poorly painted surfaces, causing them to swell so that they will not open or close properly. Veneers may come loose on the lower part of the doors.

HAILSTONES

Hailstones are distinguished from ordinary hail, which consists of small gains of frozen rain or ice-like snow. Storms that produce hailstones ranging in size from 1/2 inch to 2 or 3 inches in diameter, are believed by some authorities to be associated with cyclonic disturbances in the upper atmosphere. The stones are comprised of laminations of ice, one layer being built up on top of another. They are extremely hard, the larger ones being difficult to break with a hammer.

Hailstorms may occur during a tornado or occasionally during a hurricane. They also take place alone, or with rainstorms. They occur most often in the early spring months and are more prevalent in the central states than in other parts of the country.

The damage that can be done to buildings during a hailstorm depends on the quantity and size of the stones, and on whether rain accompanies or follows the storm. Large hailstones are capable of shattering slate, asbestos and Spanish tile roofs. They can break plate glass windows and puncture asphalt shingles and built-up roof coverings. Even the smaller hailstones are capable of perforating the asphaltic felt-type roofing. Aluminum roofing and siding, though not punctured, may be so badly pocked or dented that replacement is necessary. Wood siding also may be dented from hailstones, necessitating filling and painting, or, in the more serious cases, complete replacements.

Roofs that have been damaged by hailstones may admit rain during the storm, or later before temporary or permanent repairs can be made. As a result, interiors are subject to further damage by water.

AIRCRAFT AND VEHICLE

Damage by vehicles, as a rule, pertains to those than run on land or tracks. The vehicle may be a passenger automobile out of control colliding with the side of a building to cause minor damage to an exterior wall, a show window, or a porch. Or it may be a large trailer truck with brake failure coming down a hill and crashing into a building to cause extensive damage. Railroad freight cars occasionally strike industrial structures while being switched in or out of a siding.

Aircraft occasionally crash into buildings, causing damage ranging from minor injury to serious damage, particularly if fire ensues.

SPRINKLER LEAKAGE

Buildings that are equipped with automatic sprinklers, stand pipes or other fire protective equipment can be damaged by leakage from within the system. Sprinkler heads may operate as a result of some inherent defect, or by excessive heat. Pipes freeze; heads may be injured manually, or leakage can result from corrosion of pipes; tanks may collapse, or any number of accidents might occur to cause discharge. The main damage to buildings by water leakage from the sprinkler system is to wall and ceiling surfaces and decorations. Sometimes plaster and wall treatments in lofts, offices, and showrooms are affected. Finished wood floors may have to be sanded or replaced, depending on their condition. Asphalt, rubber and plastic floor tile is frequently loosened by water seeping between the tile and the under-floor. Electric wiring and fixtures, wet down by water, must be checked for repair or replacement, as water may have penetrated the armored cable or pipe conduit and caused corrosion or potential shorting in the switches and other outlets.

FLOODING, BACKING UP OF SEWERS ETC.

During heavy rainfalls, storm and sanitary sewers may not be adequate to carry off the surface waters fast enough to prevent backing up through drains and plumbing fixtures. Check valves installed in the sewer line will generally stop the backing up, but not all dwellings or commercial buildings have them.

Poorly drained lots or streets accumulate surface waters during heavy rains and basements become flooded when the water enters basement windows, seeps through walls, or comes in under cellar doors.

Creeks, rivers and tide waters frequently rise in some localities when there have been excessive rains, or when hurricanes cause sudden high tides.

Hydrostatic pressure will develop under proper conditions whereby ground water percolating through the soil is forced through foundation walls and cracks in concrete floors. Under each situation mentioned, water may collect in basements of buildings from a depth of a few inches to several feet. Hydrostatic pressure may be enough to heave up parts of the basement floor or collapse foundations. Drains, capped inside the building, may burst, causing the concrete floor to crack and heave. Sometimes, though rarely, footings and foundation walls can be undermined as a result of the displacement of subsoil by moving ground water.

When basements are inundated, wood flooring, plaster, trim and millwork will show but slight damage if the water recedes quickly. The longer the water remains, the greater possibility there is of serious injury. Electric wiring, outlets and fixtures, whenever exposed to water, should be checked for damage.

Furnaces that have been submerged will show more or less injury to asbestos insultation on the outside or insulation in the jacket. The fire clay bonding the brick in the fire box may soften and be so damaged that the fire box lining will need replacing. Oil burners, gas burners, and coal stokers will require cleaning and checking.

Electric motors of washing machines, clothes dryers, deep freeze boxes, refrigerators, compressors and elevators many times can be repaired by prompt removal for cleaning and baking out. Seriously damaged units may have to be replaced.

When wood beams and girders in older buildings have been under water for 24 hours or longer, they may swell, buckle and twist.

When sewage is mixed with the water, the problem of sanitation arises, and completely cleaning up a basement so affected can be costly. Also silt and mud frequently are mixed in the water. This requires considerable labor to remove, sometimes being of such depth that shoveling is required to get it out.

SONIC BOOM

The coming of the jet plane has brought with it the mechanical phenomenon of "sonic boom," which is caused by shock waves in the air generated by planes at supersonic speed. It is said that the cracking of a whip accelerates the tip of it to the speed of sound, producing the same effect. Sound travels at approximately 760 miles per hour at standard atmospheric conditions. When objects move through space at less than the speed of sound, air particles are readily pushed aside. At supersonic speeds the air particles build up in front of the moving object and create atmospheric pressure waves, that is, shock waves. Planes traveling at supersonic speeds produce several shock waves.

Scientists are not certain of the exact nature of the phenomenon, but it is generally believed that the waves attach to the plane when supersonic speed is reached, and as the speed of the plane increases they are swept back forming the cone shape of a wide funnel. When the pressure waves reach the ground, they usually have joined into one, two, or three waves. The energy of the waves dissipates as they move toward the ground and also when they pass over it because of friction and obstructions.

The boom (sound) heard by a person standing in the path of the cone is caused by the instantaneous pressure of the plane passing over. If two or three waves are created, there will be a rapid, almost overlapping, repetition of the boom.

To what extent a sonic boom can cause structural damage to a building has not yet been fully determined. There is no question that under proper conditions a sonic boom can shatter brittle window glass and also break doors or their frames if they are slightly ajar or loose-fitting at the time of the boom.

The pressures necessary to crack plaster and brick walls, damage roof structures, and heave floors, are much greater than can be created by a sonic boom; it is most unlikely that such structural damage can be traced to that source. Nuclear explosion tests indicate the necessity of a free stream of pressure of 150 to 300 pounds per square foot to cause such structural injury. The highest recorded pressure induced by sonic boom is reported to be 33 pounds per square foot—and this from a jet flying 280 feet away.

Press, radio, or television reports on the occurrence of a sonic boom frequently invite residents for miles around to inspect their properties, often for the first time. This results in numerous claims for cracks in plaster, walls, basement floors and so forth, which allegedly were caused by the boom.

When inspection is made to estimate damage by sonic boom, it should be verified, if possible, that one actually occurred by reporting to the Claims Officer of the nearest air base, and by questioning surrounding property owners. The Claims Officer will usually conduct his own investigation, if claim is filed, and his report should be obtained. When structural damage to walls, roofs and floors is claimed, it may be advisable to consult with a competent engineer, vibration or explosion expert.

EARTHQUAKE

The most frequent cause of earthquakes and earth tremors is the faulting of rock formations beneath the earth's crust. Where high mountains are in proximity to deep oceans, quakes are more numerous. Such a condition is found along the Pacific Coast in a line running from Mexico to the Aleutian Islands. This is part of a so-called earthquake belt. While earthquakes have occurred in other sections of this country, their frequency and intensity is much less than along the Pacific Coast.

The effect of quakes on buildings and structures varies with the seriousness of the disturbance, the distance away and the type of earth formation through which the vibration waves travel. Major injury such as broken windows, large masonry cracks, or the collapse of all or part of a structure can be readily determined on inspection. Minor damage, however, such as hairline cracks in plaster and masonry are easily overlooked or difficult to distinguish from normal settlement cracks or deterioration that existed before the quake. Many times it will serve all interests best to have the property surveyed by a vibration expert or a structural engineer. He can generally determine the damage and also the probably method and cost of repairs.

Earthquakes are sometimes followed by fire with the result that the damage is mixed, being partly due to vibration and partly due to fire. If earthquake damage is excluded in the policy of insurance, it will be necessary to separate fire and vibration damage. The best method of doing this is to prepare a two-column estimate which will show the cost of repairing the damage from both causes.

LANDSLIDE—MUDSLIDE

Landslides can be started on a dry hillside or a mountainside by some disturbance of the earth such as vibration. Most often, however, they occur when excessive rains follow a long dry spell. The porous soil permits water to percolate rapidly through it until it reaches an impervious layer of clay or rock. The water, following the slope of this impervious layer forms a slick medium on which the earth above begins to slip or slide.

The slide may cause cracking in walls and foundations due to earth movement, or may engulf or completely crush a building in its path.

When the soil which is sliding has been so completely saturated with water that it reaches the consistency of mud, the landslide is usually referred to as a mudslide. Mudslides also occur when the top stratum of soil is of a clay or adobe type which becomes saturated with water after a dry spell. When the underlayment is rock on a slope, the wet clay begins to slide by gravity.

The early movement of a mudslide is usually slow and almost unnoticed. The first indications may be the appearance of cracks in retaining walls, foundations, or concrete drives and patios. The lateral movement of the slide depends on many factors. It may stop with the cracking of a foundation wall or it may continue downward taking the entire building along with it. It may also occur over a period of days, weeks or months.

When a mudslide is discovered in the early stages, steps can be taken to prevent greater damage by making tests to determine the depth and stability of the soil and the direction of the slide. When this data has been developed, measures can be taken to shore the building and establish drainage ditches to draw off the water in the soil and divert it away from the property. Only men thoroughly experienced in handling mudslides should be sought for advice as the work is highly specialized.

FREEZE-UP

A building without heat in freezing weather may suffer damage from frozen water pipes, sprinklers, tanks, traps, toilet bowls, and steam or hotwater heating systems. Not only will the plumbing and heating fixtures and piping be damaged by bursting but, if not discovered when they begin to thaw out, the leakage of water from within the system can cause serious loss to floors, walls, decorations, electrical wiring and so forth (see "Water Damage," page 24).

Preparing the specifications for repairing a freeze-up loss should not be started until the water has been shut off and the premises thawed and dried out. Tests will have to be made by competent plumbing and heating contractors to establish the extent of damage to piping, fixtures, and so forth. A careful inspection of the entire building area involved is required to determine where damage begins and ends.

DAMAGE BY HUMANS

Damage to buildings by people includes vandalism, malicious mischief, riot and civil commotion, and also damage done by burglars. While the intent of the individuals may not be the same, the general nature of damage is.

Window or plate glass may be broken, decorations defaced, plaster walls gouged, doors and locks broken and so forth. Not infrequently, vacant or unoccupied buildings are entered by children or youths who cause serious destruction. In addition to wanton damage to the property, plumbing, heating, and electrical fixtures or their parts may be removed.

Prior to preparing an estimate of this type of loss, a careful investigation of the circumstances is essential. Police reports should be obtained, and neighbors, former occupants, or others familiar with the physical condition of the property are generally questioned to develop, insofar as possible, the damage claimed as against any prior condition which existed as a result of wear and tear.

COLLAPSE

Collapse losses involving roofs, particularly on commercial properties, are not uncommon. Most of these occur when the roof, usually flat, has been subject to an unusual heavy snow load, or an unusual accumulation of rain water. In either case, inadequate design is always suspect and the occasion generally requires the services of a structural or consulting engineering firm to investigate. Additionally, snowstorms and heavy rains are accompanied by a certain amount of wind. Consequently, when the peril of collapse is not covered under the insurance policy, there is frequently a tendency for the claim to be made under the peril of windstorm.

Because the person assigned to estimate the cost of repairs is many times the first or one of the first to arrive on the scene, he is in an excellent position to preserve evidence by photographs, diagrams and note-taking.

Collapse does not always involve a roof of a building. Floors of commercial buildings collapse partially or completely due to overloading. Basement walls of dwellings collapse inward due to hydrostatic pressure. Ceilings or parts of ceilings collapse for various reasons.

ICE DAMS ON ROOFS

Ice dams being their build-up in gutters at the eaves or cornice line. The gutters collect snow and melting water which, freezing, fills the gutter

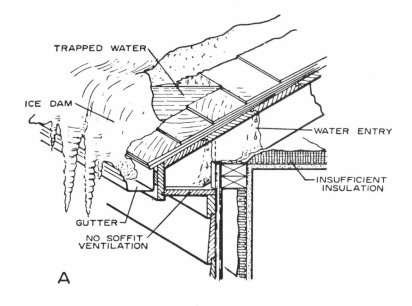

TRAPPED WATER

ICE DAM

WATER ENTRY

INSUFFICIENT INSULATION

GUTTER

NO SOFFIT VENTILATION

A

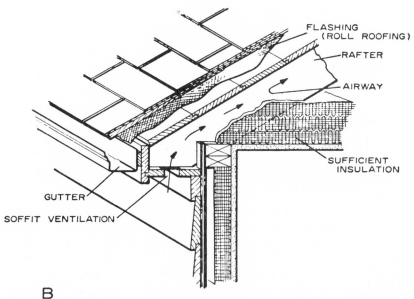

FLASHING (ROLL ROOFING)

RAFTER

AIRWAY

SUFFICIENT INSULATION

GUTTER

SOFFIT VENTILATION

B

M 134 787

Snow and ice dams: A, Ice dams often build up on the overhang of roofs and in gutters, causing melting snow water to back up under shingles and under the facia board of closed cornices. Damage to ceilings inside and to paint outside results. B, Eave protection for snow and ice dams. Lay smooth-surface 45-pound roll roofing on roof sheathing over the eaves extending upward well above the inside line of the wall.

and then snow and ice start building up on the roof from the gutter toward the ridge. Alternate freezing and thawing are factors increasing the build-up. Heat from the interior of the building melts the underside of the formation. The water cannot escape because of the frozen gutters. It eventually backs up under the shingles and enters the interior of the building causing damage to walls, ceilings and floors. An excellent illustration of how this occurs and corrective measures is shown on page 38. This is taken from "Wood-Frame House Construction," Agricultural Handbook No. 73, July 1970, U.S. Department of Agriculture, Washington, D.C.

Chapter 2.

Fundamental Considerations

PURPOSES OF BUILDING LOSS ESTIMATES

When a building or structure has been damaged by fire, windstorm, explosion or some other hazard, the owner not only wants to know how much it will cost to repair the property but, if he is insured, an estimate of the loss will form the basis of claim under his policy. The insurance carrier will want to know the amount of the damage sustained, in order to determine its liability.

There are not only variations in methods of preparing estimates, but also differences of opinion as to extent of damage and cost of making repairs. The only way to reconcile differences in estimates is to analyze and check each item. This procedure requires that the estimate itself be in complete detail, showing all measurements, and dimensions, and listing the material and labor for each item of repair or reconstruction. *The estimate becomes, therefore, not only the foundation but a fundamental device for the adjustment of a building loss.*

ESTIMATING NEW CONSTRUCTION VS. REPAIRS

The principles used to estimate the cost of restoring a damaged structure are the same as those used to estimate the cost of constructing a new building. Once the scope of the work has been established, and the specifications agreed upon, the estimate is developed from the cost of the materials and labor. The contractor's overhead charges and his profit are added, to arrive at the total cost.

The actual preparation, however, of a loss estimate differs in several respects from one made for new construction. When a builder, for example, prepares a bid for erecting a new house, he works from a set of plans and specifications. His competitors use the same plans and specifications, so that all bidders are basing their estimates on identical measurements, dimensions, materials, and work to be done. Any variations in the bids submitted will usually be the result of errors on the part of the estimator, differences of opinion as to labor costs, or the ability of one builder to obtain materials cheaper than another.

The majority of building losses are estimated without the benefit of uniformly prepared plans or specifications. Each contractor inspects the property and determines in his own mind what is necessary to repair whatever damage has been done. It is generally left up to the contractor to decide whether an item can be repaired or whether it will have to be replaced. He determines how much re-plastering is necessary, what rooms require redecorating, how much electrical wiring is damaged, or the number of squares of shingles needed to repair a roof. In addition to making his own specifications of work to be done, the person who is preparing the estimate exercises his own judgment as to the kind and quality of the materials to be used in restoration. Since there are usually no plans prepared on building losses for the use of estimators, each one takes his own measurements and makes up any sketches he feels are necessary.

The repairing of building losses will, in most instances, require the tearing out of damaged material prior to installing new materials, or will necessitate preparation of surfaces to be redecorated. Determining the cost of these labor operations is left entirely to the individual making the estimate.

This procedure of having each person who is preparing an estimate, set up his own specifications of repair, introduces a source of errors and differences which is not inherent in estimating new construction.

ESTIMATING IS NOT AN EXACT SCIENCE

To estimate, by definition, means to approximate, and it should be understood at the outset that estimating is not an exact science. *There are so many possibilities of error in making an estimate, and so many misunderstandings and variations of opinion, especially when estimating the labor, that none may be considered perfect.* It is well known that in actual practice the spread between contractors' bids on new construction will range from 10 to 30 per cent or more. This condition occurs in spite of the fact that each estimator uses identical plans and specifications. It is not unusual for even large contractors with a staff of experienced estimators to lose money on one job and make it up on another.

It is also rare that all of the detailed costs in the estimates of two contractors will be the same. The total amount for the job may check out reasonably close, but the individual items will vary up and down. The reasons for the differences in estimates are explained in Chapter 22, wherein 23 sources of error are discussed.

ACCURACY IS ESSENTIAL

While estimating is not an exact science, it nevertheless calls for the highest possible degree of accuracy in its preparation. The proper measurements, the identification of materials as to size, kind, and quality, the careful listing of items, the thoughtful consideration to labor requirements, and the double checking of all arithmetical computations are of utmost importance. Short cuts should be avoided, and outright guessing of dimensions or quantities will invariably result in errors and embarrassment.

QUALIFICATIONS OF AN ESTIMATOR

A general knowledge of materials and methods of construction is necessary in order to estimate costs. The greater knowledge a person has of building construction, the greater are the possibilities of his becoming a highly competent estimator. He need not have been a carpenter, mason, bricklayer or other journeyman. Many estimators employed by contractors have had little or no practical experience as mechanics in a building trade. They do, however, have a broad knowledge of building materials and construction which they have learned from study and observation. They are good mathematicians and detail men. They are able to visualize every operation that will be necessary to complete the work. Probably one of the most important qualifications that an estimator can have is that of making inquiry when he is uncertain or does not know. There are many excellent reference books that deal with construction methods, and no one who either undertakes to make an estimate or check one should be without good reference material.

SUBCONTRACTORS

The number of building trades which a general contractor is equipped to estimate varies. It is fairly standard practice for them to figure all of the carpentry work. On the small or medium sized jobs they may also estimate the cost of roofing, painting, plastering, concrete, or masonry. They frequently have their own men engaged in these trades and have sufficient work to keep them employed. The majority of general contractors sub-contract built-up roofing, plumbing, heating, electrical work, elevators, air conditioning, structural steel, and other mechanical or highly specialized building trades.

When a subcontractor is requested to bid on loss repairs, whoever has employed him should inspect the property with him and specify what work is to be done, and the kind and quality of materials that are to be used. He also should be instructed to prepare his estimate in detail, so that it may be subjected to analysis and audit if necessary.

Subcontractors generally add their own overhead and profit to their figures. A general contractor adds a charge for any supervision that is necessary, plus his normal profit percentage.

EXPERTS

In many cases, experts or specialists in certain building materials and operations are consulted or asked to inspect the damage and give an opinion on either the degree of damage or the cost of repairs. Damage involving building service equipment such as elevators, escalators, sprinkler systems, power plant boilers or generators, refrigerating apparatus, ventilating systems, and even such domestic appliances as electric ovens or ranges, dishwashers, dryers and air conditioners may require that either the manufacturer's representative or an expert inspect the property, before the extent of injury can be determined.

Sometimes it is prudent for the estimator to have a manufacturer's representative inspect certain specific materials such as brick of various types, marble or limestone facing and trim, cork insulation, rubber or asphalt tile flooring, or certain kinds of roofing. It is not uncommon to employ chemical or materials testing laboratories to determine whether certain materials have been damaged and, if so, to what extent repair or replacement is necessary. Among the experts frequently summoned are college professors in structural, mechanical, electrical and heating engineering. Also, professors of physics, chemistry and ceramics have been consulted on special subjects.

FUNDAMENTALS OF ESTIMATING

The principles of all estimating are identical, whether it is for new construction or for repairing damaged buildings and structures. There are five essential factors to be considered: (1) Specifications, (2) Material, (3) Labor, (4) Overhead, (5) Profit.

Specifications

Before an estimate can be properly prepared, it is necessary to know precisely what work is to be performed and what kinds of materials are to be used. As previously discussed, plans are drawn and specifications are written up for new construction projects. When estimating such jobs, identical information is made available to each estimator and the only requirements are a careful scaling of the plans and a thorough reading and understanding of the written specifications. The estimator cannot change the specifications, nor alter the measurements and dimensions on the plans.

When estimating losses, the specifications are determined at the premises by careful inspection of the property. Each part of the building that is affected by the loss is visited and all items of repair are recorded. Measurements of the building, the rooms, story heights, and any other dimensions needed for computing quantities are taken at the time of inspection. This material, when set up in orderly fashion in the finished estimate, becomes the *specifications* of repair.

When more than one person is employed to estimate the cost of restoring a damaged building, the ideal procedure is to have them make a joint inspection and survey. They can agree, at least, on the general scope of the work to be done. Unfortunately the common practice is to have them make up their specifications independent of one another. The result, in most cases, is a disagreement in opinions of the extent of repairs necessary to effect restoration. In other words, the specifications on which the cost of repairs is based are in conflict. One estimate may contemplate more plastering or painting than the other; one may provide that the floors are to be sanded; another has figured on replacing the floors; one may have included plastering of all walls while the other has included only part of them. Such differences are ultimately reconciled by reinspection and discussion, but it is far better to resolve them before any of the items in the estimate are priced. The procedure for inspecting damaged property and taking-off specifications is discussed in Chapter 3.

Estimating Materials

There are two important categories in estimating materials. One is the proper *quantity* to do the work, and the other is the proper *size and quality*.

Sometimes the quantity of a particular material is fixed by existing measurements, as when the entire floor of a room must be replaced, or an entire roof is to be recovered. There are many instances, however, when the quantity of material required is a matter of judging how much area is involved and how badly it is damaged. Where a plaster wall has suffered some water damage as a result of fire on the floor above there may be a question as to how many square yards, if any, will have to be removed and replastered. Once the extent of damage has been decided upon, the quantities of material are obtained from measurements taken in the field.

Except in cases of special agreement with the insurer and insured, the kind, size, and quality of material listed in the estimate should be the same as that which existed at the time of the loss. If the roof shingles were 235-lb. asphalt strip shingles, the estimate should contemplate replacing that type, and not a 165-lb. or 265-lb. shingle.

Describing the material as one inch by three inch pine flooring would not be sufficient. There is edge-grain (comb-grain) pine flooring, and also flat grain. There are different grades of each, although they are difficult to determine when it is charred, dirty, painted or stained. Red oak flooring is cheaper than white oak, and quartered oak is more expensive than the latter, so that any description of oak flooring should indicate whether it is red, white or quarter-sawed and, whenever possible, the grade should be noted.

The following are common examples of why proper identification of materials is essential:

(a) Douglas fir is more expensive than hemlock.
(b) Calcimine is cheaper than latex or oil base paint.
(c) Standard gauge linoleum is cheaper than heavy gauge.
(d) A 235-lb. asphalt shingle costs more than a 165-lb. asphalt shingle.
(e) A 3-ply built-up roof is cheaper than a 5-ply.
(f) Cedar shingle shakes cost more than sawed shingles.
(g) Double-strength glass costs more than single.
(h) 2-coat plaster is cheaper than 3-coat.
(i) Enamel plumbing fixtures are cheaper than porcelain.

The size or dimensions that are necessary to identify a particular material should be measured and noted. If a door is being described, it should include the number of panels (or lights of glass in a sash door), the kind of wood, the width, the height, and the thickness, otherwise it is not possible to obtain the proper price of the door from the local mill. The same rule applies to windows, cabinets, trim, and so forth. The type of hardware should be noted, as there is a substantial difference in price in the various materials, styles and grades.

Any dimension or special characteristic of the material which will affect the cost should be included in the description.

When there is a doubt as to the kind of material, and circumstances will permit, a sample should be removed and taken to a supplier for pricing. A loose shingle, a piece of flooring, a strip of wallpaper, or a piece of wood trim generally can be identified by a dealer as to its quality and cost.

The pricing of all material should be done at local sources, using lumber yards, mills, hardware stores, paint and wallpaper stores, or masonry supply houses. Where discounts are given to the trade either as a "trade" discount or a "cash" discount, they should be deducted from the cost. All customary charges for freight, cartage, or loading should be added to the cost. On large losses where materials can be ordered and delivered in carload or barge lots, the price of material should be estimated accordingly. In other words the price obtained should be the actual amount that a contractor would have to pay to have the material delivered to the job site, including all taxes.

Estimating Labor

Estimating labor consists of approximating the length of time that it will take a man, or a crew of men, to perform a specific job or operation. The cost of the labor is obtained by multiplying the number of man-hours by the hourly wage scale.

There are many factors which influence every estimator's judgment when he is considering the hours of labor required to install or apply materials or to perform specific operations. Most of these may be found within the following classifications which will serve as a check list, if the person making the estimate will ask himself these questions:

1. Are there any special or unusual conditions which the workmen will encounter?
2. Is the physical area being worked in confined, or is it open so that the workmen can move about freely?
3. Can labor-saving machines or equipment be used?
4. What experience have the men with the particular type of work they are to do?
5. Is the quality of workmanship required high or average?
6. Will the work have to be done from ladders or scaffolds?
7. Are the lighting conditions adequate?
8. Will the supervision of the men be close or loose?
9. What are the probably weather conditions under which the men will be working?

10. What is the productivity of the individual workman?
11. Are the wages that are being paid standard or substandard?
12. Is smoking permitted on the job?
13. Are there any special union regulations limiting production?
14. Is the job near to or far from a source of labor?
15. Is the supply of labor good or not good?
16. Is it necessary to protect the contents or occupants of the premises while working?

1. *Unusual Conditions.* Special working conditions of a particular job require careful analysis and visualization by the person making the estimate. The estimate may specify the painting and decorating of a room of a given size. This could be a room which was vacant and unoccupied. The work needed to prepare the wall surface might be trifling or nothing, so that all a painter had to do would be to walk in and start painting.

A second situation might involve the living room of a palatial home which is furnished and occupied. The painter would be called upon to move furniture, protect the wall-to-wall carpeting, remove and rehang pictures, sandpaper woodwork, remove and reset a radiator, or even wash the walls and woodwork prior to decorating.

A third supposition might be that this room was an office in which two or three persons worked who could not be moved out. The painter may have to work around them and also be very careful not to spatter them or the contents with paint.

Each situation presents circumstances that are peculiar to the particular job. Actually none are serious *per se,* but additional time must be given the painters where their rate of work performed is reduced by physical conditions. If it is known that a painter will average a given number of square feet an hour under conditions similar to those outlined in the first situation, then it is possible to approximate the additional time that must be allowed to meet the conditions described in the other two situations.

2. *Physical Area.* The physical area in which a person is working will affect his efficiency. In confined and cramped quarters such as closets or blind-attics, a mechanic's movements are hampered and he consequently slows down. In wide open areas, without restriction, efficiency reaches its highest point, everything else being equal.

3. *Labor-Saving Equipment.* The type of work tools and equipment that are used have an appreciable effect on the rate of speed with which work is done. The use of portable electric saws and drills, electric or pneumatic hammers, aluminum ladders, scaffold brackets, paint spray guns, electric wall paper removers, and other modern devices greatly improves working conditions and reduces cost through higher rates of production.

4. *Experience of Men.* The experience and familiarity of the workman with the job he is assigned to affects his rate of production. A roofer who puts on

asphalt shingles every day of the week, or a floorlayer who lays hardwood floors as a specialty, will turn out nearly twice as much work per hour as a carpenter who performs these tasks only occasionally. Experienced paperhangers will put on more rolls of paper per hour than a painter who has hung paper only a few times. A carpenter who has never laid out or built a winding staircase will take hours longer than a stair builder who has built many.

Workmen unfamiliar with fire loss repairs, will perform slower than men who are experienced in such lines.

There is always the new problem arising in loss repairs which confronts the most experienced journeyman such as repairing non-stock millwork, an ornamental church door, ornamental plaster, patching parquet flooring, matching paint colors inside or out, and repairing brick or stonework. Situations like these slow the workman down as he figures out the problem and performs the operation.

5. *Quality of Workmanship.* The quality of workmanship required directly affects the time required to do the work. The higher the quality, the more time will be needed, because it takes more time to do a good job. High grade painting on trim, where successive coats are hand rubbed takes more hours than 2-coat work in less expensive residences. Cabinet work, or paneling of the highest grade, requires many more hours per unit or square foot than inexpensive cabinetmaking or paneling.

6. *Work Done From Ladders.* Operating from a scaffold, or from a ladder, slows a man down compared to his rate for similar work done on the ground. He not only works with regard for his personal safety when on a scaffold or ladder, but he is also hampered by lack of freedom of movement.

7. *Lighting Conditions.* The lighting conditions in the area where work is being done will affect performance, for a man naturally works faster when he can clearly see at all times exactly what he is doing. When it is apparent that lighting conditions will be poor, or where portable extension lights have to be shifted about as the work is done, some consideration must be given to a reasonable increase in the time normally allowed.

8. *Supervision.* The type of individual who organizes and lays out the work, and who constructively supervises operations will have a decided effect on the cooperation and efficiency of the workmen. The more experienced the supervisor, and the more capable he is in handling people, the lower will be his unit labor costs, compared with those of a loosely supervised job or one which is given little or no supervision.

9. *Weather Conditions.* Weather conditions will influence the length of time it takes to perform work operations. Where repairs are to be made to the exterior of a building, the workmen will do more in pleasant weather than in cold winter months when they keep outside fires going to warm themselves, or go inside occasionally to escape the cold. They work with lighter clothing and can move about more freely. In winter months it is often necessary to provide temporary protection for materials, and allow time for shoveling away snow, especially in northern climates.

Conversely, in extremely hot humid seasons or climates, particularly some of the far southern sections, workmen slow down, especially when working directly in the sun and heat.

10. *Productivity of Workmen.* The individual productivity of workmen varies. Some mechanics do excellent work but are slow; others do mediocre work but turn out large quantities. Many persons develop a fixed rate at which they work, others are naturally lazy, talkative, or clock watchers. Construction workmen are no different from persons engaged in factories, offices, or mercantile establishments. They possess all of the human perfections and imperfections. A contractor who knows the individual characteristics of his men and estimates accordingly has an advantage over his competitor who is unfamiliar with the productiveness of the men who will do the job.

11. *Wages Paid.* The old adage that "a laborer is worthy of his hire" has a certain amount of truth to it. Where a man is aware that he is being paid less than others who are doing similar work, or less than the prevailing wage, he tends to turn out the irreducible minimum of work.

12. *Is Smoking Permitted?* Where smoking is prohibited in such occupancies as hazardous manufacturing plants or department stores, workmen who are making repairs will take time out to smoke outside of the prohibited area. Such time breaks, particularly on small jobs, can increase labor costs.

13. *Union Regulations.* Local union regulations, which limit or restrict the amount of work a man may do during a work day, naturally govern the basis for estimating labor costs. Such restrictions may stipulate the number of bricks a bricklayer can lay, or may prohibit a plasterer from taking on a second job at another location, if he has completed one small job within one work day. Other limitations may require the presence of a foreman, even if only one workman is needed, or may pertain to the use of labor-saving equipment such as paint sprayers or rollers. Where such customs are known to exist, the person making the estimate should become thoroughly acquainted with the regulations, in order to give proper consideration when preparing his figures.

14. *Source of Labor.* Where a job is located many miles from the source of labor, the estimate may have to include a provision for transportation. Workmen are not expected, in certain jurisdictions, to travel great distances to and from work on their own time or at their own expense.

15. *Supply of Labor.* Labor is a commodity and follows the laws of supply and demand. Where the supply is ample or greater than the demand, the wage rate may be lower and the production is frequently higher. Where the supply is less than the demand, the opposite is often the case.

16. *Necessary to Protect the Contents or Occupants.* When it is necessary for workmen to protect the contents or the occupants of a building, the labor costs increase depending upon the degree of care required. Painting or plastering, for example, in an office building or a retail store during working hours reduces considerably the rate at which a man works, for he must be cautious not to damage

the contents, and usually works around the occupants who are intent on performing their own regular jobs.

Labor Rates

The larger contractors, and probably the more successful smaller ones, keep cost records or time studies for each job they complete in order to determine for future reference how long it takes a man to do a particular operation.

If a roofer keeps a record of the length of time it takes to apply strip asphalt shingles to a particular roof having 20 squares, and he finds that it takes 40 man-hours, he knows that it will require an average of 2 hours per square.

$$\frac{40 \text{ man-hours}}{20 \text{ squares}} = 2 \text{ hours per square}$$

Should he encounter a similar roof with corresponding working conditions, he could be reasonably certain of his labor by multiplying the number of squares by 2 hours. If he measured the roof and found that it required 15 squares, the hours of labor required would be 2 x 15 = 30 hours. In other words, for all roofs of similar shape, pitch, and so forth, under similar working conditions, he may estimate his labor by using a unit of 2 hours per square.

In the same manner, a plumber knows that under normal conditions a man can rough-in a bathroom, or set plumbing fixtures in a certain number of hours. A bricklayer's records will note that a man can lay a particular type of brick in a particular type wall at the rate of so many hours per thousand. A contractor will know from his own experience that on new construction work the carpenters can frame a moderate price-range ranch house at the average rate of a certain number of hours per thousand board feet. A plastering contractor knows from past experience that his men can apply 3-coat work at so many hours per 100 square yards. Likewise, a painter knows that he can apply paint at the rate of so many square feet per hour under known circumstances. Contractors and subcontractors for the different building trades know or can determine by consulting time studies, approximately what the *average rate* of work will be under given conditions. When the physical conditions are altered, the rate of production is increased or decreased accordingly.

As an example, a painter may know from experience that his men can apply flat paint to the interior, smooth-plastered walls, ceilings and woodwork of a new building at the rate of about 150 square feet an hour. He estimates jobs on this basis and generally makes a fair profit. If he is called upon to estimate the cost of redecorating the interior of a dwelling where fire has discolored the painted walls and ceilings, he generally will have to give consideration to a number of conditions not contemplated on jobs where his men can apply the paint at the rate of 150 square feet per hour. Some of these considerations might be developed from the following questions:

1. Is it necessary to wash down the walls and ceilings before starting to paint?
2. Must blistered paint on the woodwork be removed and sanded?
3. Are there plaster cracks or nail holes to be filed?
4. Will the painters have to remove and replace furniture and pictures?
5. Can drop cloths be used to protect the contents instead of moving them?
6. Will the radiators have to be disconnected to paint behind them? Will radiators have to be painted?
7. Is there an unusual number of windows, and are the individual lights large or small?
8. Is the surface to be painted smooth or rough like Craftex?
9. How many closets are there and what are their sizes?
10. What type of paint will be used? Coldwater, flat paint, or enamel?
11. Can a roller be used in place of brushes?
12. How many coats and colors?
13. Are there built-in cabinets?
14. What quality of workmanship is required?
15. How badly is the room cut up?
16. Is it necessary to disconnect electric fixtures, or remove hardware?

While none of these conditions is of great importance when related to the over-all job, nevertheless several of them might add up to create a sizeable difference in cost. The estimator must properly adjust his estimate of labor to allow for the additional time that would be necessary to do the extra work encountered.

This can be done in one of two ways. He can decrease the rate of applying the paint from 150 to 125 or 100 square feet per hour, depending on his best judgment. The other way would be to maintain the normal rate of application of paint at 150 square feet per hour but to allow additional hours for the extra work anticipated in each room.

The latter method in this instance would be preferable because it is more accurate. The painter might work his labor estimate up in this manner.

Living Room 14' x 16' x 9'

Wall area 2(14' x 16')	=	60	
	x	9	
		540	sq ft
Ceiling area 14' x 16'	=	224	sq ft
Total area	=	764	sq ft
Labor painting one coat	$\dfrac{764}{150}$	= 5.1	hours

Add time to remove, paint, and
  reset two radiators .... 1.0 hours
Add time to wash down walls .... 3.0 hours
Add time to patch plaster cracks .... .5 hours
  Total hours of labor .... 9.6 hours

Conditions under which the rate of work per hour is increased or decreased must be such that the rate at which a man works is *directly* affected. In the illustrative example, had the painter found the damaged walls were rough-plastered, he would then justifiably reduce the rate at which the paint could be applied, because it takes more time to paint rough-plastered walls than it does to paint smooth-plastered walls.

Similarly, a roofer would increase or decrease the rate at which shingles could be applied depending on the type of shingle and on the roof pitch, or on how cut-up the roof is in regard to dormers, and so forth. If the roofer, however, found it necessary to erect a scaffold, this should be treated as a separate item rather than attempting to alter the rate of laying shingles to include the added cost of scaffolding.

The conditions under which the rate of applying materials is to be increased or decreased must directly affect the rate of production, otherwise labor to perform additional operations should be treated separately.

Sources For Judging Labor Necessary. There are three sources for an estimator to judge the hours of labor that will be required to perform an operation or to install material.

1. His own personal experience through having done the work.
2. Observation of others who are doing the work.
3. Time studies made by others.

Obviously, very few people have performed all the operations of constructing a building. Many estimators have had personal experience in various trades and they can rely to some extent on that background.

Contractors, superintendents, foremen and time-study engineers develop a knowledge of labor requirements, having been associated with and observed men at work on many or most phases of construction.

A third source for estimating labor is through the records of others. Numerous books have been written by contractors, builders, teachers, and architects on the subject of estimating construction costs. One or more of them will be found on the bookshelves of every large and successful builder. They provide excellent references. No one looks upon them as the final word or as an indisputable authority, but since they provide the experience of the many, and frequently state the physical conditions under which the work was performed, they do provide an excellent basis for estimating labor. (See Bibliography.)

In each of the subsequent chapters, in which estimating of particular trades is discussed, the subject of labor is treated as it relates to each specific trade. Tables are given for estimating average hourly production rates for individual operations. *They are, in general, adequate for normal working conditions and include the time required to start, finish, and clean up.* Factors which affect the production rate of workmen, as outlined in this chapter, should be given careful consideration before increasing or decreasing the hours shown in the tables.

The prevailing local hourly wage rates should always be used in determining labor costs. The wages that are paid within a given building trade vary in different parts of the country. Union wage scales may be obtained from local sources or from the U.S. Department of Labor, Bureau of Labor Statistics. Table 2-1 lists the hourly wage scales for major building trades in the principal cities as published by the U.S. Department of Labor. The basic scale is shown, and also employer contribution to insurance and pension funds, which must be added. In some areas, additional contributions are made by employers under programs for holiday, vacation and unemployment funds. It will be noted that the basic scale for carpenters, for example, ranges from $6.10 per hour in Charleston, South Carolina, to $9.86 per hour in New York City.

Overhead

Contractors usually consider two types of overhead to be included in their estimate.

1. Specific overhead, chargeable to each job for which an estimate is being prepared.

2. General overhead, not chargeable to each specific job.

Table 2-1

PRELIMINARY

Table 3. Union hourly wage rates and employer insurance, pension, vacation, and other fund payments for selected building trades in 105 cities, April 1, 1975

City	Bricklayers					Carpenters					Electricians				
	Basic rate[1]	Insurance[2]	Pension	Vacation pay	Other[3]	Basic rate[1]	Insurance[2]	Pension	Vacation pay	Other[3]	Basic rate[1]	Insurance[2]	Pension	Vacation pay	Other[3]
Albuquerque, N. Mex	*$7.560	*48¢	40¢	80¢	7¢	$7.400	65¢	50¢	†46¢	4¢	$7.812	30¢	1%+70¢	-	*7%
Atlanta, Ga	8.400	40¢	40¢	-	1¢	8.130	40¢	45¢	-	-	9.550	6%	8%	-	-
Baltimore, Md	9.000	40¢	50¢	-	10¢	8.100	60¢	54¢	-	2¢	8.800	50¢	1%+20¢	-	-
Birmingham, Ala	8.250	25¢	25¢	-	3¢	7.300	40¢	20¢	-	3¢	8.500	30¢	1%	-	-
Boise City, Idaho	*8.150	40¢	30¢	-	-	7.620	42¢	30¢	40¢	8¢	7.971	40¢	1%+25¢	-	*6%
Boston, Mass	9.050	70¢	90¢	-	9¢	9.450	60¢	50¢	-	-	10.100	50¢	*1%+$1.35	-	-
Buffalo, N.Y	9.790	-	$1.000	-	61¢	8.780	$1.250	$1.100	-	5¢	10.790	45¢	1%+70¢	-	-
Burlington, Vt	8.150	50¢	20¢	-	†1¢	*7.100	*45¢	30¢	-	-	8.100	30¢	1%	-	-
Butte, Mont	8.750	-	-	-	-	6.610	40¢	55¢	75¢	-	†7.800	35¢	†1%+25¢	-	-
Charleston, S.C	5.700	-	-	-	-	6.100	-	-	-	-	7.000	30¢	1%	-	-
Charleston, W. Va	8.500	50¢	75¢	-	1¢	9.250	25¢	25¢	-	-	8.800	30¢	1%+7¢	27¢	-
Charlotte, N.C	6.700	25¢	-	-	-	6.200	35¢	20¢	-	-	6.550	25¢	1%	-	-
Chattanooga, Tenn	8.300	30¢	30¢	-	2¢	7.330	35¢	30¢	-	2¢	*8.450	*45¢	*1%+20¢	-	-
Cheyenne, Wyo	8.600	-	-	-	-	†7.500	35¢	30¢	*20¢	†15¢	8.530	42¢	1%	-	-
Chicago, Ill	9.700	50¢	60¢	25¢	5¢	9.650	55¢	63¢	-	-	10.250	56½¢	61¢	57¢	15½¢
Cincinnati, Ohio	9.695	45¢	35¢	45¢	7½¢	9.750	45¢	55¢	-	2½¢	9.200	40¢	1%+30¢	8%	20¢
Cleveland, Ohio	9.000	75¢	65¢	$1.100	7¢	9.500	47¢	60¢	$1.000	3¢	9.760	50¢	1%+37¢	75¢	-
Columbia, S.C	(6)	(6)	(6)	(6)	(6)	5.650	-	-	-	-	6.200	40¢	1%	15¢	-
Columbus, Ohio	9.200	30¢	45¢	-	-	8.600	35¢	20¢	-	3¢	9.330	28¢	1%+60¢	-	16¢
Dallas, Tex	8.240	35¢	50¢	-	4¢	7.810	30¢	30¢	-	-	8.189	4%	1%	*4%	-
Dayton, Ohio	8.630	45¢	50¢	37¢	9¢	8.240	45¢	90¢	-	4¢	9.880	45¢	1%+75¢	-	-
Denver, Colo	8.650	45¢	60¢	25¢	25¢	8.135	48¢	60¢	40¢	-	*10.240	65¢	1%+25¢	-	-
Des Moines, Iowa	8.915	35¢	30¢	-	-	8.100	30¢	25¢	-	-	9.680	35¢	1%	-	-
Detroit, Mich	8.470	70¢	8%	11%	14¢	8.410	60¢	8%	10%	3¢	9.100	85¢	1%+80¢	9%	50¢
Duluth, Minn	[7]8.570	30¢	25¢	50¢	[7]5¢	8.350	30¢	-	30¢	-	8.890	4%	1%	11%	-
El Paso, Tex	6.130	43¢	20¢	-	3¢	6.150	30¢	-	-	-	7.450	25¢	1%	-	-
Erie, Pa	10.210	30¢	-	-	3¢	8.800	30¢	30¢	-	3¢	10.200	3%	3%	-	-

City	Painters					Plasterers				
	Basic rate[1]	Insurance[2]	Pension	Vacation pay	Other[3]	Basic rate[1]	Insurance[2]	Pension	Vacation pay	Other[3]
Albuquerque, N. Mex	*$6.870	30¢	20¢	-	2¢	*$7.100	43¢	30¢	*50¢	8¢
Atlanta, Ga	7.700	45¢	55¢	-	-	7.770	35¢	55¢	-	5¢
Baltimore, Md	7.445	85¢	45¢	-	-	8.450	45¢	30¢	-	-
Birmingham, Ala	*7.450	*40¢	-	-	-	*7.270	50¢	-	-	3¢
Boise City, Idaho	7.260	25¢	25¢	-	-	7.500	32¢	32¢	20¢	-
Boston, Mass	8.760	62¢	85¢	-	5¢	8.550	45¢	-	-	-
Buffalo, N.Y	8.655	52½¢	30¢	-	61¢	11.060	-	-	-	$1.300
Burlington, Vt	5.750	-	-	-	-	8.150	50¢	20¢	-	†1¢
Butte, Mont	6.420	25¢	10¢	-	-	7.450	35¢	25¢	-	-
Charleston, S.C	5.000	25¢	20¢	-	-	5.300	-	-	-	-
Charleston, W. Va	7.220	-	-	-	-	8.650	-	-	-	-
Charlotte, N.C	5.000	25¢	20¢	-	-	*5.650	25¢	-	-	-
Chattanooga, Tenn	5.500	-	25¢	-	-	7.450	-	-	-	-
Cheyenne, Wyo	8.510	50¢	45¢	-	-	8.690	-	-	-	18¢
Chicago, Ill	*8.100	42½¢	25¢	-	20¢	9.295	50¢	31½¢	-	12½¢
Cincinnati, Ohio	8.730	-	15¢	35¢	-	9.195	-	45¢	$1.000	22½¢
Cleveland, Ohio	9.120	48¢	39¢	50¢	3¢	9.480	-	-	$2.000	10¢
Columbia, S.C	5.000	25¢	20¢	-	-	(6)	(6)	(6)	-	(6)
Columbus, Ohio	8.600	30¢	15¢	-	1¢	8.300	45¢	20¢	-	2¢
Dallas, Tex	7.425	35¢	3¢	-	-	7.885	35¢	-	50¢	5¢
Dayton, Ohio	†8.930	40¢	40¢	50¢	-	7.750	-	-	$1.250	4¢
Denver, Colo	8.510	50¢	45¢	-	-	8.690	-	-	-	18¢
Des Moines, Iowa	8.140	-	-	25¢	-	8.335	-	-	-	-
Detroit, Mich	8.800	65¢	83¢	80¢	-	9.740	60¢	50¢	50¢	7¢
Duluth, Minn	8.260	30¢	20¢	-	-	8.350	30¢	-	45¢	-
El Paso, Tex	5.510	24¢	-	-	-	6.410	28¢	-	-	4¢
Erie, Pa	7.500	30¢	40¢	-	3¢	9.130	30¢	25¢	-	3¢

City	Plumbers					Building laborers				
	Basic rate[1]	Insurance[2]	Pension	Vacation pay	Other[3]	Basic rate[1]	Insurance[2]	Pension	Vacation pay	Other[3]
Albuquerque, N. Mex	$8.340	42¢	75¢	50¢	6¢	*$5.320	*34¢	25¢	-	4¢
Atlanta, Ga	8.950	35¢	50¢	-	6¢	5.300	20¢	†25¢	-	5¢
Baltimore, Md	8.530	50¢	45¢	-	5¢	5.750	27½¢	30¢	-	2¢
Birmingham, Ala	*9.400	*40¢	*40¢	-	4¢	5.150	15¢	30¢	-	3¢
Boise City, Idaho	8.090	37¢	40¢	70¢	6¢	5.970	45¢	45¢	20¢	10¢
Boston, Mass	*10.600	70¢	60¢	-	5¢	7.000	50¢	45¢	-	15¢
Buffalo, N.Y	10.440	53¢	52¢	-	11¢	7.955	65¢	$1.200	-	10¢
Burlington, Vt	7.650	40¢	40¢	-	-	5.600	50¢	45¢	-	†5¢
Butte, Mont	8.450	40¢	50¢	-	-	5.470	42½¢	31½¢	75¢	-
Charleston, S.C	7.500	30¢	20¢	-	-	(6)	(6)	(6)	(6)	
Charleston, W. Va	9.270	35¢	30¢	-	-	6.000	25¢	25¢	-	-
Charlotte, N.C	6.500	30¢	30¢	-	6¢	3.850	15¢	10¢	-	-
Chattanooga, Tenn	8.200	25¢	30¢	20¢	7¢	5.050	10¢	20¢	-	2¢
Cheyenne, Wyo	6.920	50¢	35¢	$1.000	-	5.200	33¢	10¢	-	5¢
Chicago, Ill	9.820	55¢	55¢	-	55¢	7.200	57¢	85¢	-	1¢
Cincinnati, Ohio	9.920	40¢	90¢	50¢	33¢	8.450	50¢	30¢	-	-
Cleveland, Ohio	9.180	60¢	70¢	$1.000	8¢	7.220	72¢	65¢	$1.000	30¢
Columbia, S.C	7.040	28¢	30¢	-	-	(6)	(6)	(6)	(6)	
Columbus, Ohio	9.570	40¢	80¢	85¢	22¢	6.450	50¢	24¢	-	4¢
Dallas, Tex	7.750	38¢	90¢	25¢	5¢	5.270	27½¢	30¢	-	4¢
Dayton, Ohio	9.610	35¢	70¢	50¢	4¢	6.790	50¢	-	-	4¢
Denver, Colo	8.600	55¢	65¢	50¢	10¢	5.000	55¢	45¢	-	12¢
Des Moines, Iowa	8.650	30¢	50¢	40¢	10¢	7.000	41¢	37½¢	-	-
Detroit, Mich	9.170	70¢	88½¢	$1.340	47¢	6.920	65¢	55¢	60¢	-
Duluth, Minn	7.860	30¢	45¢	$1.000	10¢	7.050	30¢	15¢	25¢	-
El Paso, Tex	6.650	31¢	30¢	-	-	3.840	30¢	-	-	-
Erie, Pa	9.720	30¢	50¢	-	2¢	7.300	40¢	50¢	-	3¢

See footnotes at end of table

U.S. DEPARTMENT OF LABOR
Bureau of Labor Statistics

Union Hourly Wage Scales for Major Building Trades in Principal Cities in The United States.

Table 2-1 (cont.)

Table 3. Union hourly wage rates and employer insurance, pension, vacation, and other fund payments for selected building trades in 105 cities, April 1, 1975—Continued

City	Bricklayers: Basic rate [1]	Insurance [2]	Pension	Vacation pay	Other [3]	Carpenters: Basic rate [1]	Insurance [2]	Pension	Vacation pay	Other [3]	Electricians: Basic rate [1]	Insurance [2]	Pension	Vacation pay	Other [3]
Evansville, Ind	*$8.830	*50¢	30¢	50¢	1¢	*$8.850	40¢	35¢	-	3¢	$9.600	30¢	1%	40¢	-
Fargo, N. Dak	8.250	40¢	15¢	30¢	5¢	7.010	-	30¢	[4]20¢	-	8.050	30¢	1%	4%	2%
Fresno, Calif	8.070	48¢	85¢	[4]$1.000	15¢	10.000	*72¢	*$1.150	[9]75¢	2¢	9.860	50¢	1%+70¢	[4]50¢	5¢
Grand Rapids, Mich	7.800	35¢	25¢	40¢	5¢	7.550	50¢	40¢	30¢	2¢	8.270	40¢	1%+15¢	25¢	-
Hartford, Conn	9.600	50¢	25¢	-	-	9.210	50¢	30¢	-	-	9.950	75¢	1%+45¢	-	-
Honolulu, Hawaii	7.450	65¢	75¢	-	7¢	7.430	49¢	85¢	[9]42¢	3¢	8.250	55¢	[10]1%+$1.00	5 2/5%	16 3/5%
Houston, Tex	7.840	32 1/2¢	30¢	-	-	7.770	45¢	30¢	-	-	8.304	30¢	2%	[4]4%	-
Indianapolis, Ind	8.890	30¢	20¢	50¢	7¢	8.950	47¢	50¢	50¢	2¢	8.883	3%	5%	5%	-
Jackson, Miss	6.920	-	-	-	2¢	6.350	30¢	20¢	-	3¢	7.350	-	1%+50¢	-	-
Jacksonville, Fla	7.660	30¢	40¢	-	3¢	7.300	39¢	30¢	[4]30¢	[4]14¢	8.500	30¢	1%+43¢	43¢	-
Kansas City, Mo	*9.225	35¢	35¢	*$1.000	-	*9.220	33¢	30¢	25¢	-	*9.570	23¢	1%+2¢	60¢	2/5%+20¢
Knoxville, Tenn	8.410	-	-	-	2¢	7.090	-	30¢	-	2¢	7.850	30¢	1%	-	-
Lansing, Mich	7.500	-	60¢	$2.000	10¢	9.110	50¢	40¢	65¢	2¢	9.180	40¢	1%+25¢	10%	-
Las Vegas, Nev	8.720	35¢	30¢	$1.000	8¢	8.560	45¢	60¢	80¢	2¢	11.150	63¢	1 1/5%	-	-
Little Rock, Ark	7.550	30¢	25¢	-	1¢	*7.000	35¢	25¢	-	-	7.670	30¢	1%	5%	-
Los Angeles, Calif	8.830	70¢	*90¢	[4]50¢	20¢	*8.530	*87¢	*$1.300	[9]*75¢	2¢	9.780	71¢	†1%+$1.25	[9]45¢	-
Louisville, Ky	8.880	40¢	40¢	-	5¢	8.750	30¢	30¢	-	2¢	*10.380	29¢	1%+40¢	-	-
Lubbock, Tex	7.900	-	20¢	-	-	6.850	30¢	30¢	-	-	7.150	30¢	1%	6%	-
Madison, Wis	8.200	25¢	40¢	25¢	-	8.100	25¢	15¢	25¢	-	8.880	34¢	1%	7%	-
Manchester, N.H.	7.650	45¢	40¢	-	3¢	6.890	35¢	30¢	-	1¢	9.300	25¢	1%	-	-
Memphis, Tenn	9.000	35¢	20¢	25¢	-	8.050	30¢	20¢	-	-	*8.680	35¢	1%+40¢	-	-
Miami, Fla	9.700	50¢	39¢	-	1¢	8.700	55¢	45¢	-	5¢	*10.250	4 1/2%	1%+20¢	8 1/2%	2%
Milwaukee, Wis	9.030	70¢	60¢	56¢	5¢	8.500	60¢	60¢	51¢	4¢	9.430	51¢	1%+20¢	7%	1%
Minneapolis, Minn	8.560	45 1/2¢	23¢	56¢	5¢	8.210	40¢	30¢	50¢	-	8.550	6 1/2%	4%	9%	1 1/2%
Mobile, Ala	9.040	-	90¢	-	5¢	8.360	30¢	35¢	-	5¢	*9.450	30¢	1%+25¢	12 1/2¢	-
Montgomery, Ala	†6.900	-	-	-	-	6.500	-	-	-	-	7.425	30¢	1%	-	-
Nashville, Tenn	8.000	-	-	-	-	7.650	15¢	10¢	-	2¢	7.620	35¢	1%	-	-
Newark, N.J	10.800	70¢	60¢	-	3¢	9.730	6%	6%	-	-	10.350	5%	7%	10%	([11])

City	Painters: Basic rate	Insurance	Pension	Vacation pay	Other	Plasterers: Basic rate	Insurance	Pension	Vacation pay	Other
Evansville, Ind	*$7.750	*55¢	30¢	-	-	*$9.880	-	-	50¢	-
Fargo, N. Dak	7.250	-	-	-	-	8.300	-	-	-	-
Fresno, Calif	8.820	40¢	20¢	*45¢	7¢	7.240	56¢	60¢	75¢	15¢
Grand Rapids, Mich	6.700	30¢	20¢	25¢	-	7.190	32¢	18¢	$1.000	5¢
Hartford, Conn	8.850	50¢	30¢	-	-	9.600	50¢	25¢	-	-
Honolulu, Hawaii	7.230	42¢	50¢	[9]5%	6¢	8.240	65¢	[10]$1.050	-	42¢
Houston, Tex	6.285	27 1/2¢	30¢	30¢	-	7.525	32¢	30¢	52 1/2¢	5¢
Indianapolis, Ind	7.700	20¢	20¢	40¢	-	8.600	40¢	-	$1.000	7¢
Jackson, Miss	5.950	[4]15¢	-	-	-	*6.500	25¢	*30¢	-	5¢
Jacksonville, Fla	6.900	30¢	35¢	-	-	7.410	25¢	30¢	-	5¢
Kansas City, Mo	8.740	*55¢	*70¢	-	-	*10.250	-	-	-	10¢
Knoxville, Tenn	6.750	-	30¢	-	-	7.750	-	-	-	-
Lansing, Mich	*8.715	42¢	40¢	50¢	†1¢	7.820	40¢	25¢	$1.010	6¢
Las Vegas, Nev	8.950	47¢	25¢	$1.000	-	8.980	50¢	-	90¢	15¢
Little Rock, Ark	6.350	*30¢	30¢	-	-	7.000	-	-	-	10¢
Los Angeles, Calif	9.220	40 1/2¢	40¢	40¢	20¢	8.905	63¢	$1.750	[4]55¢	15¢
Louisville, Ky	7.590	25¢	10¢	-	3¢	8.650	-	-	-	-
Lubbock, Tex	5.750	-	-	-	-	7.000	-	-	-	4¢
Madison, Wis	7.590	40¢	30¢	30¢	-	8.300	25¢	40¢	-	-
Manchester, N.H	6.425	17 1/2¢	-	-	-	7.650	45¢	40¢	-	3¢
Memphis, Tenn	7.600	30¢	30¢	-	3¢	8.600	-	-	-	28¢
Miami, Fla	*8.750	45¢	40¢	-	2¢	9.700	50¢	39¢	-	1¢
Milwaukee, Wis	8.010	45¢	60¢	50¢	2¢	7.940	75¢	60¢	66¢	3¢
Minneapolis, Minn	8.130	35¢	25¢	40¢	-	8.430	45¢	20¢	60¢	10¢
Mobile, Ala	8.500	30¢	5¢	-	-	8.860	30¢	35¢	-	5¢
Montgomery, Ala	5.550	-	20¢	-	-	5.600	-	-	-	-
Nashville, Tenn	6.500	-	40¢	-	-	7.000	30¢	20¢	-	7¢
Newark, N.J	8.400	50¢	50¢	10¢	-	10.800	70¢	60¢	-	3¢

City	Plumbers: Basic rate	Insurance	Pension	Vacation pay	Other	Building laborers: Basic rate	Insurance	Pension	Vacation pay	Other
Evansville, Ind	$8.850	40¢	60¢	10%	-	*$6.400	35¢	*35¢	-	*12¢
Fargo, N. Dak	8.100	38¢	20¢	[4]40¢	-	4.990	20¢	-	-	-
Fresno, Calif	9.780	$1.150	$1.440	[9]$1.000	10¢	6.535	80¢	$1.400	[9]90¢	10¢
Grand Rapids, Mich	8.670	38¢	35¢	92¢	5¢	5.310	40¢	30¢	55¢	5¢
Hartford, Conn	10.100	62¢	62¢	-	20¢	6.900	45¢	40¢	-	10¢
Honolulu, Hawaii	8.700	72¢	$1.050	[9]10 1/2%	-	5.710	44¢	51¢	18¢	10¢
Houston, Tex	7.310	32¢	47¢	50¢	3 1/2¢	5.600	28¢	30¢	-	-
Indianapolis, Ind	8.650	40¢	70¢	75¢	5¢	*6.400	35¢	*35¢	-	*9¢
Jackson, Miss	8.220	30¢	50¢	-	-	3.950	15¢	15¢	-	2¢
Jacksonville, Fla	*9.250	40¢	45¢	-	5¢	4.500	20¢	-	-	-
Kansas City, Mo	8.640	50¢	60¢	[4]$1.000	-	*7.500	45¢	40¢	50¢	26¢
Knoxville, Tenn	7.650	35¢	45¢	30¢	5¢	4.780	15¢	15¢	-	1¢
Lansing, Mich	9.040	60¢	50¢	60¢	4¢	6.400	40¢	30¢	55¢	1¢
Las Vegas, Nev	9.900	65¢	$1.500	$1.600	15¢	6.230	26¢	75¢	$1.000	-
Little Rock, Ark	7.650	†30¢	75¢	35¢	5¢	*4.870	15¢	20¢	-	-
Los Angeles, Calif	10.240	10%	16%	[9]13%	3/4%	6.620	75¢	$1.500	43¢	10¢
Louisville, Ky	*10.540	36¢	75¢	-	5¢	6.400	25¢	25¢	-	2¢
Lubbock, Tex	7.250	-	40¢	25¢	-	4.025	27 1/2¢	10¢	-	-
Madison, Wis	8.890	30¢	40¢	35¢	†7/10%	7.000	30¢	25¢	25¢	-
Manchester, N.H.	8.600	43¢	35¢	-	2¢	5.810	40¢	40¢	-	5¢
Memphis, Tenn	9.470	40¢	15¢	-	5¢	4.950	20¢	20¢	-	7¢
Miami, Fla	10.090	67¢	$1.050	-	8¢	6.500	60¢	30¢	-	20¢
Milwaukee, Wis	9.210	50¢	60¢	55¢	9¢	7.490	55¢	60¢	51¢	4¢
Minneapolis, Minn	7.730	38¢	45¢	$1.350	2¢	6.800	40¢	35¢	40¢	†-
Mobile, Ala	*10.150	30¢	45¢	-	-	5.420	30¢	35¢	-	5¢
Montgomery, Ala	8.300	25¢	25¢	-	-	3.450	10¢	10¢	-	-
Nashville, Tenn	7.340	40¢	40¢	-	†2¢	4.800	20¢	15¢	-	†-
Newark, N.J.	9.425	50¢	90¢	85¢	17 1/2¢	7.200	40¢	45¢	-	5¢

See footnotes at end of table.

Table 2-1 (cont.)

Table 3. Union hourly wage rates and employer insurance, pension, vacation, and other fund payments for selected building trades in 105 cities, April 1, 1975—Continued

City	Bricklayers Basic rate [1]	Insurance [2]	Pension	Vacation pay	Other [3]	Carpenters Basic rate [1]	Insurance [2]	Pension	Vacation pay	Other [3]	Electricians Basic rate [1]	Insurance [2]	Pension	Vacation pay	Other [3]
New Haven, Conn	$9.350	50¢	50¢	-	-	$9.300	50¢	30¢	-	-	*$10.060	70¢	1%+40¢	-	-
New Orleans, La	8.690	25¢	15¢	-	-	7.950	20¢	20¢	-	-	8.455	30¢	1%+20¢	*5%	-
New York, N.Y	9.640	78¢	$3.380	30¢	5¢	9.860	95¢	$1.630	55¢	5¹/₄	¹²9.500	4⁷/₅%	3¹/₅%+80¢	5⁷/₁₀%	¹³¹/₂%+40¢
Norfolk, Va	7.250	35¢	20¢	-	1¢	7.000	20¢	20¢	-	1¢	*7.500	5%	4%	-	-
Oakland, Calif	9.400	85¢	95¢	*75¢	20¢	10.000	*72¢	*$1.150	*75¢	2¢	9.370	70¢	1%+50¢	*10%	1%
Oklahoma City, Okla	8.370	45¢	35¢	-	5¢	8.000	*30¢	-	-	1¢	*7.820	30¢	1%+20¢	8%	-
Omaha, Nebr	8.375	35¢	30¢	*60¢	-	7.880	35¢	30¢	40¢	-	10.080	48¢	1%+50¢	-	¹/₈¢
Peoria, Ill	8.920	45¢	40¢	35¢	3¢	8.510	35¢	40¢	-	5¢	8.810	30¢	1%+30¢	4%	-
Philadelphia, Pa	9.500	80¢	68¢	-	-	9.270	$1.230	55¢	-	5¢	10.820	5%	5%	-	-
Phoenix, Ariz	9.130	*65¢	*$1.000	*50¢	10¢	8.285	65¢	88¹/₂¢	*25¢	-	9.495	60¢	1%+70¢	*10%	-
Pittsburgh, Pa	9.405	45¢	60¢	-	2¹/₂¢	8.540	5%	7%	-	6%	9.100	35¢	1%+20¢	60¢	2¢
Portland, Maine	*7.750	40¢	*40¢	-	*1¢	*6.850	*45¢	30¢	-	-	7.650	30¢	1%+25¢	-	-
Portland, Oreg	*9.300	45¢	45¢	*30¢	11¢	*7.890	¹⁴55¢	65¢	*25¢	3¢	8.832	45¢	1%+40¢	*8%	-
Providence, R.I	9.150	65¢	55¢	-	-	8.810	60¢	40¢	-	-	8.800	38¢	1%+25¢	-	-
Raleigh, N.C	6.700	25¢	-	-	-	6.200	35¢	20¢	-	-	6.500	25¢	1%	-	-
Reading, Pa	8.600	45¢	50¢	-	-	8.130	52¢	45¢	-	-	9.190	33¢	1%	-	-
Richmond, Va	7.350	35¢	20¢	-	¹1¢	6.600	20¢	20¢	-	-	7.450	5%	1%	-	-
Rochester, N.Y	10.225	30¢	58¢	-	5¢	9.380	55¢	59¢	-	35¹/₂¢	10.450	40¢	1%+75¢	-	†¹/₄%
Rock Island, Ill	8.350	35¢	30¢	50¢	3¢	8.310	35¢	50¢	-	3¢	8.880	32¢	5¹/₂%	-	17¢
Sacramento, Calif	9.000	73¢	72¢	*75¢	18¢	10.000	*72¢	*$1.150	*75¢	4¹/₄¢	9.145	60¢	1%+65¢	*4%	*4%
St. Louis, Mo	8.250	52¢	70¢	65¢	5¢	8.760	40¢	50¢	*50¢	5¢	8.930	5%	6¹/₂%	7%	8¹/₂%
St. Paul, Minn	8.560	45¹/₂¢	23¢	56¢	5¢	8.210	40¢	30¢	50¢	3¢	8.900	6%	2%	7%	1%
Salt Lake City, Utah	8.810	36¢	42¢	-	10¢	7.610	45¢	55¢	25¢	-	9.500	40¢	1%+25¢	-	-
San Antonio, Tex	7.870	35¢	30¢	25¢	5¢	7.120	38¢	30¢	40¢	1¢	8.030	25¢	1%	35¢	15¢
San Diego, Calif	8.500	70¢	90¢	*50¢	24¢	*8.960	56¢	85¢	50¢	5¢	8.973	48¢	1%+78¢	*10%	2¢
San Francisco, Calif	10.020	¹⁴90¢	58¢	*$1.000	20¢	10.000	*72¢	*$1.150	*75¢	†2¢	9.170	70¹/₂¢	1%+35¢	*4%	*8%
Santa Fe, N. Mex	*8.060	*48¢	40¢	80¢	7¢	7.400	65¢	40¢	60¢	4¢	7.812	30¢	1%+70¢	7%	-

City	Painters Basic rate [1]	Insurance [2]	Pension	Vacation pay	Other [3]	Plasterers Basic rate [1]	Insurance [2]	Pension	Vacation pay	Other [3]
New Haven, Conn	$8.200	35¢	40¢	-	-	$9.350	50¢	50¢	-	-
New Orleans, La	6.375	17¹/₂¢	20¢	-	-	*7.450	30¢	20¢	15¢	8¢
New York, N.Y	8.000	7¹/₂%	6¹/₂%+50¢	-	-	8.550	$1.100	65¢	90¢	11¢
Norfolk, Va	6.450	-	-	1¢	-	7.700	-	-	-	1¢
Oakland, Calif	10.020	64¢	65¢	*80¢	-	8.300	62¹/₂¢	93¢	*70¢	17¢
Oklahoma City, Okla	7.200	40¢	35¢	*35¢	-	8.800	-	-	-	-
Omaha, Nebr	7.350	-	25¢	30¢	-	8.350	35¢	30¢	-	15¢
Peoria, Ill	8.050	40¢	30¢	-	-	9.200	35¢	30¢	-	5¢
Philadelphia, Pa	8.325	37¹/₂¢	30¢	30¢	-	9.470	53¢	-	-	-
Phoenix, Ariz	*8.600	*57¢	*38¢	50¢	14¢	8.295	60¢	85¢	*75¢	18¢
Pittsburgh, Pa	8.605	50¢	20¢	-	-	8.490	37¢	70¢	-	12¢
Portland, Maine	4.650	-	-	-	-	7.000	40¢	-	-	-
Portland, Oreg	*8.110	*35¢	35¢	*20¢	16¢	†8.050	50¢	60¢	*45¢	10¢
Providence, R.I	8.150	50¢	50¢	†-	-	8.550	50¢	35¢	-	-
Raleigh, N.C	5.000	25¢	25¢	-	-	5.500	15¢	-	-	-
Reading, Pa	7.440	50¢	31¢	20¢	-	8.900	-	28¢	-	-
Richmond, Va	5.400	-	-	-	-	7.600	-	-	-	1¢
Rochester, N.Y	8.620	35¢	42¢	-	34¢	10.225	30¢	58¢	-	5¢
Rock Island, Ill	7.220	40¢	40¢	-	-	9.180	-	-	-	-
Sacramento, Calif	10.020	64¢	65¢	40¢	40¢	8.090	45¹/₂¢	75¢	*$1.000	11¢
St. Louis, Mo	8.830	30¢	30¢	21¢	-	9.100	32¹/₂¢	50¢	-	5¢
St. Paul, Minn	8.280	45¢	25¢	15¢	-	8.270	50¢	25¢	65¢	7¢
Salt Lake City, Utah	7.470	21¢	18¢	-	3¢	7.960	45¢	55¢	-	-
San Antonio, Tex	6.550	-	20¢	-	-	7.400	-	-	-	5¢
San Diego, Calif	*8.770	74¢	*80¢	*65¢	9¢	8.710	45¢	$1.250	85¢	10¢
San Francisco, Calif	9.770	74¢	80¢	40¢	40¢	8.820	68¢	$1.250	*75¢	*80¢
Santa Fe, N. Mex	*6.870	30¢	20¢	-	2¢	*7.100	43¢	50¢	*50¢	8¢

City	Plumbers Basic rate [1]	Insurance [2]	Pension	Vacation pay	Other [3]	Building laborers Basic rate [1]	Insurance [2]	Pension	Vacation pay	Other [3]
New Haven, Conn	$9.880	55¢	50¢	-	-	$6.900	45¢	40¢	-	5¢
New Orleans, La	8.400	40¢	60¢	*50¢	-	6.010	10¢	10¢	-	-
New York, N.Y	9.500	¹⁶$2.180	-	72¢	$1.440	7.850	$1.200	$1.310	-	2¢
Norfolk, Va	6.510	35¢	-	-	1¢	4.050	10¢	10¢	-	2¢
Oakland, Calif	10.800	78¢	$1.400	⁹$1.000	13¢	6.535	80¢	$1.400	*90¢	10¢
Oklahoma City, Okla	8.000	55¢	40¢	*70¢	8¢	5.500	10¢	15¢	-	1¢
Omaha, Nebr	9.760	35¢	55¢	-	30¢	5.467	35¢	30¢	-	*15¢
Peoria, Ill	8.425	24¢	60¢	50¢	5¢	7.700	30¢	35¢	-	5¢
Philadelphia, Pa	10.420	57¢	92¢	-	-	7.350	55¢	30¢	-	5¢
Phoenix, Ariz	9.390	65¢	$1.240	$1.250	12¢	6.230	60¢	65¢	*25¢	7¢
Pittsburgh, Pa	9.110	60¢	62¹/₂¢	-	14¹/₂¢	6.750	40¢	40¢	-	2¹/₂¢
Portland, Maine	8.200	¹⁴45¢	35¢	-	-	5.000	50¢	45¢	-	*10¢
Portland, Oreg	8.370	70¢	$1.000	$1.000	6¢	6.900	50¢	65¢	25¢	7¢
Providence, R.I	9.640	75¢	60¢	-	-	7.000	45¢	40¢	-	*7¹/₄¢
Raleigh, N.C	6.500	30¢	30¢	-	-	3.850	15¢	10¢	-	-
Reading, Pa	10.020	57¢	92¢	-	8¢	6.480	25¢	25¢	-	5¢
Richmond, Va	8.250	40¢	30¢	-	-	4.050	10¢	10¢	-	1¢
Rochester, N.Y	9.990	40¹/₂¢	$1.520	9¢	16¢	7.690	65¢	65¢	-	45¢
Rock Island, Ill	¹⁷8.303	¹⁷39¢	¹⁷65¢	¹⁷5%	¹⁷7¢	7.120	30¢	50¢	-	3¢
Sacramento, Calif	9.660	77¢	$1.300	*$1.300	12¢	6.535	80¢	$1.400	*90¢	10¢
St. Louis, Mo	9.455	50¢	50¢	50¢	50¢	7.675	40¢	75¢	*30¢	*13¢
St. Paul, Minn	8.320	38¢	45¢	$1.220	-	6.800	40¢	35¢	40¢	-
Salt Lake City, Utah	8.400	51¢	75¢	*25¢	9¢	6.005	25¢	25¢	30¢	4¢
San Antonio, Tex	7.910	25¢	40¢	*45¢	12¢	4.740	28¢	20¢	-	1¢
San Diego, Calif	10.240	10%	16%	*13%	1%	6.310	60¢	$1.390	50¢	15¢
San Francisco, Calif	10.140	$1.570	$1.000	42¹/₂¢	$1.490	6.535	80¢	$1.400	*$1.000	10¢
Santa Fe, N. Mex	8.340	42¢	75¢	*50¢	6¢	*5.320	*34¢	†25¢	-	4¢

See footnotes at end of table.

Table 2-1 (cont.)

Table 3. Union hourly wage rates and employer insurance, pension, vacation, and other fund payments for selected building trades in 105 cities, April 1, 1975 —Continued

City	Bricklayers					Carpenters					Electricians				
	Basic rate [1]	Employer contribution to fund				Basic rate [1]	Employer contribution to fund				Basic rate [1]	Employer contribution to fund			
		Insurance [2]	Pension	Vacation pay	Other [3]		Insurance [2]	Pension	Vacation pay	Other [3]		Insurance [2]	Pension	Vacation pay	Other [3]
Savannah, Ga	$7.250	20¢	20¢	-	1¢	$7.100	35¢	-	-	-	$7.550	†35¢	†1%+25¢	-	-
Schenectady, N.Y	9.290	40¢	35¢	-	5¢	8.700	50¢	50¢	25¢	5¢	9.850	35¢	1%+70¢	-	-
Scranton, Pa	9.000	50¢	50¢	-	5¢	8.170	40½¢	50¢	-	5¢	8.400	35¢	1%+57¢	50¢	10¢
Seattle, Wash	8.770	55¢	35¢	*25¢	13¢	8.150	50¢	60¢	*20¢	2¢	8.507	25¢	1%+40¢	*6%	-
Shreveport, La	7.450	30¢	30¢	-	-	*7.400	*20¢	25¢	-	2¢	7.632	45¢	1%	*4%	-
Sioux Falls, S. Dak	8 7.650	25¢	30¢	50¢	-	7.510	-	20¢	-	-	8.170	30¢	1%	4%	2%
South Bend, Ind	8.940	40¢	35¢	-	53¢	8.330	35¢	50¢	25¢	2¢	9.150	35¢	2½%	5%	4%
Spokane, Wash	*9.160	*55¢	*50¢	-	5¢	8.090	50¢	55¢	*25¢	2¢	†8.234	35¢	1%+40¢	*8%	-
Springfield, Mass	9.550	55¢	55¢	-	7¢	9.150	35¢	50¢	-	-	*9.470	48¢	1%+15¢	-	-
Syracuse, N.Y	9.100	61¢	70¢	-	-	9.070	55¢	60¢	-	5¢	9.750	42¢	1%+80¢	-	†9½¢
Tampa, Fla	8.350	30¢	50¢	-	4¢	7.865	30¢	20¢	-	2½¢	*9.440	4½%	2%	-	-
Toledo, Ohio	10.335	55¢	40¢	-	4¢	9.750	55¢	50¢	50¢	3¢	10.750	30¢	1%+55¢	-	-
Topeka, Kans	8 8.550	35¢	25¢	-	12¢	*7.650	35¢	25¢	-	10¢	*9.500	40¢	1%+50¢	-	-
Trenton, N.J	9.150	55¢	$1.000	-	4¢	9.200	50¢	70¢	50¢	3¢	11.350	4¢	1%+40¢	-	(11)
Tulsa, Okla	8.180	30¢	40¢	33¢	4¢	7.800	25¢	25¢	-	-	7.650	36¢	1%+25¢	30¢	-
Washington, D.C	9.550	50¢	40¢	-	5¢	8.500	35¢	39¢	-	-	9.350	35¢	1%+75¢	-	-
Wichita, Kans	8 8.320	45¢	25¢	25¢	6¢	*8.250	35¢	-	-	5¢	8.850	35¢	1%	-	-
Wilmington, Del	9.200	60¢	85¢	-	5¢	9.600	48¢	40¢	-	7¢	*9.850	*8½%	5%	-	-
Worcester, Mass	9.000	70¢	60¢	-	5¢	9.550	50¢	50¢	-	2¢	10.130	5%	1%+24¢	-	-
York, Pa	8.550	25¢	45¢	-	-	8.090	25¢	35¢	-	-	9.450	25¢	1%+15¢	-	-
Youngstown, Ohio	8.550	40¢	35¢	50¢	4¢	8.520	40¢	50¢	50¢	4¢	9.440	40¢	5%	8%	-

City	Painters					Plasterers				
	Basic rate [1]	Insurance [2]	Pension	Vacation pay	Other [3]	Basic rate [1]	Insurance [2]	Pension	Vacation pay	Other [3]
Savannah, Ga	$6.350	-	-	-	-	$5.750	-	-	-	-
Schenectady, N.Y	8.920	-	35¢	-	-	9.290	40¢	35¢	-	5¢
Scranton, Pa	7.100	60¢	-	-	10¢	8.950	-	40¢	-	5¢
Seattle, Wash	7.750	14 44¢	45¢	6	6¢	7.810	45¢	70¢	*60¢	10¢
Shreveport, La	7.250	-	-	-	-	7.450	-	-	50¢	-
Sioux Falls, S. Dak	6.070	-	-	-	-	7.530	-	-	-	-
South Bend, Ind	7.630	-	20¢	-	-	8.270	40¢	50¢	50¢	2¢
Spokane, Wash	8.460	31¢	50¢	-	6¢	8.880	45¢	-	-	-
Springfield, Mass	8.320	50¢	35¢	-	-	9.550	55¢	55¢	-	7¢
Syracuse, N.Y	8.580	60¢	45¢	-	3¢	8.710	53¢	45¢	-	2½¢
Tampa, Fla	*7.300	30¢	*30¢	-	-	*8.360	30¢	50¢	-	9½¢
Toledo, Ohio	8.490	55¢	40¢	50¢	-	9.900	55¢	-	50¢	3¢
Topeka, Kans	8.070	35¢	-	-	10¢	*8.300	-	-	-	10¢
Trenton, N.J	8.400	50¢	50¢	10¢	4¢	9.150	55¢	$1.000	-	4¢
Tulsa, Okla	7.500	-	25¢	25¢	-	7.650	-	-	-	-
Washington, D.C	8.640	41¢	18¢	-	1¢	8.500	45¢	25¢	-	†3¢
Wichita, Kans	7.470	-	20¢	-	-	*7.400	-	35¢	-	-
Wilmington, Del	7.820	60¢	25¢	-	†1¢	8.970	65¢	-	-	†7¢
Worcester, Mass	8.980	75¢	50¢	-	2¢	9.000	70¢	60¢	-	5¢
York, Pa	6.600	25¢	15¢	-	-	7.600	25¢	25¢	-	-
Youngstown, Ohio	8.265	40¢	40¢	$1.000	-	8.150	45¢	-	80¢	4¢

City	Plumbers					Building laborers				
	Basic rate [1]	Insurance [2]	Pension	Vacation pay	Other [3]	Basic rate [1]	Insurance [2]	Pension	Vacation pay	Other [3]
Savannah, Ga	*$8.750	35¢	*50¢	-	-	$3.750	15¢	10¢	-	-
Schenectady, N.Y	9.500	50¢	75¢	-	12¢	7.600	50¢	70¢	-	7¢
Scranton, Pa	9.540	35¢	50¢	-	1¢	6.800	48¢	50¢	-	5¢
Seattle, Wash	9.610	58¢	95¢	90¢	6¢	6.910	60¢	80¢	-	6¢
Shreveport, La	7.340	-	30¢	55¢	5¢	*4.500	15¢	10¢	-	3¢
Sioux Falls, S. Dak	8.010	32¢	15¢	17¢	-	6.320	-	25¢	-	-
South Bend, Ind	*8.800	38¢	5 70¢	50¢	8¢	*6.950	*40¢	*35¢	-	*12¢
Spokane, Wash	9.600	14 35¢	85¢	60¢	8¢	6.850	45¢	65¢	-	13½¢
Springfield, Mass	9.390	57¢	75¢	-	1¢	7.000	50¢	45¢	-	15¢
Syracuse, N.Y	9.100	51¢	57¢	-	†5¢	7.600	55¢	55¢	-	5¢
Tampa, Fla	†8.910	†45¢	40¢	75¢	5¢	5.850	22½¢	20¢	-	2½¢
Toledo, Ohio	9.235	55¢	75¢	$1.000	46¢	8.740	50¢	30¢	20¢	3¢
Topeka, Kans	9.750	30¢	30¢	-	-	*6.400	30¢	35¢	-	20¢
Trenton, N.J	9.250	57¢	80¢	$1.140	10¢	7.000	40¢	55¢	-	5¢
Tulsa, Okla	8.720	35¢	40¢	*25¢	-	5.600	25¢	20¢	-	-
Washington, D.C	9.430	58¢	55¢	-	3¢	7.230	38¢	45¢	-	-
Wichita, Kans	*9.530	*59¢	*80¢	-	-	*5.950	45¢	25¢	-	20¢
Wilmington, Del	9.140	78¢	*70¢	9%	†5¢	7.000	50¢	40¢	-	7¢
Worcester, Mass	9.930	65¢	45¢	-	-	7.100	40¢	45¢	-	5¢
York, Pa	9.300	30¢	40¢	-	-	5.550	35¢	25¢	-	-
Youngstown, Ohio	9.010	35¢	40¢	$1.250	8¢	6.540	50¢	30¢	-	$1.040

[1] These rates represent the minimum wage rates (excluding holiday and vacation payments regularly made or credited to the worker each pay period) agreed upon through collective bargaining between employers and trade unions.

[2] Includes life insurance, hospitalization, and other types of health and welfare benefits; excludes payments into holiday, vacation, and unemployment funds, when such programs have been negotiated.

[3] Includes all other nonlegally required employer contributions, except those for apprenticeship fund payments, as indicated in individual agreements.

[4] Part of the negotiated rate; not included in base rate shown.

[5] Part of the basic rate transferred to insurance, pension, and/or vacation plans.

[6] No union rate in effect on survey date.

[7] Part of basic rate transferred to promotional fund.

[8] Rate in effect prior to April 1, 1975; on that date, a new rate was in negotiation.

[9] Part of the negotiated rate; not included in base rate shown. Includes contributions for vacations and holidays; separate data are not available.

[10] Includes contribution for supplemental annuity plan.

[11] In addition to legally required contributions, employer pays employee's share of Temporary Disability Insurance.

[12] Based on minimum daily rate for a 5-hour day. Time and one-half the hourly rate is paid for the 6th and 7th hours worked in any day. If a journeyman electrician is employed less than 7 hours, he receives $10.860 an hour.

[13] Employer also pays employee's share of FICA.

[14] Includes contribution for dental insurance.

[15] Includes Rock Island and Moline, Illinois and Davenport, Iowa.

[16] Includes contribution for insurance and pension; separate data are not available.

[17] The following rates are applicable to plumbers in Davenport, Iowa: Basic rate, $8.750; insurance, 40 cents; pension, 60 cents; vacation, 25 cents; and other, 7 cents.

* Represents either a newly negotiated or deferred increase which became effective during the quarter.

† Revision of data previously reported.

NOTE: Information on employer contributions to insurance (welfare), pension funds, vacation pay, and other fund payments, as provided in labor-management contracts, is presented as cents-per-hour or as percent of basic rate; in actual practice, however, some employer payments are calculated on the basis of total hours worked, or gross payroll. These variations in method of computation are not indicated in the above tabulation. Payments directly to the worker each pay period for or in lieu of benefits are footnoted.

Some contracts also provide for employer contributions to an apprenticeship fund. Information on payments to this fund was not collected.

Overhead Charged To Specific Jobs:

Foreman and/or Superintendent wages
Watchman wages
Rubbish removal, clean up costs
Temporary buildings and structures
    Office (trailer, etc.)
    Tool and/or storage sheds
    Toilet facilities
    Fences, guardrails, etc.
    Signs
Building permits
Utilities
    Temporary heat, light and power
    Temporary water
    Temporary telephone
    Out of town jobs
        Travel cost
        Premium labor rates
        Board and room
Plant
    Rental of machines and equipment
    Scaffolding built on premises or rented
    Maintenance of machinery and equipment
    Depreciation
Legal expense
Insurance
    Public liability insurance
    Fire and Extended Cover insurance
    Workmen's Compensation
Taxes
    Social Security taxes
    Federal Unemployment taxes
    State Unemployment taxes
    Sales taxes on materials
Performance Bonds

Overhead Not Chargeable To Specific Jobs:

Contractor's office, shop, etc.
Rental
Depreciation

Heat, light and power for office, shop, etc.
Stationery, postage, supplies
Business machines
Property and liability insurance
Telephone
Other
Employees
   Executives
   Clerks
   Estimators
   Draftsmen
Travel
Advertising
Donations
Legal retainers

Overhead is not a theoretical charge but one subject to analysis. A contractor engaged in small to medium size jobs may operate out of an office in his home, his wife may do his clerical work, answer the phone, and keep his books of account. He may store equipment and surplus materials on his premises. His general overhead would be trifling in dollar-cost contrasted with a contractor doing heavy construction work with a separate office, shop, estimators, draftsmen, clerks, etc. Yet when the annual overhead expenses of each of these contractors is related *percentage-wise* to their income, it is surprising how, in most instances, the percentage to be charged against each job estimated, ranges from 8 to 10 per cent.

Profit

The profit percentage a contractor charges against the overall *job cost* varies with the size of the job and the competition. As the dollar amount of the job increases, the percentage profit a contractor is content with decreases particularly when the bidding is competitive.

Small jobs of a few hundred dollars or more may well justify 10 to 20 per cent profit. In other words a $300 to $400 repair job cost could justify a profit of $60 to $80 whereas on a job cost of $5,000 or more the contractor would be satisfied with 10 per cent. If it is competitive bidding the contractor will lower his profit percentage. Unfortunately, most of the estimating on repair or replacement of buildings covered by property insurance is not competitive. So the generally accepted profit allowed is 10 per cent of the actual cost of repairs before general overhead.

OSHA (Occupational Safety and Health Administration)

The U.S. Department of Labor has published safety and health regulations for the construction industry pursuant to the Williams-Steiger Occupational Safety and Health Act of 1970.

These rules or regulations, while mostly applicable to heavy construction, are enforceable as respects small contractors and builders of residential properties. Authorized representatives of the U.S. Labor Department have the right of entry to any construction site to investigate compliance with the regulations or may designate the services, personnel, or facilities of any State or Federal agency to inspect. Many states have adopted the Plan and have primary responsibility on a reimbursable basis by the Government.

A complaint of the smaller contractors and builders is that the cost of compliance increases their operating overhead anywhere from 2 to 5 percent with no visible benefits in many instances. Nevertheless OSHA should be taken into consideration in estimating, especially where substantial demolition and removal of debris is involved. Federal Register, Part 2, Vol. 37, No. 243 can be obtained from the Department of Labor. This document outlines the regulations. A brief resume´of *some* of the rules are stated below for illustrative purposes. Note particularly Regulations pertaining to Demoliton (Subpart T).

Basically, no contractor or subcontractor shall require any laborer or mechanic to work in surroundings which are *unsanitary, hazardous, or dangerous to his health or safety*. It is the responsibility of the employer to initiate and maintain such programs necessary to comply.

The prime contractor and subcontractor(s) may agree on who shall furnish (for example) toilet facilities, first aid services, fire protection or drinking water, but such agreement will not relieve either one of his legal responsibility.

Machines or power tools not in compliance with the Regulations must be tagged unsafe, locked, or removed from the premises.

Employers must instruct employees in avoidance of unsafe conditions to avoid exposure to illness or injury. If harmful animals or plants (poison ivy?) are present, the employee is to be instructed regarding the hazards and in any first-aid. The same rules apply where employees are required to handle flammable liquids, gases, or toxic materials.

Employers must insure availability of medical personnel for advice and consultation on matters of occupational health. Provisions shall be made for prompt medical attention in case of serious injury.

If no infirmary, clinic, hospital, or physician is reasonably accessible, a person certified in first-aid shall be available at the work site.

First-aid supplies shall be easily accessible. The first-aid kit must contain approved materials in a waterproof container and contents must be checked by the employer before being sent to the job and *at least weekly.*

Equipment must be furnished to transport any injured person to a physician or hospital, or else communication must be available for ambulance service.

Telephone numbers of physicians, hospitals, or ambulances must be conspicuously posted.

An adequate supply of potable water must be supplied in all places of employment. Containers shall be capable of being tightly closed and supplied with a tap. (No dipping of water from containers.) Container is to be clearly marked. A common drinking cup is prohibited. Where single service cups are supplied, a sanitary container for unused and one for used cups is to be provided.

Toilets are to be provided in accord with the Regulations.

Protection against the effects of noise-exposure shall be provided when sound levels exceed those set forth in a Table of "Permissible Noise Exposures." The protective devices inserted in the ears must be fitted individually by competent persons. Cotton is not acceptable.

All construction areas must be lighted to minimum intensities listed in a Table designating "foot-candles" for specific areas.

Hardhats (usually supplied by employer) are to be worn by employees working in areas where there is possible danger of head injury from impact, falling objects, or electrical shock or burns. Goggles and face protection are to be provided employees who are engaged in operations where machines or equipment present possible eye or face injury.

Fire fighting equipment shall be furnished, maintained, conspicuously located, and periodically inspected. A fire alarm system (telephone, siren, etc.) shall be established so employees and local fire department can be alerted. Alarm code and reporting instructions shall be conspicuously posted at entrance.

Employers shall not permit the use of unsafe hand tools.

Scaffolding requirements are extensive. They must be equipped with guardrails and also toeboards on all open sides and ends of platforms over 6 feet above ground or floor. If persons are permitted to work or pass under the scafford a 1/2" mesh screen must be installed between guardrail and toeboard.

Subpart T—Demolition

(1926.850) Preparatory operations.

(a) Prior to permitting employees to start demolition operations, an engineering survey shall be made, by a competent person, of the structure to determine the condition of the framing, floors, and walls, and possibility of unplanned collapse of any portion of the structure. Any adjacent structure where employees may be exposed shall also be similarly checked. The employer shall have in writing evidence that such a survey has been performed.

(b) When employees are required to work within a structure to be demolished which has been damaged by fire, flood, explosion, or other cause, the walls or floor shall be shored or braced.

(c) All electric, gas, water, steam, sewer, and other service lines shall be shut off, capped, or otherwise controlled, outside the building line before demolition work is started. In each case, any utility company which is involved shall be notified in advance.

(f) Where a hazard exists from fragmentation of glass, such hazards shall be removed.

(g) Where a hazard exists to employees falling through wall openings, the opening shall be protected to a height of approximately 42 inches.

(h) When debris is dropped through holes in the floor without the use of chutes, the area onto which the material is dropped shall be completely enclosed with barricades not less than 42 inches high and not less than six feet back from the projected edge of the opening above. Signs, warning of the hazard of falling materials shall be posted at each level. Removal shall not be permitted in this lower area until debris handling ceases above.

(1926.851) Stairs, passageways, and ladders.

(a) All stairs, passageways, ladders and incidental equipment thereto, which are covered by this section, shall be periodically inspected and maintained in a clean safe condition.

(1926.852) Chutes.

(a) No material shall be dropped to any point lying outside the exterior walls of the structure unless the area is effectively protected.

(b) A substantial gate shall be installed in each chute at or near the discharge end. A competent employee shall be assigned to control the operation of the gate, and the backing and loading of trucks.

(1926.854) Removal of walls, masonry sections, and chimneys.

(a) Masonry walls, or other sections of masonry, shall not be permitted to fall upon the floors of the building in such masses as to exceed the safe carrying capacities of the floors.

(b) No wall section, which is more than one story in height, shall be permitted to stand alone without lateral bracing, unless such wall was originally designed and constructed to stand without such lateral support, and is in a condition safe enough to be self-supporting. All walls shall be left in a stable condition at the end of each shift.

The word *shall* means mandatory, the penalties can be severe for non-compliance so even small contractors and subcontractors find it expedient to attend and/or have a foreman or superintendent attend training facilities set up by the Department. This can be expensive in time and travel for those long distances from such centers.

LOCAL CUSTOMS AND PRACTICES

One who estimates building losses or who checks estimates on a nationwide basis will need to adjust his thinking to local customs and practices within the trade. Such customs will show variations in the four areas:

1. Nomenclature
2. Construction methods and materials.
3. Estimating methods.
4. Agreements of unit costs.

1. A wood sill for a frame building may be termed simply a sill in one section, or a sole-plate, shoe, or mudsill in other sections. Roof boards, or roof sheathing, is known as decking or roof decking in some sections. A leader pipe running from the gutters to the ground is frequently called a down-spout, and sub-flooring may be known as under floor, or a porch may be referred to as a piazza, stoop, or veranda. A stud in some rural areas is termed a *scantling,* and a ridgepole is variously called a *ridgeplate* or *ridge-piece.* Edge grain, comb grain, and vertical grain refer, in different areas, to the same type of flooring.

2. Construction methods vary in many sections of the country. Some of these variations may be due to differences in climate. A northerner traveling in the southern states for the first time may be surprised not to see chimneys on most of the dwellings. Frame dwellings in the South are frequently built with single flooring and no exterior wall sheathing underneath the weather boards, brick veneer or stucco finish. The method of applying stucco in the South, by placing chicken wire directly over the studs, is almost unheard of in northern climates. Light roof framing on low-pitched roofs, common in southern climates, would be inadequate in northern areas where snow loads must be considered in designing roof construction. Shallow footings or none at all, such as are found in frost-free climates, would violate the building codes in northern cities where depths of 3 to 4 feet are required.

It is essential for an estimator to be familiar with local construction methods so that, when a building is completely destroyed, he will be able to figure the reproduction cost as it existed.

3. The different customs used in estimating are many and varied throughout the country. There may be a tendency toward standardization in local areas, but even here there will be found contractors who employ different methods to estimate the same thing. One estimator may figure the milling and cutting waste of 1'' x 6'' tongue and groove boards at one-sixth, while another will use 20 per cent. One paperhanger, in computing the number of rolls of paper he needed for a room, may allow 20 per cent waste; another paperhanger may determine the waste by allowing one extra roll for two window openings, two extra rolls for three windows, and so forth. In calculating the quantity of wall sheathing, bevel siding, and similar materials, some contractors deduct the area of all window and door openings; others leave them in or deduct one half the area. In figuring interior plaster, or exterior stucco, there are differences in opinion as to whether or not openings should be deducted.

Many painters, when estimating the cost of painting a room, will measure the area of the walls and the ceiling, ignoring the window or door openings, and multiply the area by a unit cost. In other sections of the country the same method will be used, but in addition a flat charge will be made for each door, window, and for the base board, on the theory that a different kind of paint or different color may be used on these units. Some painters estimate a job by the hours required, and figure a flat rate per hour for material, labor, overhead and profit. All of these systems seem to work, but whether they all are accurate is subject to analysis. The person who is called upon to make estimates in different parts of the country learns the various customs and practices, in order to discuss details intelligently when the occasion arises.

4. In some of the larger cities, contractors or estimators for the company, and those customarily figuring losses for insureds, tacitly have a general understanding or agreement on the approximate unit costs for such items as ordinary framings, roofing, painting, plastering, electric outlets, flooring, and so forth. Such unit costs are for the purpose of reaching an agreement in order that losses may be adjusted. They are not necessarily representative of a unit cost that might be used in a *competitive* estimate not involving the adjustment of an insurance loss.

EFFECTS OF ECONOMICAL FACTORS

As in other businesses, economic factors play a dominant role in construction job cost. Building material costs during the early part of the historic "Depression" of the 1930's were at rock bottom. Merchantable framing lumber was retailed in the neighborhood of $20 per 1,000 board feet. Carpenters in a metropolitan area were happy to get a dollar an hour for an eight hour day with no fringe benefits. The same quality framing lumber in 1975, as a result of inflationary prices, retailed at $170 to $200 per 1,000 board feet. Carpenters in the metropolitan area were less than content to receive $9.42 per hour with fringe benefits of $2.43 making a total of $11.85.

When materials are scarce, and prices change upward each month, contractors are reluctant to take jobs on a firm bid basis. More and more jobs are let on a cost-plus-basis with a condition that the total cost will not exceed a specified (safe) figure. If a job *is* taken on a fixed contract it can usually be assumed the figure is generous in order for the contractor to protect himself against rising costs of labor or material, or both, before the work is completed. An understandable situation.

DEMOLITION AND REMOVAL OF DEBRIS

Before new materials can be installed in a structure that has been damaged, all of the burned, wet, broken or otherwise injured materials must be removed and eventually disposed of, either on the premises by burning, or off of the premises by carting away to a public or private dump.

Damaged brickwork, lath and plaster, steel, framing lumber, roof shingles, finished flooring, and so forth, when not obliterated, will have to be removed to make ready for new materials. Generally, this work is done by common or unskilled labor if a substantial amount of demolition is involved. When a few rafters, floor joists or studs are to be taken out and replaced, a carpenter normally would do it. If a moderate amount of finished flooring, trim, electric wiring, plumbing pipes and fixtures are to be removed, the building trades journeyman who will install the new work generally takes care of the demolition. Much will depend on local working conditions and labor customs. In different sections of the country this operation appears in building loss estimates under one of the following various designations: Tear out and cart; demolition and trucking; debris; demolition; removal of debris; debris removal; and drayage. These terms all refer to the same operation, which is tearing out the damaged material and disposing of it.

There are no tables, charts, or formulas for estimating the cost of this item. It is usually all labor and expense, plus any specialized equipment or material that must be purchased or rented to do the work. Experience, analysis, the ability to visualize each step, plus good judgment are needed to estimate the cost.

In most cases it will be found at the end of an estimate in the form of a lump sum amount. Many estimators make an educated guess at the overall cost of the item, rather than trying to analyze each operation. Such overall guesses are usually on the high side and, in fact, it is not uncommon for contractors to use the "debris" item to bury contingencies.

In some sections of the country, it is an accepted practice to add a few cents to the square foot price of shingles to allow for the cost of removing and carting away the old ones. Other unit prices will sometimes be increased to include removing and carting of the debris based on the cost per square foot, square yard, or cubic foot.

Because it is an operation precedent to reconstruction, and is almost always a labor, trucking, and expense item, it should be carefully studied, just as any other construction cost is analyzed. The only way in which this can be done is to visualize each individual operation and estimate the number of man-hours it may take to do the work. If oak or finished pine floor, subfloor, plaster, or brickwork must be removed, the estimated number of hours for each should be noted. If wood joists, studding, or rafters are to be removed, the operation is visualized and the probable number of hours is noted. After recording the estimated labor for each type of work to be performed, a summary should be made up by building trades, or by floor, room, or some other suitable classification. In this way omissions are frequently discovered and it also provides a sound basis for double-checking and for any later discussion during the adjustment.

Illustrative Example
Demolition and Removal of Debris

Remove 6 square wood shingles	9 hours
Remove roof boards 20' x 30'	6 hours
Remove 8 rafters	4 hours
Remove 200 sq yds plaster and wood lath	10 hours
Tear out 150 sq ft 1" x 3" flooring	8 hours
Total labor	37
	@ $5.00
	$185.00
Loading and carting to public dump	75.00
Total	$260.00

It should be remembered that many times *undamaged* materials may have to be torn out because they are contiguous to materials that are damaged. Badly charred attic rafters or floor joists are a typical example where the roofing, in the case of charred rafters, is undamaged, and the finished floor, in the case of charred joists, is undamaged. The cost of removing the undamaged finished floor and roofing must be taken into account, as well as the cost of replacing them.

Machinery and equipment of a tenant may have to be moved and later replaced, in order to make the repairs to a floor. Charred joists under a bathroom floor may require the removal and replacement of all of the tile floor and plumbing fixtures above, to make proper repairs. Where a tenant has abandoned the contents debris, the cost of moving any portion of it to repair the building may have to be included in the estimate.

When subcontractors are employed to estimate for a particular building trade, a clear understanding should be reached as to who will estimate the cost of removing the debris and carting it away. In this way omissions or duplications can be eliminated.

If the debris cannot be satisfactorily disposed of on the premises by burning, or by dumping, an item of trucking must be added. When a contractor has his own truck and driver, he generally knows what the proper charge should be. If he is required to hire a truck and driver, the local rate per day multiplied by the number of estimated days the truck is in service will determine the cost. The capacity of the truck, the time to load, and the estimated number of loads that can be hauled per day are considerations to be taken into account.

When preparing an estimate of the cost of removal of debris the answers to the following questions will aid in determining the probable cost.

1. What kind of material is to be removed?
   (a) How is it installed, or attached to the structure?
2. Is there contents debris either blocking or mixed in with the building debris?
   (a) Will the cost of removing contents debris be borne by the tenant?
       (or)
   (b) Must the cost be included in the building debris item?
3. What degree of care must be used in removing the debris?
   (a) To save and preserve it for reuse.
   (b) To avoid damage to other property.
   (c) To avoid injury to workmen.
   (d) To avoid injury to occupants.
   (e) To avoid injury to passersby.
4. Are protective barricades necessary?
   (a) Partitioning, sidewalk fences, canopies.
5. Is special equipment needed?
   (a) Scaffolding, debris chutes, etc.
   (b) Bulldozers.
   (c) Cranes or derricks.
   (d) Pneumatic compressor and tools.
   (e) Electric drills or hammers.
   (f) Acetylene torches.

6. What type of labor will be used?
   (a) Common.
   (b) Skilled.
7. Is the damaged area easily accessible?
   (a) Are elevators, stairways, and so forth, intact?
   (b) Is the working space confined, or in open?
8. Can trucks load and park readily in the area?
9. How close is the building to a public dump, or disposal site?
   (a) Is there a charge for dumping?
10. Can the debris be burned on the premises?
11. Is shoring of floors or walls necessary for removal of debris?
12. Can any material be sold for salvage?
    (a) As scrap metal.
    (b) As second-hand building materials.
    (c) If contents have been abandoned, do they have any value?
13. Are specialists or subcontractors needed to perform any particular operation?
14. Must the tenants' property be shifted or moved?
    (a) Machinery and equipment.
    (b) Furniture and fixtures.
    (c) Stock.
15. Are there special ordinances relating to wrecking? Special permits?

ESTIMATING SALVAGE

There are two kinds of salvage to be considered in estimating building losses.

1. Units or materials that, though undamaged, must be temporarily removed and later will be reinstalled.
2. Units or materials that are damaged or partially damaged, and which are to be removed and sold for salvage.

The first category would include such items as trim or millwork removed and stored during replastering, also service units, i.e., oil burners, plumbing and electrical fixtures, refrigerators, laundry equipment, which have to be removed mainly for protection while repairs are being made. Considerable care and skill is required to remove many such items so as not to damage them.

In the second category, salvage to be sold for value, there are very few building materials to be considered. Structural steel, and copper cable, wire or roofing in quantities, provide fair salvage depending on the scrap market, and the cost of getting it to the market. Common brick at one time was considered good salvage when labor wage rates were low enough to warrant the cost of cleaning and handling. In some sections of the country, second-hand bricks are now premium-priced as veneer for new homes. This architectural oddity is limited and confined to certain geographical areas.

Partially damaged service units of heating or plumbing fixtures bring small salvage, if any. Rough and finished carpentry salvage for resale is extremely rare.

KINDS OF ESTIMATES

There are several kinds of estimates, each of which has its special place in approximating the reproductive cost of buildings, depending on the degree of accuracy desired. The following are those most commonly used:

1. Cubic foot estimate
2. Square foot estimate
3. Semi-detailed, or panel method
4. The sectional, or bay method
5. Original cost method
6. Detailed estimate

All but the *detailed estimate* are used, primarily, to obtain the total replacement cost of a building rather than to determine partial repairs.

Cubic Foot Estimate

When the new-replacement cost of a building is desired on an approximate level only, one of the best methods used is the *cubic foot system*. The cost of building a dwelling, church, factory and so forth is either known or estimated in detail. The cost is divided by the total number of cubic feet contained within the outer walls, roof, and lowest floor of the structure. The quotient is the cost per cubic foot for the particular building under consideration.

Illustrative Example

$$\frac{\text{New replacement cost}}{\text{Number of cubic feet}} = \frac{\$33,000}{30,000} = \$1.10 \text{ per cubic foot}$$

The cost per cubic foot may then be applied to similar buildings by measuring the cubic foot volume and multiplying by the unit cost.

Though the theory behind cubic foot estimating is sound, there are a number of inherent weaknesses in the method which, unless understood and compensated for, will diminish its reliability. Very few buildings are identical in specifications, except in some housing developments, and it is therefore important to recognize construction and architectural differences between the building from which the unit was developed and the one to which it is being applied. The cubic foot cost of identical buildings decreases as the size (cubic volume) increases and vice versa. The ratio of the length to the width of a specific type of structure affects the cubic foot cost. The number and the height of stories in buildings of apparently similar construction and layout changes the cost per cubic foot. Any special or unusual interior finishes, built-in features, heating or air conditioning systems add to the difficulty of determining the proper cubic foot cost to be applied. The cost of porches, breezeways, and patios, which are not included in the volume computation, will have to be estimated and added.

While cubic foot appraisals sometimes come within 10 per cent of the actual replacement cost, it must be considered a coincidence rather than a rule. Their main value is in quickly and inexpensively approximating building replacement costs. A builder in competition would not contract to build strictly on a cubic foot estimate.

There are several books and manuals published, with tables, charts and pictures, which are excellent guides for selecting an appropriate cost per cubic foot or square foot.

Measuring Cubic Foot Volume

The proper method of calculating the cubic foot volume of a building depends solely on the method used by the person who developed the cost per cubic foot.

A recognized procedure is to compute the square foot area of the building using the outside dimensions and multiplying that area by the height in feet from the underside of the lowest floor (the basement if there is one) to a point halfway between the ceiling and the ridge or weather surface of the roof. Figure 2-1 shows several shapes of buildings and how the height is usually measured. The principal variation in methods used is that some systems measure from the top of the basement or lowest floor rather than from its underside.

Square Foot Estimating

The replacement cost of buildings can also be approximated by measuring the square foot area of each floor in a building and multiplying by a predetermined cost per square foot. The unit is developed, as in the cubic foot method, from known replacement costs which are divided by the square foot floor area.

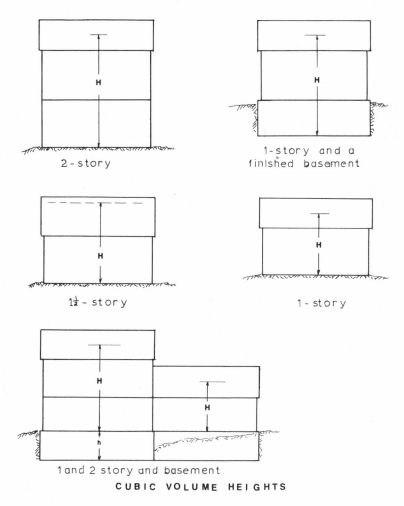

2-story

1-story and a
finished basement

$1\frac{1}{2}$ - story

1- story

1 and 2 story and basement

CUBIC VOLUME HEIGHTS

Figure 2-1

Recommended heights (H) to use in computing the cubic-foot volume of a building.

The square foot method is more reliable when applied to buildings in which the construction and layout of the various floors are similar such as schools, hotels, office buildings and hospitals. Most of the weaknesses of the cubic foot method are inherent also in the square foot method.

Measuring Square Foot Area

The area included in calculating the square foot cost of a building includes the first floor area, using outside dimensions, and similar livable areas above the first floor. Open attics and basements are not included but rather priced separately as the costs of those areas are much lower.

The cost of partitions, floor and roof construction can be determined in the same manner. Interior and exterior doors and windows are counted and priced according to the individual cost for material and labor to install. Electric outlets and fixtures may be counted and priced per unit for labor and material. Built-in features, chimneys, fire places, stairways, heating and plumbing are added, generally in lump sum amounts. To complete the estimate, overhead and profit are added.

This method, in the hands of an experienced and thoughtful estimator, ranks next to the detailed method for accuracy. It is shorter than preparing a detailed estimate, but takes more time than the cubic foot or square foot system. It is especially suitable for office or factory buildings, apartment houses and warehouses.

The Sectional or Bay Method

A number of types of buildings are built in bays or sections of practically identical construction. By detail-estimating the new replacement cost of a single bay, the entire cost of the building can be computed by multiplying by the number of bays. End walls must be added, and any dissimilarities in construction of one bay from others should be noted and taken into consideration. Such things as partitions, stairways, window or door openings, special equipment or services are added.

This method is useful in estimating foundry type buildings, lumber sheds (either the open-front type or closed), barns, and also some warehouses and factories. Inspection of a building will quickly disclose whether the method should be employed. It is equally as accurate as the semi-detailed method and, for many buildings, is much faster.

Original Cost Method

When the original construction cost of a building is known or available, it can be brought up to date by applying a reliable cost index. The cost indexes are published by the American Appraisal Company, Engineering News-Record, publishers of appraisal books and construction trade journals. Building cost indexes are based on average conditions and indicate the percentage change in construction costs from one year to another, using a particular year, usually 1913, as 100 per cent.

When using this method, it is important to know whether the original cost of the building included excavations, grading, foundations, footings, and so forth. If the cost of the lot is included, its approximate value at the time the building was built must be deducted before applying the cost index. If any additions have been made to the building, outside of maintenance items, since it was built, proper allowance should be made for them.

This method of estimating is approximate, the degree of accuracy being contingent on how recently the building was built, the dependability of the original cost figures, the reliability of the cost index, and the experience and judgment of the person applying it.

The Detailed Estimate

The detailed estimate is an itemization of all of the material and labor costs required to repair or replace a building or structure. It includes everything that a contractor will be required to provide and do to restore the property. It is the most accurate and dependable method of estimating because of its detailed build-up. It is also the most useful and satisfactory kind of an estimate for negotiating the adjustment of a building loss. Its main advantages over the other types are that:

(a) It identifies the room or area involved.
(b) The sizes, dimensions, and measurements are noted.
(c) The kind and quality of material are shown.
(d) The work to be performed is specified.
(e) The quantities of material are shown.
(f) The hours of labor allowances for each operation are shown.
(g) The material prices can be verified.
(h) The wage scales can be verified.
(i) All arithmetical calculations can be double-checked.
(j) It is orderly and therefore easy to follow.
(k) Comparison with other estimates is made easy.
(l) Differences are readily spotted.
(m) It provides a sound basis for discussion of details.

A detailed estimate is divided into two parts. The first part consists of methodically listing each operation and all of the materials that are necessary to complete the repairs or replacement of a structure. It follows a logical sequence of reconstruction. The second part of a detailed estimate consists of computing the cost of the materials and the cost of labor for each of the items listed. The total cost of the work is obtained by adding together the cost of the materials and the cost of the labor. When a building or structure has been totally destroyed, the listing of work and materials may be developed from plans or more often from sketches. The majority of losses, however, are partial and therefore the itemization is based on notes taken in the field at the scene of the loss.

Chapter 3.

Fieldwork and Notes

The first and most important step in estimating building losses is the meticulous taking of measurements and notes in the field, because they form the basis for computing quantities, and pricing all materials and labor. The degree of care and thoughtfulness used in itemizing each operation is a major control of the accuracy of the final cost. The taking of notes is simply the process of identifying and listing all of the items of construction that are to be repaired or replaced, together with a word or statement to describe the work to be done.

EQUIPMENT FOR ESTIMATING

The estimator should carry a standard wood, metal, or plastic, 6-foot folding rule, and a suitable notebook. A strong flashlight is recommended for inspecting darkened rooms, basements, closets, or other poorly lighted areas. Some estimators use a 6- or 8-foot steel tape, and many carry a 25- or 50-foot steel tape for longer measurements. A small awl, carried in the brief case, is convenient for securing one end of a long tape when working alone. It can be stuck into a wood surface or brick-joint. Many estimators prefer using a clip-board and pad in place of a notebook; some also carry a camera in their car, as well as a collapsible aluminum ladder; others are experimenting with tape recorders to take notes.

PRELIMINARY SURVEY

When the estimator arrives at the place of loss, before taking any notes, his first consideration is to obtain an over-all mental picture of extent of damage, the scope of the work involved in making the estimate, and any specific or unusual problems with which he will be confronted. To do this, he should make a general preliminary inspection of the premises.

If the damage is slight and confined to a small area, there is usually no need for an extensive survey, for a brief examination of the surroundings will disclose all of the information required to begin taking notes. If the damage is wide-spread and serious, a careful reconnaissance should be made in order to formulate a plan of procedure. During the preliminary survey a great many observations will be made. Some may suggest a special method, pattern, or sequence for taking-off the loss. Some will disclose problems that require special treatment. Most of these observations will include the following:

A. The Physical Characteristics of the Building
1. Its use and occupancy.
    (a) Private or multiple dwelling
    (b) Offices
    (c) Hotel, motel, etc.
    (d) Manufacturing
    (e) Mercantile, institutional
    (f) Other occupancy
2. Number of stories, basement, sub-basement.
3. General plan of building and layout of interior.
4. Type of construction.
    (a) Frame
    (b) Brick, cement block
    (c) Brick-veneer
    (d) Mill
    (e) Fire-resistive
    (f) Metal clad on wood or steel frame
    (g) Non-classified
5. Special architectural or construction features.
6. Quality of materials and workmanship.
7. Age of building.
8. Physical condition of building prior to loss.
9. Obsolescent features.
10. Structural defects prior to the loss.
11. Unrepaired damage prior to the loss.
12. Damage resulting from an uninsured peril.
13. Building code violations.

B. Need for Emergency or Temporary Work
    1. Removal of building or contents debris.
      (a) For safety
      (b) For accessibility
    2. Removal of building equipment.
      (a) For its protection
      (b) For immediate repair
    3. Drainage of plumbing, sprinkler system, and heating facilities to prevent freezing.
    4. Shut off utilities—gas, water, electric.
    5. Install temporary or permanent roof.
    6. Temporary heat required to dry out building.
    7. Close openings in walls, windows, roof.
    8. Pump out basement, elevator pits, etc.
    9. Mop up water on floors—spread sawdust.
    10. Install temporary lights to prepare estimate.
C. Accessibility for Inspection
    1. Need for ladders or special equipment.
    2. Need to obtain tenants' keys for admission.
D. Need for Plans, Specifications, Photographs or Other Evidence
    1. Availability of plans or specifications.
    2. Need for their preparation by the estimator.
    3. Need for photographs.
    4. Obtain description of construction and lay-out from tenants.
    5. Obtain description of construction and lay-out from owner.
    6. Obtain description of construction and lay-out from neighbors.
    7. Need for preserving physical evidence.
    8. Need for consulting manufacturers of special or unusual materials or equipment.
E. General Extent of Damage
    1. General area or part of building affected.
      (a) Parts in general needing repair
      (b) Parts in general needing replacement
    2. Have any repairs or replacements been made?
    3. Allow waiting period before deciding on damage to
      (a) Finish floors, asphalt tile, etc.
      (b) Plaster or wallboards
      (c) Paint and wallpaper, etc.
F. Question Ownership or Interest of Tenants and Landlord
    1. Improvements by tenant.
      (a) Store fronts, painting, etc.
    2. Equipment used by tenant.
      (a) Air conditioning, heating, light fixtures, etc.
      (b) Duct work, power wiring, etc.
G. Need to Engage Experts

   H. Need to Employ Subcontractors

   I. Specific Problems Due To Location

      1. Congested or business area.

         (a) Accessible for trucks loading and unloading.

         (b) Need for sidewalk canopy.

         (c) Need for debris chute.

      2. Proximity to labor and material sources.

      3. Existence of union regulations.

      4. Need to conform to special building ordinances.

      5. Possibility of condemnation of part or all of building.

   J. General Area or Parts of Building *Not* Involved

ORDER IN TAKING NOTES

Following the preliminary survey of the damage, a methodical take off of the loss is begun. The starting place on fire repairs may be at the point of origin or area of the most severe damage. It may be either on the exterior, or interior of the building. When taking notes of damage by windstorm or hail, it is customary to start with the exterior and then go to the interior. Where the damage is caused by sprinkler leakage, the obvious place to begin taking notes is on the floor where the sprinkler leakage occurred and work downward. The conditions encountered in each instance will disclose the most likely place to commence taking the measurements and notes.

Experience has taught that progressing from one room to the next, and from one floor to the next, is a satisfactory plan to follow. One should never take off part of the loss in one area, move to another, and then return to the first area. Nor should one take notes on the interior for a while, go to the exterior, and then return to the interior and continue note taking. Not only will the finished estimate appear disorderly, but there is always a possibility of omissions and duplications. Insofar as it is practical, the work should be completed in one area, or one room on a floor, before moving to the next area or room on that floor.

Many estimators follow a rather fixed order of listing damage. This is particularly adaptable where a number of rooms are involved. A typical pattern for taking notes in a room might be as follows:

   1. Rough carpentry items.

      (a) Studding

      (b) Floor joists

      (c) Rough flooring

   2. Finish carpentry items.

      (a) Windows

    (b) Doors

    (c) Trim

    (d) Finish floor

3. Plastering (or wall treatment).
4. Painting and decoration.
5. Plumbing pipes or fixtures.
6. Heating pipes or fixtures.
7. Electric outlets and fixtures.
8. Demolition and removal of debris.
9. Special or miscellaneous items.

This order of listing becomes fixed in the mind of an experienced estimator, and he tends to follow it automatically, always alert, however, to any special items.

There are no rules for the order of taking notes other than, whatever the procedure, they should conform to a logical, easy-to-follow sequence. The following outlines provide suggested systems for taking measurements and notes on various types of losses:

FIRE LOSSES

1. *Partial Losses* (*All parts of structure not involved*):
   (a) Start taking measurements and notes at point of origin, or the area of most serious destruction, and follow the course of the fire proceeding from room to room, floor to floor, to the portion of building not affected. Go to exterior and record necessary repairs on all sides, and to roof.
   (b) Start taking measurements and notes at basement or lowest floor level. Proceed upward, going from room to room, floor to floor, to the highest floor involved. Go to exterior, recording necessary repairs on all sides and to roof.
   (c) Start at roof, attic, or highest floor involved. Proceed downward, going from room to room, floor to floor, to lowest floor involved. Go to exterior, recording necessary repairs to all sides and to roof.
2. *Total Losses* (*or substantially all parts involved*):
   (a) Start taking measurements and notes at basement or lowest floor level. Proceed upward room by room, floor by floor, to highest floor. Go to exterior and record damage to all sides and roof.
   (b) Start taking measurements and notes, at basement or lowest floor, by building trades, for entire structure, i.e. masonry, framing, sheathing, flooring, roofing, plastering, painting, plumbing, heating, electric, and so forth.

WINDSTORM AND HAIL LOSSES

1. *Partial Losses (all parts of structure not involved)*:
    (a) Start taking measurements and notes at the exterior, recording necessary repairs on roof and all sides. Proceed to the interior, working from highest floor to lowest floor involved, going from room to room on each floor.
2. *Total Losses (or substantially all parts involved)*:
    Proceed as in 2(a) and 2(b) under "Fire Losses."

EXPLOSION, BLASTING OR SONIC BOOM LOSSES

1. *Partial Losses (all parts of structure not involved)*:
    (a) Start taking notes and measurements at lowest floor involved and proceed upward room by room, floor by floor, to highest floor involved. Go to exterior, recording necessary repairs to all sides and roof.
2. *Total Losses (or substantially all parts involved)*:
    (a) Proceed as in 2(a) or 2(b) under "Fire Losses."

SMOKE AND FURNACE BOX EXPLOSION LOSSES

(a) Start taking measurements and notes at furnace room. Proceed from room to room, floor to floor, to highest floor involved. Go to exterior to note any damage.

VEHICLE DAMAGE LOSSES

(a) Start taking measurements and notes at point of collision. Complete exterior and proceed to interior, going from lowest floor involved to the highest floor involved.

LIGHTNING LOSSES

(a) Start taking measurements and notes at point where lightning struck. Complete examination of all sides and roof of structure. Go to interior and note damage beginning at lowest floor involved to highest floor involved, or follow path of lightning where it entered building.

SPRINKLER LEAKAGE AND WATER DAMAGE LOSSES

(a) Start taking measurements and notes at floor on which the leakage originated. Proceed downward room by room, floor by floor, to lowest floor involved.

FLOOD, WAVE-WASH, BACKING UP OF SEWER LOSSES

(a) Start taking measurements and notes at basement or lowest floor. Proceed, room by room, to floors above that are involved.

Illustrative Example of Note Taking from Roof Downward

RE: DONALD DART
             12 E. 12 RD.
             DARBY, ILL.
             FIRE: 6/22/58.

ROOF

             3-2" x 6" x 18' rafters
             Repair 4" x 4" plate
             5' x 18' 1" x 6" roof boards
             3 square 210# asphalt shingles
             Repair cornice
             Rehang gutter

ATTIC BEDROOM 16' x 18' x 7' 6"
             New sheet rock clg.
             Paint room 2 coats
             Sand and shellac floor
             2 electric outlets, 1 fixture

SECOND FLOOR—FRONT B.R. 12' x 14' x 8'
             2 lights 16" x 26" d.s. glass
             Paint 2 coats
             1 electric outlet and fixture

FIRST FLOOR—LIVING RM. 16' x 21' x 9'
             Paint clg. only—1 coat

KITCHEN 6' x 8' x 9'
             Paint clg.—1 coat

DINING RM. 11' x 12' x 9'
             Paint room—2 coats

EXTERIOR
             Paint cornice and gutter to match

DEMOLITION AND CARTAGE

Illustrative Example of Note Taking from Basement Upward

RE: R.G. SMITH
             106 2ND. ST.
             BELTON, ILL.
             FIRE: 2/20/58.

BASEMENT 28' x 36' x 7'
             Replace Timkin Oil Burner
             Overhaul controls
             6-2" x 10" x 12' First Floor Joists
             5 sets bridging
             Whitewash walls

Paint stairway to First Floor
2 electric outlets—1 fixture

FIRST FLOOR

Kitchen 8' x 10' x 9'
4 lights 8" x 10" d.s. glass
Repair rear door lock
2 electric base outlets
Paint 2 coats
Dining Rm. 10' x 12' x 9'
Replace 1" x 6" underfloor 10' x 12'
Replace 1" x 3" white oak floor 10' x 12'
Sand and shellac entire floor
New 6"—2-member oak base board
Paint 2 coats
Living Rm. 14' x 18' x 9'
Paint 2 coats
Stairhall to Second 4' x 14' + 4' x 6' x 9'
Paint, including stairs

SECOND FLOOR—S/E. BEDROOM 10' x 12' x 8'

Paint 2 coats
Other rooms show no damage

TEAR OUT AND CART DEBRIS

DESCRIPTIVE NOTES

The actual recording of notes in the field should be brief, simple, and readily understandable to anyone reviewing them, even though they may never have visited the scene of the loss. They should convey the following information:

1. The *location on the premises* where each item of repair or replacement is to be made.
2. All *measurements* required to compute the quantities of material, unless the quantity is determined in the field, in which case it should be noted.
3. The *kind, size, and quality of material* to be used.
4. A descriptive word or statement of the *work to be performed.*

1. Location on the Premises. The notes should show in what building the repairs are contemplated. They should identify the floor, and the room, or part of structure involved, as: *Main Residence, Second Floor, Master Bedroom,* or *Factory Warehouse, First Floor-Rear, Shipping Office;* or *Dormitory Building, Third Floor, South Wall.*

Notes made in this manner will make it clear to anyone the precise location to which the estimator is referring.

2. *Measurements.* Since very few loss estimates are prepared from plans from which measurements can be scaled or read, it is necessary to show in notes all measurements that are necessary to compute areas and quantities. The size of each room to be plastered, painted, or papered, or where floors are to be laid or scraped, should be indicated, giving the length, width, and height. The measurements of roof areas to be repaired or recovered, walls to be repaired or replaced, and other measurements from which quantities are computed, should be noted.

Illustrative Example

Main residence
 Second floor
  Master bedroom 14′ x 16′ x 8′
(or)
Repair wall plaster 6′ x 8′
(or)
Repair 210-lb. asphalt shingles 16′ x 20′
(or)
Replace 8″ common brick wall 12′ x 32′
(or)
Repair 1″ x 3″ oak floor 6′ x 10′
(or)
Main residence
 Southwest side
  Replace bevel siding 10′ x 16′

In each instance, measurements that are noted as illustrated will provide the basis for computing the quantity of material needed, the labor to apply it, and the total cost of the operation.

3. *Kind, Size and Quality of Material.* The cost of a particular kind of material varies with its size and quality and it is impossible to price it without adequate information. Almost all materials used in the construction of a building are available in different grades, and the field notes should include sufficient data to properly identify them.

Illustrative Example
Install 4 lights double-strength glass 18″ x 24″.
(or)
Replace 1 pair 3′ x 4′ x 1 3/8″ —8-light pine sash.
(or)
Plaster ceiling 2-coat sand-finish on gypsum lath.
(or)
Apply Certigrade No. 3 Black Label—16″ wood shingles.

4. *Work To Be Done.* Estimators use descriptive verb-words or their abbreviations to tell what work or operation is to be performed. They should be clear and informative. The following are examples of those most commonly used.

Build	Install	Plaster
Cart or truck	New (this or that)	Rebuild
Decorate	Paint	Refinish
Erect	Paper	Remove
Hang	Patch	Repair
Replace	Sand	Tear out
Reset	Scrape	Wash down
Resurface	Straighten	Whitewash
Rewire	Take down	

Much time can be saved by becoming familiar with these words and by accumulating others. They should be employed when recording notes. A few abbreviations are frequently useful, such as

Ref. . . . . Refinish
Rem. . . . . Remove
Rep. . . . . Repair
P&D. . . . . Paint and decorate
T&C . . . . Tear out and cart
Pl. . . . . Plaster
WW. Whitewash

MEASUREMENTS AND DIMENSIONS

In measuring the length, width, and height of rooms, and in measuring surface area, except where scale plans or sketches are being made, it is considered sufficiently accurate to measure to the nearest whole foot or half foot. A room with actual dimensions of 12'2" x 18' 5" x 8' 2" may be recorded in the notes as 12' x 18' 6" x 8".

Dimensions taken for building materials, however, which are sold in specified lengths, widths, or thicknesses, should always be scaled accurately. The sizes of doors and windows, and their thicknesses, make a considerable difference in the price at the lumber yard. Baseboard and finish flooring are sold in different widths. Glass comes in single, double, semi-plate and plate thicknesses. Wood shingles are made in 16-, 18-, and 24-inch lengths. The dimensions of trim material vary for interior millwork, exterior cornices, water table, and corner boards, and they must be scaled accurately.

SKETCHES AND PLANS

Line drawings to record outside measurements, story heights, roof shape and pitch, interior layout of rooms or service facilities, can be extremely useful. On serious or total losses, they may provide a basis for eventually drawing formal plans. They are also a means for obtaining or double-checking quantities, locating windows, doors, and special building fixtures or equipment.

Sketches showing the details of certain construction, or architectural features, may also be useful later when the amount of material and labor is being estimated. A line sketch of a cornice construction or a framing detail of a building, takes only a few moments to make, but it may well serve to

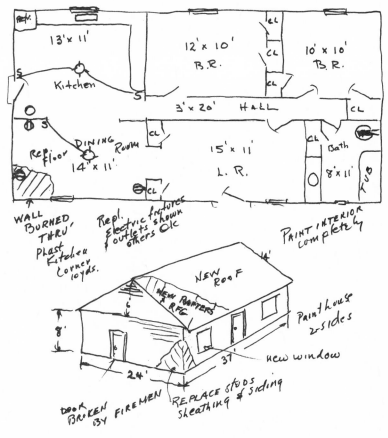

Figure 3-1

A typical "field sketch" to record data for later use.

avoid a return trip to the premises should the written notes be inadequate, or if memory fails. As a matter of practice, many experienced estimators tape the outside of a building and draw a line sketch showing the dimensions before they enter the premises on all but the smaller losses.

When a loss is very serious, or a constructive total loss, plans of some kind will almost always be necessary to prepare an estimate. In some cases original plans may be obtained from the local Building Department, from the owner, from the architect, or contractor who built the building. In the majority of cases, however, the plans are either not available or incomplete, and new ones must be prepared. Very often the person who makes the estimate draws such plans.

Much of the data for plans can be obtained at the scene of the loss. It usually is necessary to get a complete description of the building from persons who have been closely associated with it. As many persons as necessary should be interviewed to verify details. The following persons or firms are suggested as likely sources of information:

> Owner or members of the family.
> Friends of the owners who have visited premises.
> Tenants or employees of tenants.
> Superintendent of building.
> Employees of owner.
> City inspector.
> Insurance inspector.
> Watchman.
> Caretaker.
> Builder of property.
> Original architect.
> Realty management firm.
> Former appraiser of property.
> Neighbors.
> Real estate sales people familiar with the property.
> Maintenance or repair firms who have worked in the building.
> Any persons possessing photographs taken before the loss.

The questioning of persons concerning layout and construction is best done at the premises in order that their memories are refreshed by inspection and examination of the remaining portion of the building. It is well to have a list of questions ready to save time and to make certain that nothing is omitted. The list should follow the sequence of actual restoration from the ground, or foundation, to the roof. Circumstances in each case will govern the precise pattern of questioning to follow, but if the estimator will cover the following major points, he will include most of the construction details of the average building.

I WALLS

(A) *Basic Construction*
1. Wood frame
   (a) Studded
   (b) Post and girder
   (c) Log
2. Steel frame
3. Masonry
   (a) Stone
   (b) Brick
   (c) Cement block
   (d) Hollow tile
   (e) Concrete

(B) *Exterior Finish*
1. Sheathing
   (a) Wood
   (b) Composition
2. Siding
   (a) Wood
   (b) Shingles
   (c) Metal
3. Paint
4. Stucco
5. Brick veneer
6. Stone veneer

(C) *Interior Finish*
1. Plaster and lath
2. Composition board
3. Metal
4. Wood
5. Tile
6. Paint and paper

II FLOORS

(A) *Basic Construction*
1. Wood frame
   (a) Joisted
   (b) Post and girder
2. Steel
3. Masonry
   (a) Concrete on earth
   (b) Reinforced concrete
   (c) Masonry Arch

(B) *Finish*
1. Rough flooring
2. Finish flooring
3. Paint, varnish, etc.
4. Cement
5. Stone
6. Terrazza
7. Tile
8. Tile, linoleum, etc.
9. Insulation

III ROOF

(A) *Basic Construction*
1. Wood frame
   (a) Post and girder
   (b) Rafters
   (c) Trusses
2. Steel
3. Masonry
   (a) Reinforced concrete

(B) *Finish*
1. Roof decking
2. Shingles
3. Roll roofing
4. Built up
5. Metal
6. Insulation

IV EQUIPMENT AND SERVICE UNITS

(A) *Millwork*
1. Windows, doors
2. Trim
3. Stairways
4. Built-in facilities
   (a) Cabinets, etc.
   (b) Mantles
   (c) Bookcases

(B) *Heating and Cooking*
1. Fireplaces
2. Central heating system
3. Floor or space units
4. Ranges
5. Hot water heaters

(C) *Plumbing*
   1. Toilets
   2. Bathtubs and shower stalls
   3. Sinks
(D) *Miscellaneous*
   1. Sprinkler systems

  2. Fire escapes
  3. Incinerators
  4. Refrigeration
  5. Air conditioning
  6. Elevators and escalators

EMERGENCY OR TEMPORARY REPAIRS

After a loss has occurred, there are usually measures which have to be taken promptly to protect and preserve the property from further damage by the elements, trespassers, corrosion, dampness, water, and so forth. Most of the steps to be taken will fall into one or more of the following five classifications:

1. *Protection Against the Elements*
    - (a) Close up openings in roofs and walls. Board up windows and doors. Permanently glaze windows and sash. Temporarily or permanently repair roof and skylights.
    - (b) Drain plumbing and heating pipes, boilers, etc.
    - (c) Provide temporary heat.
    - (d) Install temporary partitions or barricades.
2. *Protection Against Water and Dampness*
    - (a) Air and ventilate premises.
    - (b) Mop floors, scatter sawdust, bore holes in floors.
    - (c) Shut off water.
    - (d) Pump out basement.
    - (e) Clean, oil, dry equipment.
3. *Protection Against Trespassers*
    - (a) Close openings in walls and roof.
    - (b) Provide locks for doors and windows.
    - (c) Provide watchman.
4. *Protection* Against *Unsafe Conditions*
    - (a) Take down hazardous masonry walls and chimneys.
    - (b) Brace.
    - (c) Remove excessive weight of debris or equipment from weak floors.
    - (d) Shore up weak floors or roof structures.
    - (e) Remove contents that swell in water (newsprint, cottonseed, etc.).
    - (f) Shut off gas and electric.
5. *Provide Emergency Facilities*
    - (a) Temporary electric lighting.
    - (b) Temporary electric power.
    - (c) Temporary repairs to elevators.

When he arrives at the premises to inspect the damage and take off the loss, the estimator may discover that part or all of the temporary protection and emergency repairs has been completed. In that case, his notes should be taken in sufficient detail to check any bills or details which may be submitted later.

If none of the temporary or emergency work has been started, the estimator should include in his notes the specifications he believes necessary.

DAMAGE THAT EXISTED PRIOR TO THE OCCURRENCE OF THE LOSS

When a person is inspecting a building loss and making his notes, his objective, in the majority of cases, is to determine the extent of damage which has been caused by a specific hazard or peril covered by the policy. It is necessary, therefore, to be alert to any conditions which indicate a need for repair, but which are in no way related to the peril involved. Buildings that are poorly maintained often contain defects which may consist only of deteriorated paint, broken glass or cracks in the plaster, or they may be far more serious, including such things as rotted timbers, sagged roofs, or bulged masonry walls. Frequently, these conditions are in such close proximity to the damage caused by the peril involved that they are inadvertently included in the estimate of loss. Many times, in fact, it is difficult to separate the two.

The more common defects and types of deterioration will be evident by the following:

1. Sagging roofs.
2. Sagging floors.
3. Cracks in exterior brick walls.
4. Cracks in exterior stucco.
5. Cracks in foundations or basement floors.
6. Cracks in interior plaster.
7. Stained or soiled wallpaper and paint.
8. Efflorescence on interior plaster.
9. Rust and corrosion on metal building materials, equipment or apparatus.
10. Rotted timbers or other woodwork.

These defective conditions may be the result of one or more of the following common causes:

1. Earth movement or subsidence, due to improper drainage, poor fill, shrinkage from drought, tunneling operations below, etc.

2. Poor foundation design—not below frost line, too narrow a footing, improper reinforcing or concrete materials and mixtures.
3. Inadequate size of timbers or their spacing to safely support load intended.
4. Shrinkage of wood timbers and floors.
5. Overloading by tenants.
6. Poor mixture or materials in stucco or masonry mortar.
7. Poor mixture in wall plaster.
8. Plaster or stucco improperly applied.
9. Inadequate bracing and bridging; or ties to veneer brick.
10. Fungus or insect injury to wood supports.
11. Seepage of rain through brickwork or other masonry.
12. Leakage of rain or melted snow through roofing, behind flashing, around windows and doors, or directly down chimney.
13. Hydrostatic pressure under basement floor, or backing up of storm and sanitary sewers.
14. Abuse by use, especially by tenants.
15. Vibration.

An experienced estimator is generally able to recognize most of these conditions and will comment in his notes concerning their existence, their probable cause and cost to repair.

DAMAGE FROM A PREVIOUS LOSS NOT REPAIRED

A property owner who has suffered damage to his building does not always make full repairs, or make repairs on the same basis allowed in the insurance adjustment. If the damage is not structural, even though structural members are involved (as replacement of charred rafters, beams and girders), and if it can be covered up, or is not in an objectionable location, the repairs may consist of patching. In some instances no repairs are made at all.

When a subsequent loss occurs that takes in part or all of the area involved in the previous loss, the overlapping may be completely overlooked by the person making the inspection. It is not easy to distinguish old from new damage in every case. The more serious the second occurrence, especially if it is a fire damage, the more difficult it becomes. Old char maintains the same appearance over many years, except that it may collect dirt, dust and spider webbing. These are signs to look for. Another indication of a prior loss is new-looking or smoked wood cleats, or nailers, on top of actually charred wood. Whitewash is frequently used to cover and kill the odor in char. When whitewashed charred timbers or under-floor are observed, some inquiries should be made about a previous fire. The most likely sources for checking whether there has been a previous loss and also to learn what allowances for damage were made in the adjustment are:

1. Present owner.
2. Former owner.
3. Tenants in the area involved.
4. Superintendent or his assistants.
5. Present insurer's loss department.
6. Former insurer's loss department.
7. Adjuster who handled former loss.
8. Contractor who estimated former loss.
9. Builder who repaired former loss.
10. Real estate management firm.
11. Any of the Company Bureaus through which certain losses clear.
12. Fire department.
13. Insurance agent or broker.

If definite data is found, or reliable statements are available which will disclose the details of the previous loss, a comparison should be made between that information and the actual repairs made. In this manner duplications may be avoided.

VIOLATIONS OF LOCAL BUILDING CODES OR ORDINANCES

Most municipalities have building codes and ordinances which are designed to regulate the alteration, repair, and construction of buildings. They also stipulate certain materials and construction for buildings that may be damaged by fire, lightning, windstorm, and so forth. For example, knob and tube wiring, the use of BX, or armored cable of less than 12 gauge is frequently not permitted in new construction. The code often provides that if an existing structure is damaged, and non-standard wiring is involved, then either the part damaged, or *all* of the wiring in the building must be replaced in accordance with the provisions of the building code.

Building codes and ordinances may also apply to chimney construction, fire-resistive shingles in place of wood shingles, and sizes of timbers and other matters relating to safe construction.

The field notes should indicate any violations, and the estimate should include only the *repairs* necessary to restore what existed at the time of loss, because the standard fire policy is limited to repairing and replacing with like kind and quality, without any increase of loss "occasioned by ordinance or law regulating construction or repair of buildings."

REPAIRS MADE PRIOR TO INSPECTION

It is not uncommon for the estimator to find that repairs were partially or completely made prior to his inspection. When such repairs are minor, and readily explained as to the scope, there is no particular problem. The estimator obtains a copy of the estimate or invoice for making the repairs and prepares his specifications and costs, frequently having the owner or owner's representative along to answer any questions.

If, however, the repairs are extensive, the estimator may need considerable time and ingenuity to establish with any degree of accuracy what occurred and develop a set of specifications on which to base his estimate.

His principal sources of information will be:

1. The insured or his representative.
2. The general contractor's estimate or records of payroll and materials used.
3. Subcontractor's records, estimates, etc.
4. Tenants in the area (if any).
5. Other insurers if they are involved.
6. Any plans, pictures, building permits, etc.

When the estimator has satisfied himself that there was a fire, windstorm, sprinkler leakage, or whatever, and he has a general idea of the part or parts of the structure affected, he can, by using his imagination and the information furnished him, develop an estimate. He has one advantage in that the materials used, as to kind and quality, are subject to verification. He may not be able to establish with certainty that particular rooms required redecorating or the walls and ceilings needed re-plastering. Judgment and the credibility of persons interviewed will be his guide.

CONSTRUCTIVE TOTAL LOSSES

Estimators are many times confronted with the question of whether the building involved is a constructive total loss. Where valued policy laws are in effect, it is an extremely important question if the amount of insurance is in excess of the cost of rebuilding. Even where valued policy laws are not in effect it is important if a reasonable part of the structure remains upon which to begin rebuilding.

The general rule applied by the courts seems to be:

> Whether a building is an actual total loss (by fire) depends upon whether a reasonable prudent owner, uninsured, desiring to rebuild, would have used the remnant for restoring the building.

ESTIMATING LOSSES CAUSED BY MIXED PERILS

Occasionally damage is caused by two perils—one covered, the other not covered. For example, during hurricanes along the seacoast, properties close to the shore are frequently damaged by wind (a covered peril), and by wave-wash (an excluded peril). Similarly, if a fire policy excludes "explosion" and one occurs followed by fire, the damage done by the explosion is not covered. Or, if a fire policy does not cover earthquake, and one occurs, only ensuing fire damage is covered.

Damages by mixed perils such as those described, require the estimator to resort to a three-column estimate in which he lists the repairs to damage covered, the damage not covered, and the damage which is questionable as to its being covered. (See Estimating in Catastrophes, Chapter 5)

However, if a covered peril such a fire operates and during the fire there is an explosion, all of the damage is covered because fire is the proximate cause.

Chapter 4.

Pricing and Completing the Estimate

After the damaged property has been inspected and all the measurements and specifications for repair have been compiled, the final step to complete the estimate and obtain the total cost of repairs, is the pricing. It consists of computing the cost of the materials required for each item listed, and estimating the number of hours and cost of the labor necessary to install the material or perform the operation indicated. Material costs are based on the local prices, delivered to the job. Labor costs are based on prevailing local wage scales for the building trades involved.

PRICING METHODS

As explained under "Fundamental Considerations" (Chapter 2), there are different customs and practices among contractors for figuring the cost of material and labor. Whatever procedure is followed, if it is to be considered reliable, it must take into consideration these two factors.

1. Quantity of Material x Price = Cost of Material
2. Hours of Labor Required x Wage Scale = Cost of Labor

There are three general methods of pricing and setting up material and labor costs in an estimate.

1. The material is priced in detail for each item, and the labor cost is bulked in a lump-sum amount.
2. The material and labor are individually priced for *each* item, and shown separately.
3. A unit cost is applied to each item to obtain a single sum for the combined cost of material and labor.

METHOD 1: MATERIAL PRICED, LABOR LUMPED

Many of the smaller contractors and builders employ this method. All of the materials are listed for each item of repair, quantities are extended and properly priced. The cost of labor is then determined by approximating the hours or days required to perform all of the work. Where minor repairs are involved, the method can be fairly accurate, but when it is used on losses where extensive repairs are required, it can take on the character of *guessing* rather than *estimating*. Carpentry work is often figured in this manner.

Illustrative Example

Framing	Dimensions	FBM			
42 Joists	2" x 12" x 16'	1,344			
48 Studs	2" x 4" x 9'	288			
19 Plates	2" x 4" x 12'	152			
12 Joists	2" x 8" x 16'	256			
2 (Girder)	2" x 12" x 14'	56			
2 (Girder)	2" x 12" x 16'	64			
38 Studs	2" x 4" x 8'	202			
18 Plates	2" x 4" x 12'	144			
15 Rafters	2" x 4" x 18'	180			
8 Rafters	2" x 6" x 18'	144			
Blocking	4" x 6" x 8'	16			
		2,846	@ .160	$	455.36
Roof Sheathing					
1,000 FBM (1" x 6")			@ .150		150.00
Sub-Flooring					
1,200 FBM (1" x 6")			@ .150		180.00
Finish Floor					
1,500 FBM (1" x 3" C.G. Pine)			@ .300		450.00
				Material:	$1,235.36
Carpenter Labor (25 days @ $48.00)					1,200.00
Material and labor:					$2,435.36

It should be noted that some contractors who use this method may set up the labor cost in the finished estimate as a lump sum, but *actually* they have worked out the labor cost for each individual item which can be verified from their working papers.

METHOD 2: MATERIAL AND LABOR PRICED AND SHOWN SEPARATELY

In this method the material cost is computed for each item and extended to a "Material Cost" column. Labor for each item is determined and the cost is extended to a "Labor Cost" column. A third column indicates the "Total" cost for both material and labor.

Illustrative Example

Item			Material Cost	Labor Cost	Total Cost
Framing					
12 Studs	2″ x 4″ x 10′	80 FBM			
16 Joists	2″ x 6″ x 16′	256			
20 Rafters	2″ x 6″ x 14′	280			
		616 FBM	@ $ .160	$ 98.56	
20 Carpenter hours			@ 6.00		$120.00	$218.56
Sheathing					
864 FBM			@ .150	129.60	
18 Carpenter hours			@ 6.00		108.00	237.60
					$456.16

The form in which the estimate is set up varies with individual preferences of contractors. Printed forms are available from firms like Frank R. Walker, Chicago, Illinois, that specialize in contractors' stationery. Some of them provide columns for both quantities and unit prices for material and labor. Examples of this and other forms appear on pages 98, 99 and 100.

METHOD 3: UNIT COSTS FOR MATERIAL AND LABOR COMBINED

Considerable estimating, especially for new construction, is done by the *unit cost* method. A unit cost is the combined cost of the material and labor needed to install a unit of material.

If, for example, a roofer knows that he can lay 235-lb. asphalt strip shingles on a plain one-story roof of a dwelling at the average rate of 1 square every 2 hours, he can estimate the cost of doing a roof in two ways. If the roof requires 20 squares of shingles, the roofer can estimate the time

Form 4-1

PRACTICAL STANDARD Forms for Contractors FORM 514 MFD. IN U.S.A. FRANK R. WALKER CO., PUBLISHERS, CHICAGO												

GENERAL ESTIMATE

BUILDING _____ ESTIMATE NO. _____

LOCATION _____ SHEET NO. _____

ARCHITECTS_____ ESTIMATOR _____

SUBJECT_____ CHECKER _____

DATE _____

DESCRIPTION OF WORK	NO PIECES	DIMENSIONS		EXTENSIONS	EXTENSIONS	TOTAL ESTIMATED QUANTITY	UNIT PRICE M'T'L	TOTAL ESTIMATED MATERIAL COST	UNIT PRICE	TOTAL ESTIMATE LABOR COST

Form 4-2

FORM 420 5 74

GAB General Adjustment Bureau, Inc **Detailed Estimate**

DATE _____

BRANCH _____ OFFICE TELEPHONE NO _____

OWNER	FILE NUMBER	BUILDING	CONTENTS
LOCATION	TELEPHONE NO	$ _____ ESTIMATED REPLACEMENT COST	$ _____ ESTIMATED REPLACEMENT COST
POLICY NO	CLAIM NO	DATE OF LOSS	TYPE OF LOSS	$ _____ LESS DEPRECIATION	$ _____ LESS DEPRECIATION
TYPE OF BUILDING	BUILDING AGE	TOTAL SQ FT	$ _____ ACTUAL CASH VALUE	$ _____ ACTUAL CASH VALUE
ROOF TYPE	TYPE OF CONSTRUCTION	NO ROOMS	NO BATHS		

DESCRIPTION	CHECK APPLICABLE ITEM	BUILDING → / CONTENTS →	LABOR HOURS / AGE	LABOR RATE / ITEM COST	① TOTAL LABOR / DEPRECIATION	② MATERIALS / ③ ACTUAL CASH VALUE
(CONTINUE ON FORM 448)			TOTAL			

TOTAL LABOR (BUILDING)	①	$
TOTAL MATERIALS (BUILDING)	②	$
TOTAL REPAIRS (BUILDING)		$
LESS DEPRECIATION		$
ACTUAL CASH VALUE (CONTENTS)	③	$
LESS DEDUCTIBLE OR CO-INSURANCE PENALTY		$
ORDERED BY (APPRAISALS ONLY)	TOTAL CLAIM	$

AGREED TO BY (APPRAISALS ONLY) _____ PREPARED BY _____

Form 4-3

Kemper
INSURANCE

BUILDING REPAIR ESTIMATE

	CLAIM NO.

INSURED

LOCATION OF PROPERTY

TYPE OF LOSS	LOSS DATE	TYPE OF BUILDING

REPLACEMENT VALUE	ACTUAL CASH VALUE	ESTIMATOR	DATE OF ESTIMATE

WHEN COMPLETING THE DATA BELOW, TRY TO DEVELOP AND USE "UNIT PRICES"; OTHERWISE, ESTIMATE "MATERIAL COST" AND "LABOR COST" SEPARATELY.

DESCRIPTION OF WORK REQUIRED (IDENTIFY MATERIALS — SHOW ROOM OR AREA MEASUREMENT)	QUANTITY	UNIT PRICE (WHEN USED)	MATERIAL COST	LABOR COST	TOTAL COST

EK432 2-73 1M PADS

PRINTED IN U.S.A.

required by multiplying 20 by 2 and obtain 40 hours. If the price of the shingles is $9.00 and his scale of wages is $6.00 per hour, his estimate would be

20 squares shingles	@ $9.00 =	$180.00
40 hours labor	@ 6.00 =	240.00
Cost of material and labor	. . . .	$420.00

The unit cost method of estimating the shingling would take this form.

UNIT COST OF 1 SQUARE SHINGLES

1 square shingles	@ $9.00 =	$ 9.00
2 hours labor	@ 6.00 =	12.00
Cost per square	. . . .	$21.00

Since there are 20 squares of shingles to be laid, the total cost of the material and labor is $21.00 x 20 = $420.00.

The advantage of this method of estimating is immediately apparent. The unit cost may be reduced to the cost per square foot by dividing $21.00 by 100 (1 square = 100 square feet). In this example the cost per square foot is $.21, and, with that unit cost in mind, all the roofer is required to do in order to estimate similar jobs is to measure the roof, obtain the number of square feet needed and multiply by $.21. Whether the unit cost per square or per square foot is used is a matter of choice. If the roof he is estimating requires 3,000 square feet of shingles (30 squares) the estimate would be 3,000 x $.21 = $630.00 or 30 x $21.00 = $630.00.

This is the principle of developing unit costs. It can be used for most of the building materials, and also for pure labor operations such as removing shingles from a roof, washing down walls or brickwork, and so forth.

DEVELOPING UNIT COSTS

A unit cost may be developed by starting out with a quantity of material for which the average labor rate of application or installation is known. The quantities used for *developing* the unit cost for typical materials are:

Concrete forms	100 sq ft	(square feet)
Concrete	1 cu yd	(cubic yard)
Framing lumber	1,000 FBM	(feet board measure)
Sheathing)		
Subflooring)		

Roof decking)	1,000 FBM	
Wood siding)		
Wood flooring)		
Wallboards	100 sq ft	
Wood trim	100 lin ft	(lineal feet)
Cement block	100 block	
Brickwork	1,000 brick	
Lath and plaster	100 sq yds	(square yards)
Painting	1 gal	(gallon)
Roofing	1 square	(100 square feet)

The quantities generally used to *express* unit costs are shown in Table 4-1.

Table 4-1
QUANTITIES GENERALLY USED IN ESTIMATING MATERIALS & UNIT COSTS

Kind of Material	*Quantity*
Acoustical tile	sq. ft.
Aluminum siding	sq. ft.
Asbestos shingles	square
Asbestos siding	square
Baseboard, mouldings, etc.	lin. ft.
Beam, wood	b/f FBM
Beam, steel	lb. or ton
Block, concrete or cinder	ea. or per 100
Boards	b/f FBM
Bolts	lb.
Brick, clay	per M (1000)
Brick, glass	ea. or sq. ft.
Building paper (or felt)	roll or sq. ft.
Cabinets, kitchen	ea. or lin. ft.
Carpeting	sq. yd.
Casing, door & window	lin. ft. or side
Celotex	sq. ft.
Cement, bagged	bag
Chimney flue lining	ea. or lin. ft.
Cinder block	ea. or per 100
Composition shingles	bdl. or sq.
Composition sheathing	per sq. ft.
Concrete block	ea. or per 100
Concrete forms	sq. ft.

Kind of Material	*Quantity*
Concrete; poured	cu. ft. or cu. yd.
Coping	lin. ft.
Cornice	lin. ft.
Counter top (vinyl, linoleum, Formica)	sq. ft. or lin. ft.
Dimension lumber	b/f FBM
Doors	each
Disposal units	each
Door hardware	each or set
Downspouts	lin. ft.
Electrical fixtures	each
Electrical wiring	lin. ft. or outlet
Excavating	cu. yd. or cu. ft.
Fixtures, electrical	each
Fixtures, plumbing	each
Flagstones	each or sq. ft.
Flashing, metal	lin. ft.
Flooring, wood	b/f FBM or sq. ft.
Flue liners	each or lin. ft.
Floor covering, carpet & linoleum	sq. yd.
Framing, wood	b/f FBM
Furring, grounds	per 100 lin. ft.
Girders, wood	b/f FBM
Girders, steel	lb. or ton
Glass, window	each
Glass, plate	sq. in. or sq. ft.
Glass block	each
Guttering	lin. ft.
Gravel or stone	cu. yd. or ton
Hardware, door & window	each or set
Hollow tile	each or per 100
Insulation	sq. ft.
Joist, wood	b/f FBM
Joist, metal	each or lb.
Kitchen cabinets	lin. ft. or unit
Knobs, door and cabinet	each or set
Lath, metal or gypsum	sq. yd.
Lath, wood	bdl. or sq. yd.
Linoleum	sq. yd.

Table 4-1 (Continued)
QUANTITIES GENERALLY USED IN ESTIMATING MATERIALS
& UNIT COSTS

Kind of Material	*Quantity*
Marlite	sq. ft.
Masonite	sq. ft.
Metal lath	sq. yd.
Metal roofing	per sq. or sq. ft.
Mortar, lime	bag or cu. yd.
Mouldings, wood	lin. ft.
Nails	lb.
Paint	qt. or gal.
Painting	sq. yd. or sq. ft.
Paper, building	roll or sq. ft.
Pipe, water or gas	lin. ft.
Plaster	bag or lb.
Plastering	sq. yd.
Plates, wood	b/f FBM
Plate glass	sq. ft.
Plywood	sq. ft.
Putty	lb.
Rafters, wood	b/f FBM
Registers, heat	each
Reinforcing rod	lb.
Ridge tin	lin. ft.
Roofing, metal	sq. ft. or sq.
Roofing, shingles (wood or composition)	bdl. or sq.
Roofing, roll	roll or sq.
Roofing, built-up	sq.
Roof sheathing, wood	b/f FBM
Roof sheathing, gypsum panels	sq. ft.
Roof sheathing, plywood	sq. ft.
Sand	cu. yd. or ton
Sanding floors	sq. yd. or sq. ft.
Sash, window	each
Screen, window & door	each
Screen wire	sq. ft.
Screws	lb., doz., or gross
Sheathing, boards	b/f FBM
Sheathing, plywood	sq. ft.
Sheet rock panels	sq. ft.
Shelving, boards	b/f or lin. ft.

Kind of Material	*Quantity*
Shingles, asbestos, wood & composition	bdl. or sq.
Shutters, window	each or pair
Siding, metal & asbestos	sq. ft.
Sills, stone or concrete	cu. ft.
Sills, wood	b/f FBM
Skylight glass	sq. ft. or unit
Space heaters	each
Stairs, metal	assembly
Stairs, wood	assembly or per tread & riser
Siding, wood	b/f FBM
Steel beams	each or ton
Stone, decorative	CWT (100 lb.)
Storm doors and windows	each
Structural steel	per lb. or ton
Stucco work	sq. yd.
Studding	b/f FBM, or per stud
Tanks, fuel & water	gal. capacity
Terrazzo floors	sq. ft.
Tile, floor & wall	each or sq. ft.
Tile, building	each or sq. ft.
Tinwork, gutter, ridge & valley	lin. ft.
Toilet bowls & tanks	each
Trim, door & windows	lin. ft. or set
Underlay, felt	roll or sq. ft.
Underlay, plywood	sq. ft.
Valley tin	lin. ft.
Varnish	qt. or gal.
Varnishing	sq. ft. or sq. yd.
Venetian blinds	each or sq. ft.
Ventilators, roof	each or sq. ft.
Vinyl, tile	each or sq. ft.
Vinyl, roll	sq. ft. or sq. yd.
Wallpaper	per roll or sq. ft.
Waterproofing	100 sq. ft. or per sq. yd.
Weather stripping	per opening
Wood paneling	per sq. ft.
Wood siding	sq. ft. (or FBM)

The following basic formula is used for developing unit costs. The local price of material is inserted, and the hours of labor are multiplied by the prevailing hourly wage scale.

Basic Formula

Quantity of material   @ $ = $
Hours of labor   @   =
Unit cost of material and labor = $

Illustrative Example
Framing Lumber

1,000 FBM lumber	@ $160.00 =	$160.00
10 Lbs nails	@ .20 =	2.00
25 Hours labor	@ 6.00 =	150.00
Unit cost per 1,000 FBM	=	$312.00

$$\text{Unit cost per FBM} = \frac{312}{1,000} = \$.312$$

One-coat Interior Paint
(Assume 1 gal paint covers 450 sq ft)

1 Gal paint	@ $7.00	= $ 7.00
3 Hours labor	@ 6.00	= 18.00
Cost per 450 sq. ft.		= $25.00

$$\text{Unit cost per sq. ft.} = \frac{\$25.00}{450} = \$.056$$
Unit cost per sq. ft.

An estimate that is priced by the unit cost method might appear in final form as follows:

Repairs To Garage

Framing
10 – 2″ x 4″ x 12′ studs	80 FBM	
6 – 2″ x 6″ x 16′ rafters	96 FBM	
	176 FBM @ .31 . . .	$ 54.56

Roof Boards
20′ x 24′	480 sq. ft.	
Waste 1/6	80 sq. ft.	
	*560 FBM @ .25 . . .	140.00

Bevel Siding
10′ x 16′	160 sq. ft.	
Waste 1/3	54 sq. ft.	
	214 FBM @ .50 . . .	107.00

Exterior Paint
2,400 sq. ft. (2 coats)	@ .14 . . .	336.00
Total material and labor		$637.56
Add overhead and profit 20%		127.51
		$765.07

*When waste is added to the square foot area, the total becomes board feet.

A VARIATION FOR SETTING UP UNIT COSTS

Many contractors, rather than using a unit cost of the combined material and labor, prefer to use a separate unit cost for each. The basic principle is the same. Using again the example of the roofer who estimates an average of 2 hours per square to apply 235-lb. asphalt strip shingles, his unit cost for material and labor may be shown individually. The price of the shingles is $9.00 per square or $.09 per square foot. His wage scale is $6.00 per hour and the cost per square for labor is 2 x $6.00 or $12.00, which is equivalent to $.12 per square foot. The estimate for 2,000 square feet (20 squares) may be set up in this manner.

	Material Unit Cost		Labor Unit Cost		Total
2,000 sq. ft. shingles @ $ .09	$180.00	$ .12	$240.00		$420.00
(or)					
20 squares of shingles @ 9.00	180.00	12.00	240.00		420.00

CHARACTERISTICS OF UNIT COSTS

Although theoretically a unit cost should consist only of the cost of the materials and the labor cost to install them, there are variations in practice. Some contractors include overhead and profit. Others include removal of damaged material preparatory to installing the new. Still others may include waste. In checking a unit cost, if the percentage of overhead and profit is known, it can be deducted. But since demolition or tearing out and waste are difficult to extract from a unit, it is recommended that they be treated separately in estimating.

It is recommended that waste *be added to the quantity of material* to be installed. Demolition, except for a few materials such as shingles, is best estimated under the general debris item. A unit cost should contemplate that the debris and damaged material have been removed preparatory to reconstruction, and that the conditions under which new materials are to be installed are comparable, to a large extent, to those normally encountered in *new construction* work.

While unit costs are developed from such quantities as 1,000 bricks or 1,000 FBM of lumber, they can be applied readily to smaller quantities of material where the working conditions are fairly normal. For example, a unit cost for finished 1" x 3" flooring developed from 1,000 FBM can be applied to the quantity of flooring required in a room 10 x 10 feet:

$$10' \times 10' = 100 \text{ sq ft}$$
$$\text{Waste } \frac{1}{3} = \underline{34}$$
$$134 \text{ FBM}$$

It would not be proper, however, to apply such a unit cost to a 2' x 8' closet where the area worked in is confined and difficult as compared to an open area contemplated in estimating the production rate for the 1,000 FBM. Here, as in all estimating, judgment must be exercised. The unit price can be increased to compensate for the additional labor required or the material and the time needed to lay the floor can be estimated separately.

Unit Costs Not Applicable to Patch Jobs

Unit costs, as a general rule, cannot be applied to small patch jobs requiring only a few yards of plaster, or several square feet of flooring or roofing. The reason is that the labor cost in patch work is very large compared to the material cost. It will distort the combined material and labor unit cost, making it difficult, if not impossible, to judge one that is adequate. In such cases the cost of the material should be obtained, and the hours of labor figured separately for the entire patch. In other words, the estimate should be made on a time and material basis. A roofer may spend an entire day patching in a square of shingles. If the shingles cost $10 a square, and labor is $6.00 an hour, the total cost per square would be:

1 square of shingles	$10.00
8 hours labor @ $6.00	48.00
Total cost per square	$58.00

This is considerably different from a unit cost developed for average straight work involving an entire roof.

1 square of shingles	$10.00
2 hours labor @ $6.00	12.00
Total unit cost per square	$22.00

Extracting Labor from Unit Costs

Estimators will frequently increase a unit cost to take care of more difficult working conditions than those contemplated when originally computing it. There is no difference in the amount of material in the unit, therefore *the entire increase must be charged to labor.* In an analysis of unit costs this characteristic must be constantly remembered, otherwise any indiscriminate increasing of unit costs *will automatically raise the hours of labor* to unrealistic quantities. If a predetermined unit cost for asphalt roof shingles is 21 cents a square foot, but is raised to 25 cents, the additional 4 cents a square foot is entirely labor.

1 square of shingles	$ 9.00	
2 hours labor @ $6.00	= $12.00	
Total	= $21.00	= $.21 Per sq ft
1 square of shingles	$ 9.00	
2 ⅔ hours labor @ $6.00	= 16.00	
Total	= $25.00	= $.25 Per sq ft

For ordinary work, two hours of labor per square of asphalt shingles should be adequate. Two and two-thirds hours per square for average work is excessive unless there are many gables or the work is unusually difficult.

It is a simple matter of arithmetic to separate the material and the labor when applying unit costs to check the actual number of hours of labor in an estimate.

Assume that the unit cost of asphalt shingles is $21 per square, and apply this unit to a roof requiring 30 squares of shingles. The total cost would be $21 x 30 = $630. If the labor in the unit cost is based on a roofer laying shingles at the rate of 2 hours per square, then the hours of labor to apply the 30 squares would be 60 hours (2 hours per square x 30 squares).

If the unit cost of framing for a house is based on carpenters framing lumber at a rate of 25 hours per thousand, the total number of hours required to frame the house would be calculated by multiplying the number of thousand board feet by 25 hours. Assume that the total framing lumber in a house is 5,700 FBM. The number of hours of carpenter labor would be 5.7 x 25 = 142.5 hours.

This is a convenient way to compare hours for hours when one estimator is using a unit cost method, while another is using material cost plus a flat number of hours for a single operation or for the entire job.

REFINEMENT OF A UNIT COST

The question is often asked as to the number of decimal places to which a unit cost should be carried. The cost of framing lumber might be computed at $213. per 1,000 FBM. This would be 21.3 cents a board foot. Whether the unit is made 21 cents, 21.5 cents or 22 cents is not too important and the choice depends on a number of factors. Except in highly competitive estimating *units are used in whole cents*. If the unit cost for roll-roofing came out to 6.1 cents, the estimator might give considerable thought before he raised it to 7 cents, which is an increase of 6.1 or 14.7 per cent. He would more likely use 6 cents or $6.10 per square. In contrast, if the unit cost of oak flooring is computed at 44.2 cents, raising it to 45 cents would represent an increase of 44.2 or less than 2 per cent.

Judgment should be based on all of the circumstances of the liberality of the hour allowance in the unit as originally computed, of the competitive advantage of lowering the unit, and the people with whom the estimator is dealing.

MINIMUM CHARGES FOR SMALL JOBS

Many times small repair jobs are subject to minimum charges as to labor and material, either combined or individually. For example, one cubic yard of ready-mix concrete to be delivered over several miles may cost the same as two yards. One lite of glass may cost as much as two or three. Several shingles to be replaced; a few hours of painting; checking electric appliances, a furnace, plumbing, etc., often are subject to minimum charges. The reason may involve the distance to be travelled, or, the fact that the best part of a day has been taken up and it is too late to go to another job. High overhead may also be a consideration in small jobs. (See also Union Regulations, p. 50).

These are important considerations for the estimator particularly in periods or in areas where builders and subcontractors are extremely busy. Small jobs can be an annoyance because of having to take a workman away from a large job for a few hours or days.

In inflationary periods, the cost of making an estimate is frequently billed to the customer if the job is not given to the firm or individual who made the estimate.

THE FINISHED ESTIMATE

Since the real purpose and usefulness of an estimate of damage is to provide the groundwork for discussing the adjustment of a loss, the final form should set forth every item of repair or replacement. The work to be done should be so itemized and clearly described, that even the uninitiated would be able to understand it. Quantities and the cost of labor and materials should be indicated.

Bulking, grouping, or lumping two, three, or several items into one sum, only tends to confuse and introduce unnecessary areas of misunderstanding. It raises differences of interpretation as to exactly what the estimator meant or intended. Such practices are without good reason. An estimator, if he endeavors to be as accurate as possible, must break down each operation of repair. The final estimate should reflect such a breakdown. Each item should lend itself to analysis because most of the differences in estimates can be traced to errors in measurements, arithmetic, material costs, labor allowances, or judgment.

Follow Field Notes

Generally, when field notes have been taken off as suggested in Chapter 3, there is a logical starting point, and continuity flows throughout these notes. The sizes of rooms, descriptive specifications, dimensions, and measurements need only be transcribed to the final estimate. Areas and quantities can be computed, priced, and extended.

Frequently, it may be desirable or convenient to keep certain types of work or trades together. Framing, for example, may involve the roof structure, some of the sidewalls, and perhaps a few of the floor joists in the rooms below. These items may be pulled out of the notes and listed under one heading, "Framing," in the final estimate. When subcontractors are used on such trades as plastering, painting, electrical or steelwork, the work contemplated by each should be specified in his bid or estimate in detail to permit checking. A subcontractor's estimate in a lump sum amount has little or no value.

Bulking Labor

This practice is not recommended in the final form. It is meaningless to anyone except possibly the person who prepared the figures. Not many estimators can guess at the overall number of hours that a job of any size will take. Many very capable estimators show labor in bulk, but they obtain the final labor figure by a careful analysis of each operation. Such analysis is usually in their notes or worksheets. Because building loss estimates require explicit details for negotiating, the labor as well as the material should be itemized for each operation.

Showing Unit Costs

The final form need not necessarily show individual unit costs because they may be computed by dividing the cost of the item by the quantity. For simplicity, however, it is desirable that unit costs be shown for each item.

Overhead and Profit

It is also recommended for uniformity that contractor's overhead and profit be added at the end of the entire estimate. Unit costs, or labor and material costs, for each item are better shown on a "net" basis; that is, one that excludes overhead and profit. The cost of each item should be made up solely of the material and the labor necessary to perform the operation or install the material. One exception to this recommendation may be made in catastrophe losses where unit costs, which sometimes include overhead and profit, are agreed to by the local contractors' association and representatives of the insurance companies (See Chapter 5).

Other exceptions may be made where the pricing of certain items is based on a unit cost quotation of a subcontractor. This may be a roofer, painter, mason, electrician or tiler who quotes a price per unit which the estimator uses, and which includes the subcontractor's overhead and profit.

Recapitulation

Estimators who figure a great many building losses recognize the value of recapping their estimates by trades. A comparison of the recapitulation of two estimates will quickly reveal whether the differences are in one or two items, or are scattered throughout. Recapping an estimate is done after the entire estimate is completed and priced. The amounts of money shown for every item in a particular trade as carpentry, painting, plastering, or electric are added together. The various trades are listed on a separate sheet entitled "Recapitulation" and the total is shown opposite each trade. This should be done with great care, with a check-off of each item as it is listed from the estimate. The total of the recapitulation must always equal the total for the detailed estimate.

CUSTOMARY FORMS USED

One of the most acceptable methods for setting up the finished estimate is to list each item of repair. The quantities are multipled by a combined unit cost for material and labor. It is a popular form that is used in many sections of the United States and Canada, being especially favored in larger cities where there is a higher incidence of loss, and where experience has well demonstrated its time-saving usefulness in discussing the details.

Another method that is used which serves the same purpose, is to list the items, and show the cost of the materials and the labor separately. In this method, generally a third column is used to indicate the total cost of the material and labor for each listed item.

Individual estimators use variations of these two methods, according to local custom and practice. These approaches are entirely satisfactory, provided they are subject to analytical discussion and also provided they do not contain bulked items or lumped dollar amounts with no apparent explanation.

The following loss estimates have been selected from a number submitted by contractors with considerable experience in estimating building losses and in making repairs.

Form 4-4

LOUIS PASQUALUCCI AND SON, INC.
GENERAL CONTRACTORS
QUINCY, MASS.

GEORGE A. PASQUALUCCI, A.S.A.
40 SUMNER STREET
P.O. BOX 514
QUINCY, MASS. 02169
472-1452

BUILDING APPRAISALS

SENIOR MEMBER
AMERICAN SOCIETY OF APPRAISERS

January 1, 1975

Lumbermens Mutual Casualty Company
196 Main Street
Brockton, Massachusetts

Attention of Mr. Charles Gillon

> Re: Insured: Vicky L. Smith
> Claimant: Martin P. Jones
> 112 Elm Street, Boston
> Damage to foundation by vehicle
> Date of Loss: July 3, 1974

Dear Mr. Gillon:

At your request, I have examined the above captioned reference and submit the following report to restore the property to its pre-loss condition:

	LABOR	MATERIAL	TOTAL
FRONT PORCH OPEN TYPE 25'0" x 7'9" x 8'8" HT. TO CEILING LINE			
TEMPORARY REPAIRS			
Shore up roof - second hand lumber as requred 2 x 10 1-12'8" 2 x 6 1 - 12'10" 1/6 1/5 Allowance - Lump Sum		52.50	52.50
DEMOLITION			
Floor complete, framing, underpinning Corner posts - 3 No. Rail and balustrades 2 sections - Total	388.80	96.00	484.80
SHORING & BRACING			
Jack up porch entire length and re-align. Note: approximately 9¼" out of square	110.55	45.00	155.55

Form 4-4 (cont.)

LOUIS PASQUALUCCI & SON. INC. - GENERAL CONTRACTORS   Re: Claimant: Martin P. Jones -2-

	LABOR	MATERIAL	TOTAL
REPAIRS			
Rebuild brick pier - 4 No.			
10" x 10" x 2'4" high -			
320 brick	112.00	64.00	176.00
Framing - 2 x 4 2/8 11 b.f.	2.64	3.52	6.16
2/10 14 b.f.	3.36	4.48	7.84
Lattice			
Wood 1 3/8" - Dimensions			
2'8" ht. x 18'3" x 7'9"			
Total	61.20	26.20	87.40
Trim			
Base 1 x 5 2/10 2/8 36 l.f.	10.80	9.00	19.80
Fascia 1 x 8 2/10, 2/8 36 b.f.	10.80	14.40	25.20
Moulding 1½" scotia 2/10 2/8			
36 l.f.	5.40	5.04	10.44
Corner boards 1 x 7 1/6 6 l.f.	1.80	2.40	4.20
1 x 6 1/6 6 l.f.	1.80	1.80	3.60
Flooring			
4" fir - matched T&G			
Entire porch 270 s.f.	121.50	199.80	321.30
Corner posts - remove and reset			
3 No. 7" x 7" x 7'0"	55.35		55.35
Railings			
Rails and balustrades			
Reset - 2 sections	38.70		38.70
PAINTING			
All new work - 2 coats			
Existing to match - 1 coat			
Trim, floor, rails, lattice			
work only - Total		240.00	240.00
	924.70	764.14	1,688.84
3% Mass Sales Tax on Material			22.92
33% of Labor - includes taxes			
benefits, insurance, etc.			305.15
			2,016.91
10% Overhead			201.69
			2,218.60
10% Profit			221.86
			2,440.46

Respectfully submitted,

George Pasqualucci

George A. Pasqualucci, A.S.A.

GP:ja

Form 4-5

TELEPHONE 751-3816

GUNN & GUNN *Construction Company*

161 N. W. 52 STREET, MIAMI, FLORIDA 33127

January 1, 1975

Universal Insurance Company
111 S. W. 8th Street
Miami, Florida

Re: Mr. John Doe
     000 S. W. 86th Avenue
     Dade County, Florida

Gentlemen:

We propose to furnish labor and materials to repair fire damage to the one story C.B.S. single family residence as shown in the following:

Kitchen 8'6"x9'x8'6"

Rocklath & plaster ceiling & drop soffit over cabinets		18 yds @	6.00 yd	$108.00
Replace 1/16" vinyl asbestos floor tile		76 sf @	.90 sf	68.40
" 23 size glass jalousie window complete		1 @		39.70
" 3620 painted wall cabinet over exhaust hood		@		65.52
" 108" formica form top with 4" backsplash		@		112.50
" tarnished wall & base cabinet hinges & pulls		@		38.71
" 36" Nutone hood & fan reusing existing ductwork		@		84.20
Labor removing & setting cabinets & form top		@		78.00
Plumbing & electrical connections for sink & cook top		@		56.00
Replace wallpaper on drop soffit-minimum job		@		50.00
Paint new plaster ceiling	2 coats	76 sf @	.14 sf	10.64
Clean, seal & paint plaster walls	3 "	297 sf @	.26 sf	77.22
Enamel doorsides & trim	2 "	2 @	7.00 ea	14.00
Clean, prepare & refinish wall & base cabinets		66 sf @	2.85 sf	188.10

Dining Room 9'x10'x8'6"

Sand & refinish oak floor		90 sf @	.37 sf	33.30
Refinish, retape & record venetian blinds		26 sf @	.98 sf	25.48
Clean & paint 3 plaster walls & ceiling	2 coats	336 sf @	.16 sf	53.76
Enamel doorside & trim	2 "	1 @		7.00
" baseboard & shoe molding	2 "	29 lf @	.13 lf	3.77
Clean jalousie windows & screens		2 @	4.50 ea	9.00
Clean ceiling light fixture		1 @		2.50

Living Room 15'6"x17'x8'6"

Sand & refinish oak floor		263 sf @	.37 sf	97.31
Refinish, retape & record venetian blinds		52 sf @	.98 sf	50.96
Clean & paint plaster walls & cei ling	2 coats	815 sf @	.16 sf	130.40
Enamel doorsides & trim	2 "	3 @	7.00 ea	21.00
" baseboard & shoe molding	2 "	65 lf @	.13 lf	8.45
Clean jalousie windows & screens		4 @	4.50 ea	18.00

Form 4-5 (cont.)

Page #2, John Doe

Hallway 3'x18'x8'6"

Clean & paint plaster walls & ceiling	2 coats		411 sf @	.16 sf	65.76
Enamel louvre doorsides & trim	2 "		2 @	9.50 ea	19.00
" doorsides & trim	2 "		3 @	7.00 ea	21.00
" baseboard & shoe molding	2 "		42 lf @	.13 lf	5.46
" attic scuttle hole cover,frame & trim	2 "		1 @		4.50
Clean wax & buff oak floor			54 sf @	.18 sf	9.72

Hall Closet 2'6"x5'x8'6"

Clean & paint plaster walls & ceiling	2 coats		140 sf @	.16 sf	22.40
Enamel doorside & trim	2 "		1 @		7.00
" baseboard & shoe molding	2 "		15 lf @	.13 lf	1.95
" shelves & hang rod	2 "		3 @	1.50 ea	4.50
Clean, wax & buff oak floor			13 sf @	.18 sf	2.34

Bathroom 5'x8'x8'6"

Clean & paint plaster walls & ceiling	2 coats		161 sf @	.16 sf	25.76
Enamel doorside & trim	2 "		1 @		7.00
Clean tile floor, wainscoat, fixtures, medicine cabinet, window & screen			@		35.00
Clean light fixture			1 @		1.50

Sub Contractors & General Conditions
Electrical- kitchen,rewire circuit to hood fan,
check all outlets, switches, recepticals to
electric panel- connect new hood fan, furnish
& install 8" drum type electric ceiling fixture   @   122.00

Job cleanup, hauling debris to dump, scaffolding
rental and building repairs permit   @   75.00
                                                                                      ‾‾‾‾‾‾‾
                                                                                      1881.81
Contractor's Overhead 10%   188.18
                                                                                      ‾‾‾‾‾‾‾
                                                                                      2069.99
Contractor's Profit 10%   206.99
                                                                   Total $2276.98

Thank you for the opportunity of submitting this proposal.

Sincerely,

GUNN & GUNN CONSTRUCTION COMPANY

JOHN GUNN

JG/lg

Form 4-6

DEWEY C. WHITE

WHITE REPAIR & CONTRACTING CO.

———— GENERAL CONTRACTORS • RESIDENTIAL - COMMERCIAL ————

FIRE DAMAGE REPAIRS A SPECIALTY

January 9, 1975   REAR - 2125 PIEDMONT RD., N.E.
ATLANTA, GEORGIA 30324

Kemper Insurance Group
1401 Peachtree Street, N.E.
Atlanta, Georgia

Re: John Doe   TRinity 2-1020
1122 Moreland Place, S. E.
Atlanta, Georgia

Attn: Mr. John F. Hardman

Dear Sir:

We propose to furnish labor and materials to accomplish the following work
at the above address.

Tear out and haul away all charred materials and debris	145.00

Roof:

Replace 6 pieces of 2" x 6" x 18'0" roof rafters 108 bd. ft. @325.00 per M	32.40
Replace 1 piece of 2" x 8" x 10'0" ridge rafters 13 bd. ft. @325.00 per M	4.23
Replace 260 bd. ft. of 1" x 6" roof decking @325.00 per M	84.50
Replace 8 pieces of 2" x 4" x 16'0" roof braces 85 bd. ft. @325.00 per M	27.63
Replace 4 squares of 15 lb. roof felt @ 3.25 per sq.	13.00
Install new roof covering on low side of split level only, where roof structure is being replaced matching existing color - 4 squares @24.00 per sq.	96.00
Spray paint and deodorize attic area complete with aluminum paint	135.00
Replace 1400 sq. ft. of 4" rockwool insulation in attic area complete to eliminate all smoke odors.	168.00
Rebuild 2'0" x 21'0" of overhang, soffit, facia and frieze board across rear of house @ 2.25 per lin. ft.	47.25
Replace 21 lin. ft. of O-G type gutter on rear of house @ 1.10 per lin. ft.	23.10
Sand blast all smoked brick veneer on rear of house	70.00

Interior:

Kitchen: 11'0" x 22'0" x 8'0"

Replace 4 pieces of 2" x 6" x 12'0" ceiling joists 48 bd. ft. @325.00 per M	15.60
Replace one 3'0" x 3'2" aluminum window unit complete	38.25
Replace one 2'8" x 6'8" rear exterior door unit complete with weather stripping and hardware	57.25
Replace 38 pieces of 2" x 4" x 8'0" wall studs and plates 203 bd. ft. @325.00 Per M	65.98
Replace one 2'6" x 6'8" cased opening complete	19.75
Replace one 2'6" x 6'8" interior door unit complete	39.00
Replace 80 sq. ft. of 5/8" plywood underlayment where damaged only @ 38¢ per sq. ft.	30.40
Replace one 4'6" x 3'2" aluminum window unit complete	58.60
Apply sheetrock wallboard to ceiling only 242 sq. ft. @ 30¢ per sq. ft.	72.60
Replace 1/4" prefinished birch paneling on sidewalls complete 352 sq. ft. @ 40¢ per sq. ft.	140.80
Replace one 4'0" x 6'8" cased opening complete	21.50
Replace 2 member base mold 66 lin.ft. @ 60¢ per lin. ft.	39.60
Replace 4 sides of door and window trim @ 6.50 ea.	26.00
Replace 4" crown mold complete 66 lin.ft. @ 40¢ per lin. ft.	26.40
Install inlaid linoleum tile floor covering complete 242 sq.ft. @ 80¢ per sq. ft.	193.60
Replace 8 lin.ft. of 3'6" high birch wall cabinets complete @ 30.00 per lin. ft.	160.00
Replace 3 lin. ft. of 2'0" high birch wall cabinets @ 16.00 per lin. ft.	48.00
Replace one 2'6" x 8'0" high birch oven cabinet	88.00

Form 4-6 (cont.)

DEWEY C. WHITE

WHITE REPAIR & CONTRACTING CO.

———————————— GENERAL CONTRACTORS • RESIDENTIAL - COMMERCIAL ————————————
FIRE DAMAGE REPAIRS A SPECIALTY

January 9, 1975   REAR - 2125 PIEDMONT RD., N.E.
                                   ATLANTA, GEORGIA 30324
Page 2   Re: John Doe   TRinity 2-1020

Replace 11 lin. ft. of 3'0" high birch base cabinets complete @ 19.00 per lin.ft.	209.00
Replace one 2'0" x 11'0" inlaid formica counter top complete with 4" back-splash @ 10.50 per lin. ft.	115.50
Replace one 3'0" wide x 5'0" bar double thick complete with formica top	140.00
Replace one 36" range hood complete with vent	79.50
Replace one Westinghouse extra-wide oven stove with glass in door, pink color	238.25
Replace one double pink sink complete with hardware	46.50
Replace one pink surface unit Westinghouse brand	108.50
Stain (7) openings @2.00 ea.	14.00
Paint ceiling only complete 242 sq. ft. @ 12¢ per sq. ft.	29.04
Stain and varnish all wall and base cabinets	78.00

Living Room: 12'0" x 22'0" x 8'0"

Replace 7 sides of door and window trim @ 6.50 ea.	45.50
Replace one 11" x 48" window glass	7.50
Replace two 11" x 30" window glass	12.00
Sand, strip and varnish (4) openings @ 3.00 ea.	12.00
Clean and paint walls and ceiling complete 808 sq. ft. @ 14¢ per sq. ft.	113.12
Machine sand and refinish hardwood floor 264 sq. ft. @ 24¢ per sq. ft.	63.36
Replace one heat thermostat control	29.00
Repair front door	16.00

Hallway: 6'0" x 6'0" and 3'0" x 6'0" x 8'0"

Replace three 2'0" x 6'8" interior door units leading to bedrooms only @39.00 ea.	117.00
Replace 5 sides of door trim @6.50 ea.	32.50
Replace attic entrance door	4.00
Repair all damaged sheetrock on sidewalls and ceiling as needed	28.50
Clean and paint walls and ceiling complete 390 sq. ft. @ 14¢ per sq. ft.	54.60
Sand and stain (5) openings complete @ 6.00 ea.	30.00
Machine sand and refinish (6) step-treads and risers	12.00
Machine sand and refinish hall floor 6'0" x 6'0" 36 sq. ft. @ 24¢ per sq. ft.	8.64

Electric:

Replace all wiring where damaged by fire or water. Replace all light fixtures where damaged only	400.00
Labor & Material	$4,044.45
20% Overhead & Profit	808.89
	$4,853.34

Prices quoted include Workmens' Compensation, Public Liability Insurance and Taxes required by Federal and State authorities pertaining to Old Age Benefits.

Trusting that we may serve you, we are

Very Truly Yours,

WHITE REPAIR AND CONTRACTING COMPANY

Dewey C. White

BY: Dewey C. White

Form 4-7

N. Frank Vought Inc.

BUILDERS AND APPRAISERS · SINCE 1882

G. RAYMOND EISENHAUER, JR.
ROBT. B. ROGERS III
EUGENE P. COHEN

26 EAST 1st STREET
MT. VERNON, N. Y. 10550
MO 8-2420
IN NEW YORK TA 3-3600

January 2, 1975

Tri-City Adjustment Company
126 - 9th Avenue
New York, New York 10032

RE: FIRE LOSS TO BUILDING - 1980 Union Street
                                Bronx, New York
     Assured: Blank Realty Corp.
     Date of Fire: December 16, 1974

Gentlemen:

our estimate of the cost to repair and make good the damage
caused by fire at the above mentioned location, using materials
of a like kind and quality as before the fire, is as follows:

FIRST FLOOR (10' Ceiling)
CANDY STORE (19'x30')
White Plains Road Side

Replace one 4'6" x 4' double hung rear window with wide jambs and trim		$175.00
Remove and reset iron window guards		50.00
Plaster ceiling and repair side walls	1,255' @ $1.00	1255.00
Wire lath and plaster two columns		150.00
Test and repair gas piping and risers		110.00
Repair and scrape flooring	600' @ 45¢	270.00
Remove and reset radiator		12.00
Ten ceiling outlets		
Four double wall receptacles)		
Three switches)		
Entry outlet and fixture)		
Fuse box)		480.00
Check and repair water lines and branches		65.00
One plate glass - 34x72)		
One plate glass - 60x72)		
Two plate glass in doors - 24x62)		
Two D T Transom glass - 60x22)		
Two D T Transom glass - 34x22)		
Two D T Transom glass - 32x18)		
Two D T glass - 26x48 in moveable sash)		306.00
One D T glass - 28x36)		
Replace one window sash and trim		60.00
Repair store front mouldings		125.00
Three exterior outlets over show windows		60.00
Repair metal cornice (Fire damage only)		75.00
Replace framing and metal panel under South show window		95.00
Paint Store Front and new rear window (No interior Painting)		145.00

Rear Passage

One door complete		85.00
Scrape and repair flooring		30.00
Plastering and patching	150' @ $1.00	150.00

(cont'd)

Form 4-7 (cont.)

N. Frank Vought Inc.

RE: FIRE LOSS TO BUILDING - 1980 Union Street
Bronx, New York
Assured: Blank Realty Corp.
Date of Fire: December 16, 1974

Ceiling outlet and fixture)	
Wall outlet and receptacle)	
One switch)	$ 80.00
Repair and ease two doors	24.00
Paint	85.00
Storeroom and Lavatory	
Repair and glaze one window	32.00
VACANT STORE	
Paint	160.00
UPHOLSTERY STORE	
Touch up painting at door and	
Passage	50.00
BASEMENT	
Test and check building wiring	
under Candy Store	75.00
EXTENSION ROOF	
Repair roof boarding	45.00
Repair rubberoid roofing and	
flashing	75.00
GENERAL	
Demolition, debris removal and	
cleaning up	275.00
	$4679.00
Builders Overhead and Profit	
15%	701.85
	$5380.85

NOTE:
If painting of store is part of Building
add $345.00 net to our estimate

-We Do Not Include-
1) Emergency or temporary repairs.
2) Store fixtures.
3) Kitchen equipment and fixtures
4) New York State sales tax

APPROXIMATE CUBIC CONTENTS -   957,500'
Unit Cost   - $1.50
APPROXIMATE REPLACEMENT COST- $1,436.250.00

Yours very truly,

N. FRANK VOUGHT, INC.

GRE:MS

Chapter 5.

Estimating in Catastrophes

When a peril such as fire, explosion, hailstorm, windstorm, earthquake, flood, or freeze-up, results in a number of claims under insurance policies greatly in excess of normal, and requiring the recruiting of additional adjusters, roofers, etc., it is generally classified as a catastrophe in the industry. Damage may be confined to one city or town, as in a tornado, or it may extend over several states as did hurricane "Camille" in 1969. It came in from the Gulf of Mexico at Mississippi and Louisiana, traveled north-northeast and left the continent at Virginia and Maryland.

Conflagrations in cities today are infrequent due to better construction and improved fire fighting services. There are still brush fires and forest fires which, burning out of control, destroy buildings and structures that lie in their paths. There are occasional violent explosions in chemical and other industrial plants which result in damage to adjacent structures and frequently to buildings at considerable distance.

The most common catastrophes confronting insurance companies are caused by windstorms occurring over a wide area, hailstorms, flooding, hurricanes and tornadoes. (See Causes of Loss—Chapter 1). The number of claims arising from a single catastrophe may be a few thousand or as many as a million as, for example, when hurricanes of severity travel over several states before dissipating.

CATASTROPHE "PLANS"

As a result of their experience in numerous catastrophes since the 1930's the insurance companies have formulated *catastrophe plans* to better organize and supervise the handling of claims. The "Objective" stated in the "Plan for Handling Catastrophe Losses" adopted by the American Insurance Association, very clearly and briefly outlines the basic problem:

Following the occurrence of a major catastrophe, conditions in the area involved present a serious handicap to the orderly and efficient adjustment of losses. The sudden necessity for handling expeditiously a multitude of claims places a severe strain upon regular adjusting facilities and there is a tendency to use personnel without due consideration of qualifications, and to close losses without proper inspection of the properties involved. Combined with the fact that the labor and material markets experience excessive demands at the same time, the orderly adjustment of losses, vital to the interests of the public as well as to our business, requires that we establish a routine for the efficient handling of such claims.

Catastrophe plans have been adopted also by the Property Loss Research Bureau, The National Association of Independent Insurance Adjusters, several state and regional agents associations, and also by companies that maintain their own staff of adjusters. All of the plans are flexible and designed for cooperative action with each other.

The purpose of each plan is to move adjusters, builders, experts, supervisory personnel, supplies and equipment into a stricken area as soon as practical following the disaster. By so doing, damaged property can be quickly protected from the elements, uniform procedures can be put into operation, claim handling and processing is fast-tracked, and claimants can be provided with ready funds through the prompt payment of claims.

CONDITIONS TO EXPECT

An adjuster or a builder who has been assigned to a catastrophe area for his first time will usually be frustrated in the early stages by the confusion that exists. The degree of the confusion will depend a great deal upon the violence of the disaster and the area that it covers.

Residents are often in shock during the first few days. Their main concern is naturally for the well-being of other members of the family, or for friends who may have been injured or made homeless. After a tornado, hurricane, or flood, people are occupied in searching for articles of personal property that have been blown or washed away from the premises.

There is cleaning up to be done. Many will have moved temporarily or permanently to other quarters, and they may be difficult to locate. When found, they may be too upset and distracted to discuss their damage calmly.

Following the more serious catastrophes, the National Guard, Army, or State Police are in charge of the town or city, patroling, issuing emergency passes only, or prohibiting admission entirely to certain areas until order has been restored and the hazard of looting has been eliminated.

Communications, power and transportation may be partially or completely out of service. Many times the drinking water becomes contaminated. Hotel, motel and restaurant facilities are inadequate to accommodate the influx of the vast number of people.

The adjuster or builder generally finds that he is expected to handle the same number of claims in a day that he normally handles in a week, and it is not uncommon for him to work from early morning to nightfall. He is constantly under such pressure from insureds, agents and his own organization that it is almost impossible to devote the necessary time to each loss to prepare a careful, orderly, estimate of damage. There is a tendency under such pressure to grow careless, and take short cuts, and to make guesses and compromises in specifications in order to move on to the next loss.

Knowing, or anticipating many or all of these conditions, the adjuster or builder is well advised to sacrifice a small amount of speed in order to perform as efficiently as possible. He should not be influenced to compete with some other adjuster or builder who is out to set a record of fast claim handling. It is not uncommon to find it necessary to reopen losses that have been handled in haste because of overpayment or under-payment— particularly the latter.

INSPECTING PROPERTIES

The procedures that been outlined under "Fieldwork and Notes" in Chapter 3, should be followed in catastrophe losses just as in any other loss. The methods of estimating, the damageability of materials, and methods of repair have been discussed in the previous chapters. However, because of the special problems in catastrophes, the following suggestions are offered:

1. A set of sample policy forms used in the area should be acquired and carried in the brief case. The form used on the particular risk should be known and its conditions well understood in order that the estimate of

damage will include all of the property covered but will not encompass excluded property or perils.

2. The inspection of the premises is always best made in the presence of the insured or his authorized representative. The insured may then point out the damaged portions of the property, and the occasion presents an opportunity to discuss specifications and cost of repairs.

3. If temporary repairs to roofs, walls, or windows have not been made, provisions for immediate attention should be made. There are numerous instances of subsequent damage by rain to the interior of the building long after the loss has been agreed upon. Frequently it is necessary to reopen and readjust.

4. Although no damage is claimed for the interior of the building it should nevertheless be carefully inspected for such things as wet or stained walls and floors, and cracks in plaster. The insured or tenant often overlooks damage of this type, and may discover them at a later date and make supplemental claim.

5. A roof that has been exposed to strong winds or hail is best inspected from the roof rather than from the ground by eye or with field glasses. Asphalt shingles may appear undamaged from the ground but on close examination may be broken or cracked underneath the tabs where they have been flapped up and then laid back again. Perforations by hail are not always visible from the ground, and if necessary a ladder should be obtained for close inspection on the roof.

6. It is good practice to walk around the exterior of the building to look for any damage that might have been overlooked by the insured or his builder.

7. Outbuildings, yard fixtures and other outside property that is covered under the policy should be inspected.

8. Previous damage such as that done by hail or wind, which has never been repaired should not be included in the estimate. (See Chapter 3.)

MIXED DAMAGE—THE FOUR COLUMN ESTIMATE

In some catastrophes part of the damage may be caused by a peril covered under the policy while other damage has been caused by an excluded peril. For example, when hurricanes strike coastal states such as in the Gulf of Mexico or along the Atlantic seaboard, properties adjacent to the coast may suffer damage from both wind, a covered peril, and from wavewash, tidewater or other flooding which is excluded. The estimator is then required to make a careful and reasonable separation of the covered and uncovered damage.

A sample of an estimate sheet used in such cases is shown on page 126. This is a four column estimate showing total cost of repairs; estimate to repair the water damage; estimate to repair debatable items; and the estimate to repair wind damage.

NATIONAL FLOOD INSURERS ASSOCIATION(NFIA)

This Association, with headquarters at 160 Water Street, New York, 10038, is a joint project of the insurance companies and the Federal government to provide flood insurance in specific parts of the country subject to flooding and wavewash.

Servicing companies for underwriting and for claim handling are designated. A comprehensive claim manual is furnished by NFIA and while the servicing company is responsible for and has authority over investigation and evaluation of flood claims it is responsible to the NFIA.

Because flooding is generally excluded under most policies and flooding and wavewash frequently occur in hurricanes along coastal areas, the NFIA requires the use of separate adjusters to handle their losses and they provide specific worksheet forms for both contents and buildings. These require separation of wind and water damage and provide space for "grey area" damage. The worksheet form for building losses is shown on page 127.

The NFIA requires adjusters to be able to prepare estimates of their own of at least $1,000 and to be able to check contractors' estimates of at least $2,500.

ADVISORY PRICE LISTS

One of the first steps taken by the administrators of a catastrophe plan is to arrange for a conference with local builders and contractors associations to discuss and agree when possible on unit costs for various building items. These unit costs are published and then used as a guide by the adjusters and builders in estimating the losses in the area. If no agreement can be reached, an outstanding, competent contractor is asked to prepare such a list which is used for the same purpose.

These advisory lists are necessary in a catastrophe because most of the claims are handled by adjusters who take off the specifications and apply the published unit costs. There are never enough constractors available to appraise the damage to each property. The unit costs agreed to are generally "full" and no flexibility is provided for in the list for grade and quality of material.

Form 5-1

General Adjustment Bureau, Inc.

SPECIAL WAVEWASH STATEMENT

TYPE OF LOSS	GAB FILE NO.
DATE OF LOSS	ADJUSTER

INSURED _____

LOCATION _____

ITEM NO. _____ TYPE OF CONSTRUCTION _____

ADDITIONS _____ DIMENSIONS _____ SQ. FT. AREA _____

NO. ROOMS _____ NO. BATHS _____ AGE _____ GENERAL CONDITION _____

EST. REPLACEMENT COST $ _____ DEPRECIATION $ _____ A.C.V. $ _____

DETAILS OF LOSS	TOTAL REPAIR OR REPL. COST	WATER	DEBATABLE	WIND
FOUNDATION				
DIMENSION LUMBER				
EXTERIOR SIDING				
SHEATHING				
ROOFING				
INT. WALLS & CEILINGS				
FLOORING				
MILLWORK				
HARDWARE				
INTERIOR PAINT				
EXTERIOR PAINT				
ELECTRICAL				
PLUMBING - AIR COND.				
DEMOLITION - DEBRIS				
OTHER				
TOTALS	$	$	$	$

COMPROMISE OF DEBATABLE TOTAL ADDED	$
TOTAL COST OF WIND REPAIRS OR REPLACEMENTS	$
LESS DEPRECIATION $ _____, DEDUCTIBLE $ _____, COINSURANCE $ _____	$
INSURANCE $ _____ AND INSURED CLAIMS	$

APPLICATION OF DEDUCTIBLE OR COINSURANCE CLAUSE:

_____

_____

_____

_____

_____

610 - (1-67)

Form 5-2

NATIONAL FLOOD INSURERS ASSOCIATION

WORKSHEET - BUILDING

Date of Report

Insured and Location	Policy No.	Co. Claim Number

Adjusting Firm and Location	File No.	Date of Loss

Measure Dimensions and Draw Diagram of Ground Floor Area. Attach Snapshot.	Previous Flood Loss Record

Previous Flood Loss Record

Prev. Loss	Date of Loss	Amount Paid
Yes ☐ No ☐		

Type of Bldg.	Building Age

Building Dimensions	Total Sq. Ft.	No. Rms.

No. Baths	Interior Wall Construction	Exterior Wall Construction

Estimated Repl. Cost _____

Less Depreciation _____

Actual Cash Value _____

Quantity	Detailed Description	Full Cost Repair	Wind Loss	NFIA Loss	Grey Area		
					Total Grey	Total E.C.	Total NFIA

NFIA-13 (Ed. 8-74)

Chapter 6.

Arithmetic for Estimating

A sound working knowledge of elementary arithmetic and simple geometry is necessary to calculate quantities and costs in estimating building losses. Unless a person is regularly or frequently called upon to add, subtract, multiply, and divide whole numbers, fractions, and decimals, the principles are forgotten, mistakes are easily made, and estimates become unreliable. A knowledge of the formulas used in calculating the areas of simple figures, and the volumes of ordinary solids, is of equal importance.

FRACTIONS

A fraction is part of a whole number, for example: 1/2, 1/3, 1/4, 1/8. The number above the line is called the *numerator;* the number below the line is called the *denominator.* The line itself signifies that the top number is to be divided by the bottom number

$$\frac{1}{2} = \frac{\text{numerator}}{\text{denominator}}$$

The fraction 3/4 means that we are dealing with three parts of a whole unit which has been divided in four equal parts. It indicates that 3 is to be divided by 4. When the division indicated by a fraction is performed, the result is the *decimal equivalent* of the fraction.

$$\frac{3}{4} = .75, \text{ or } \frac{1}{2} = .5$$

Table 6-1
COMMON FRACTIONS WITH DECIMAL EQUIVALENTS

1/64	0.015625	1/7	.1428
1/32	.3125	1/6	.1666
1/16	.0625	1/5	.2
1/12	.0833	1/4	.25
1/10	.1	1/2	.333
1/9	.1111	1/3	.5
1/8	.125	1	1.0

Any fraction less than a whole unit as 1/2, 1/6, 4/5, and so forth is called a *proper fraction*.

A fraction equal to or greater than a whole unit, as 3/3, 5/2 or 8/6, is called an *improper fraction*.

Any number that includes a whole number and a fraction as 1 1/2 is referred to as a *mixed number.*

A mixed number may be converted to a fraction by multiplying the denominator by the whole number and adding the numerator. This result is placed over the denominator of the fraction.

Illustrative Example
3 ½ = 2 x 3 + 1 = 7/2
6 ⅔ = 3 x 6 + 2 = 20/3

A fraction may be reduced to its lowest terms by dividing both the numerator and the denominator by a common number, or numbers, until they no longer can be divided by a common number.

Illustrative Example
12/30 ÷ 3/3 = 4/10
4/10 ÷ 2/2 = 2/5

In this example both the top and bottom are first divided by 3. The top and bottom of the resultant fraction 4/10 is then divided by 2 reducing the fraction to its lowest terms. The same result is obtained if the numerator and denominator of 12/30 are divided by 6.

Adding and Subtracting Fractions

Fractions may be added or subtracted only if they have a common (the same) denominator.

Illustrative Example

Add:		*Subtract:*	
3/4	= 3/4	3/4	= 3/4
+ 1/2	= 2/4	− 1/2	= 2/4
	5/4		1/4

When adding or subtracting fractions, the numerators only are added or subtracted, and the result is placed over the common denominator.

Multiplication of Fractions

To multiply two fractions, first multiply the numerators, then multiply the denominators. Reduce the resultant fraction to its lowest terms by dividing numerator and denominator by a common number.

Illustrative Example

$3/4 \times 1/3 = 3/4 \times 1/3 = 3/12 = 1/4$

(or by cancellation)

$\cancel{3}/4 \times 1/\cancel{3} = 1/4$

To multiply mixed numbers, first reduce them to improper fractions.

$1\ 1/2 \times 3\ 1/4 =$

$3/2 \times 13/4 = 39/8 = 4\ 7/8$

Division of Fractions

To divide one fraction by another, invert the divisor and proceed as in multiplication.

Illustrative Example

Divide:

4/7 by 1/2 =

$4/7 \times 2/1 = 8/7 = 1\ 1/7$

Expressed another way

$4/7 \div 1/2 = 4/7 \times 2/1 = 8/7 = 1\ 1/7$

Cancellation is useful in solving problems involving the multiplication or division of two or more fractions (it cannot be used for addition or subtraction).

Illustrative Example

Multiply

2/4 x 3/8 X 5/10 x 4/9

(or)

$$\frac{3 \times 8 \times 4 \times 2}{1 \quad 12}$$

$$\frac{\cancel{3} \times 8 \times \cancel{4} \times 2}{\cancel{12}} = 16$$

$$\frac{\cancel{3}}{1}$$

$$\overset{1}{\cancel{2}}/4 \times \overset{1}{\cancel{3}}/\cancel{8} \times \overset{1}{\cancel{5}}/\cancel{10} \times \overset{1}{\cancel{4}}/\cancel{9} = 1/24$$
$$\qquad 2 \qquad 2 \qquad 3$$

SOME APPLICATION OF FRACTIONS IN ESTIMATING

Board Feet—To compute the quantity of board feet in lumber, divide the number of pieces times the dimensions by 12.

Illustrative Example

To obtain the number of board feet in 10 pieces of 2″ x 4″ that are 12 feet long.

$$\frac{10 \times 2 \times 4 \times \cancel{12}}{\cancel{12}} = 80 \text{ FBM}$$

Number of Rafters, Joists or Studs—To find the number of joists, rafters or studs required in a given space, divide the length of the space in feet by the distance in feet, center to center between joists, rafters or studs. Add one unit for the end.

Illustrative Example

Find the number of rafters placed 16 inches on center in a flat roof 36 feet long:

$$(16 \text{ inches} = 4/3 \text{ feet})$$

$$36/\underset{3}{\underline{4}} = \overset{9}{\cancel{36}} \times 3/\cancel{4} = 27 \text{ plus } 1 = 28 \text{ rafters}$$

Obtain the number of studs placed 20 inches on center in a partition 30 feet long:

$$(20 \text{ inches} = 5/3 \text{ feet})$$

$$30/\underset{3}{\underline{5}} = \overset{6}{\cancel{30}} \times 3/\cancel{5} = 18 \text{ plus } 1 = 19 \text{ studs}$$

SURFACE MEASUREMENT

A *polygon* is a plane figure bounded by straight-line sides. A three-sided polygon is called a *triangle;* a four-sided polygon is a *quadrilateral.* The figures following illustrate various types of polygons.

In discussing the properties of any polygon in a vertical plane, the *base* is the bottom side, and the *height* or altitude is the perpendicular distance from the highest point to the base. The symbol for the *base* is "b" and the symbol for the height is "h."

The *area* of a polygon is the measurement of its surface without regard to its thickness. In estimating buildings, the calculation of areas of polygons has to do principally with the surfaces of floors, ceilings, interior and exterior walls, and of roof surfaces.

The *perimeter* of a polygon is the sum of the length of its sides. It is the distance around the figure.

RECTANGLES

The area of a rectangle equals the base multipled by the height. If the rectangle is horizontal, as the floor or ceiling of a room, the area equals the length (l) multipled by the width (w).

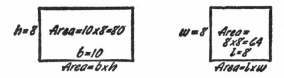

RHOMBOIDS, PARALLELOGRAMS

The area of a rhomboid equals the base multipled by the height. All opposite sides of a rhomboid are parallel.

Area = b x h
Area = 10 x 8 = 80

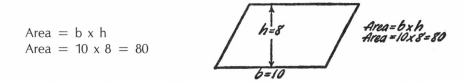

TRAPEZOIDS

The area of a trapezoid equals one-half of the sum of the top and bottom multipled by the height. Only the top and bottom sides are parallel.

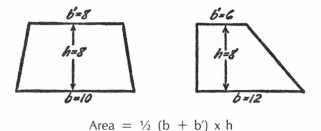

Area = ½ (b + b') x h

$$\text{Area} = \frac{8 + 10}{2} \times 8 = 72 \qquad \text{Area} = \frac{6 + 12}{2} \times 8 = 72$$

IRREGULAR POLYGONS

The areas of trapeziums or irregular polygons can only be determined by breaking the figures up into triangles, or regular polygons, and obtaining the areas separately.

Illustrative Example

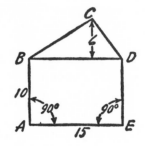

Area ABDE =   10 x 15 = 150
Area BCD = ½(6 x 15) = 45
                              Total   195

TRIANGLES

The area of any triangle equals one-half of the base times the height. (The height is the perpendicular distance from the base to the highest point.)

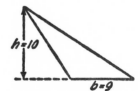

$$\text{Area} = \frac{12 \times 10}{2} = 60$$

$$\text{Area} = \frac{9 \times 10}{2} = 45$$

Hero's Formula for Area of a Triangle

The area of any triangle may also be computed by the following formula where "s" equals one-half of the sum of the sides of the triangle.

Illustrative Example

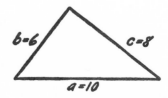

$$s = ½ (10 + 6 + 8) = 12$$

Area $= \sqrt{s(s-a)\ (s-b)\ (s-c)}$
Area $= \sqrt{12(12-10)\ (12-6)\ (12-8)}$
       $= \sqrt{12(2)\ (6)\ (4)}$
       $= \sqrt{576}$
       $= 24$

This formula is advantageous where no angular or other measurements of the triangle are available except the length of the three sides.

Rule of Pythagoras for Right Triangles

The square of the length of the hypotenuse is equal to the sum of the squares of the lengths of the other two sides.

Illustrative Example

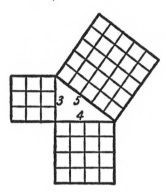

$$5^2 = 3^2 + 4^2$$
$$25 = 9 + 16$$
$$25 = 25$$

By means of this formula if any two sides of a right triangle are known, the other side can be computed.

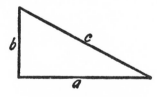

$$c = \sqrt{a^2 + b^2}$$
$$a = \sqrt{c^2 - b^2}$$
$$b = \sqrt{c^2 - a^2}$$

Illustrative Example
$a = 3 \quad b = 4 \quad c = 5$
Known a and b—find c.
$c = \sqrt{3^2 + 4^2}$
$ = \sqrt{9 + 16}$
$ = \sqrt{25}$
$ = 5$

Illustrative Example
$a = 3 \quad b = 4 \quad c = 5$
Known: c and b—find a.
$a = \sqrt{5^2 - 4^2}$
$ = \sqrt{25 - 16}$
$ = \sqrt{9}$
$ = 3$

Illustrative Example
$a = 3 \quad b = 4 \quad c = 5$
Known: c and a—find b.
$b = \sqrt{5^2 - 3^2}$
$ = \sqrt{25 - 9}$
$ = \sqrt{16}$
$ = 4$

Extracting the Square Root of a Number

(A) Extract the square root of 1764.   (B) Extract the square root of 576.

```
        17'64 (42 answer)                          5'76 (24 answer)
82  16                                    44  4
      164                                      176
      164                                      176
```

Steps to Take:

1. Begin at right side of number and point off the digits in pairs, ending at left side with one digit (if there is an odd number) or with a pair (if there is an even number of digits).
2. Find the largest square equal to or less than the first digit or pair of digits in the number (*16* in Example A, *4* in Example B). The square root of *that* number is the first digit in your answer (*4* in Example A, *2* in Example B).
3. Subtract this square as shown.
4. Bring down the next two digits in the number and place alongside the remainder.
5. Double the root already found (*4* in Example A, *2* in Example B) and place out on left side as shown.
6. Divide this *doubled root* into the first two digits of your remainder (*8* into *16* in Example A, *4* into *17* in Example B). The quotient is placed alongside the dividend and also it becomes the next digit in the answer.
7. Multiply the quotient obtained by the entire dividend (*2* times *82* in Example A, *4* times *44* in Example B).
8. The process continues in this manner for larger numbers.

Square Root from Table

Table 6-2 gives the root and the square beginning with number ten. The nearest root shown may be used or for greater accuracy interpolate.

TABLE 6-2
SQUARE ROOTS OF NUMBERS

Number	Square Root	Number	Square Root	Number	Square Root	Number	Square Root
100	10.00	150	12.25	200	14.14	250	15.81
101	10.05	151	12.29	201	14.18	251	15.84
102	10.10	152	12.33	202	14.21	252	15.87
103	10.15	153	12.37	203	14.25	253	15.91
104	10.20	154	12.41	204	14.28	254	15.94
105	10.25	155	12.45	205	14.32	255	15.97
106	10.30	156	12.49	206	14.35	256	16.00

Number	Square Root	Number	Square Root	Number	Square Root	Number	Square Root
107	10.34	157	12.53	207	14.39	257	16.03
108	10.39	158	12.57	208	14.42	258	16.06
109	10.44	159	12.61	209	14.46	259	16.09
110	10.49	160	12.65	210	14.49	260	16.12
111	10.54	161	12.69	211	14.53	261	16.16
112	10.58	162	12.73	212	14.56	262	16.19
113	10.63	163	12.77	213	14.59	263	16.22
114	10.68	164	12.81	214	14.63	264	16.25
115	10.72	165	12.85	215	14.66	265	16.28
116	10.77	166	12.88	216	14.70	266	16.31
117	10.82	167	12.92	217	14.73	267	16.34
118	10.86	168	12.96	218	14.76	268	16.37
119	10.91	169	13.00	219	14.80	269	16.40
120	10.95	170	13.04	220	14.83	270	16.43
121	11.00	171	13.08	221	14.87	271	16.46
122	11.05	172	13.11	222	14.90	272	16.49
123	11.09	173	13.15	223	14.93	273	16.52
124	11.14	174	13.19	224	14.97	274	16.55
125	11.18	175	13.23	225	15.00	275	16.58
126	11.23	176	13.27	226	15.03	276	16.61
127	11.27	177	13.30	227	15.07	277	16.64
128	11.31	178	13.34	228	15.10	278	16.67
129	11.36	179	13.38	229	15.13	279	16.70
130	11.40	180	13.42	230	15.17	280	16.73
131	11.45	181	13.45	231	15.20	281	16.76
132	11.49	182	13.49	232	15.23	282	16.79
133	11.53	183	13.53	233	15.26	283	16.82
134	11.58	184	13.56	234	15.30	284	16.85
135	11.62	185	13.60	235	15.33	285	16.88
136	11.66	186	13.64	236	15.36	286	16.91
137	11.70	187	13.67	237	15.39	287	16.94
138	11.75	188	13.71	238	15.43	288	16.97
139	11.79	189	13.75	239	15.46	289	17.00
140	11.83	190	13.78	240	15.49	290	17.03
141	11.87	191	13.82	241	15.52	291	17.06
142	11.92	192	13.86	242	15.56	292	17.09
143	11.96	193	13.89	243	15.59	293	17.12
144	12.00	194	13.93	244	15.62	294	17.15
145	12.04	195	13.96	245	15.65	295	17.18
146	12.08	196	14.00	246	15.68	296	17.20
147	12.12	197	14.04	247	15.72	297	17.23
148	12.17	198	14.07	248	15.75	298	17.26
149	12.21	199	14.11	249	15.78	299	17.29

TABLE 6-2
SQUARE ROOTS OF NUMBERS

Number	Square Root	Number	Square Root	Number	Square Root	Number	Square Root
300	17.32	350	18.71	400	20.00	450	21.21
301	17.35	351	18.73	401	20.02	451	21.24
302	17.38	352	18.76	402	20.05	452	21.26
303	17.41	353	18.79	403	20.07	453	21.28
304	17.44	354	18.81	404	20.10	454	21.31
305	17.46	355	18.84	405	20.12	455	21.33
306	17.49	356	18.87	406	20.15	456	21.35
307	17.52	357	18.89	407	20.17	457	21.38
308	17.55	358	18.92	408	20.20	458	21.40
309	17.58	359	18.95	409	20.22	459	21.42
310	17.61	360	18.97	410	20.25	460	21.45
311	17.64	361	19.00	411	20.27	461	21.47
312	17.66	362	19.03	412	20.30	462	21.49
313	17.69	363	19.05	413	20.32	463	21.52
314	17.72	364	19.08	414	20.35	464	21.54
315	17.75	365	19.10	415	20.37	465	21.56
316	17.78	366	19.13	416	20.40	466	21.59
317	17.80	367	19.16	417	20.42	467	21.61
318	17.83	368	19.18	418	20.45	468	21.63
319	17.86	369	19.21	419	20.47	469	21.66
320	17.89	370	19.24	420	20.49	470	21.68
321	17.92	371	19.26	421	20.52	471	21.70
322	17.94	372	19.29	422	20.54	472	21.73
323	17.97	373	19.31	423	20.57	473	31.75
324	18.00	374	19.34	424	20.59	474	21.77
325	18.03	375	19.36	425	20.62	475	21.79
326	18.06	376	19.39	426	20.64	476	21.82
327	18.08	377	19.42	427	20.66	477	21.84
328	18.11	378	19.44	428	20.69	478	21.86
329	18.14	379	19.47	429	20.71	479	21.89
330	18.17	380	19.49	430	20.74	480	21.91
331	18.19	381	19.52	431	20.76	481	21.93
332	18.22	382	19.54	432	20.78	482	21.95
333	18.25	383	19.57	433	20.81	483	21.98
334	18.28	384	19.60	434	20.83	484	22.00
335	18.30	385	19.62	435	20.86	485	22.02
336	18.33	386	19.65	436	20.88	486	22.05
337	18.36	387	19.67	437	20.90	487	22.07
338	18.38	388	19.70	438	20.93	488	22.09
339	18.41	389	19.72	439	20.95	489	22.11
340	18.44	390	19.75	440	20.98	490	22.14

Number	Square Root	Number	Square Root	Number	Square Root	Number	Square Root
341	18.47	391	19.77	441	21.00	491	22.16
342	18.49	392	19.80	442	21.02	492	22.18
343	18.52	393	19.82	443	21.05	493	22.20
344	18.55	394	19.85	444	21.07	494	22.23
345	18.57	395	19.87	445	21.10	495	22.25
346	18.60	396	19.90	446	21.12	496	22.27
347	18.63	397	19.92	447	21.14	497	22.29
348	18.65	398	19.95	448	21.17	498	22.32
349	18.68	399	19.97	449	21.19	499	22.34
						500	22.36

APPLICATION OF SURFACE MEASUREMENT TO ESTIMATING

Measurement of floor and ceiling areas

The floor or ceiling area of a room is obtained by multiplying the length by the width.

Illustrative Example

Area = 14′ x 16′ = 224 sq ft

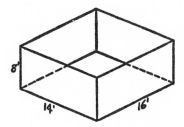

Measurement of sidewall areas of rooms

The sidewall area of a room including the window or door openings is equal to the perimeter times the height from the floor to the ceiling.

Illustrative Example
Perimeter = 2(14′ + 16′) = 60 lin ft
Area   =   60′ x 8′   = 480 sq ft

Another method is to obtain the area of the walls separately and add them together to get the total wall area.

2 walls = 2(16′ x 8′) = 256 sq ft
2 walls = 2(14′ x 8′) = 224 sq ft
480 sq ft

Measurement of exterior wall areas

The exterior areas of buildings may be obtained either by computing each wall area separately and adding them together, or by taking the perimeter of a building and multiplying by the height of the wall. Gable ends, dormers, extensions, and so forth, should be figured separately and added.

Illustrative Example

Perimeter = 2(20' + 30') = 100 lin ft

$$\frac{\times 10}{1,000 \text{ sq ft}}$$

Plus Gables = $\frac{2(8' \times 20')}{2}$ = $\underline{160}$

Total 1,160 sq ft

Measurement of Rafter Lengths

When the *rise* of a roof is known, and also the *run* (one half the span), the length of the rafter may be calculated from the "Rule of Pythagoras." (See also Table 13-2, page 316.)

Rise = 6 ft
Span = 16 ft
Run = $\frac{16}{2}$ = 8 ft
Rafter AC or BC = $\sqrt{6^2 + 8^2}$
= $\sqrt{36 + 64}$
= $\sqrt{100}$
= 10 ft

Note: To this length must be added any existing overhang.

PROPERTIES OF CIRCLES

The *circumference* of a circle is the distance around its exterior.

The *diameter* is a line from one side of the circumference to the opposite side and passing through the center of the circle.

The *radius* of a circle is a line from the center of the circle to any point on the circumference, and, therefore equals one-half the diameter.

In any circle the circumference when divided by the diameter will always equal the same number called pi with symbol π. Its value is approximately 3.1416, or 3 1/7. In other words, the circumference of a circle is about 3 1/7 times as long as its diameter.

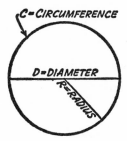

$$\frac{\text{Circumference}}{\text{Diameter}} = \pi \begin{aligned} &= 3.1416 \\ &= 3\ 1/7 \end{aligned}$$

The circumference of a circle equals π times the diameter. This is expressed as $C = \pi D$, or $C = \pi\,2R$ (also 3 1/7 x diameter).

The diameter of a circle equals the circumference divided by π. This is expressed as $D = \dfrac{C}{\pi}$

The area of a circle in terms of its radius is π times the radius squared. This is expressed as $A = \pi R^2$ *or* 3 1/7 x R^2.

Illustrative Example

The diameter of a sprinkler tank is 20 feet. What is the area of its base?

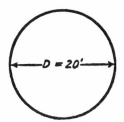

$$
\begin{aligned}
A &= \pi R2^2 \\
&= 3.1416 \times 10^2 \\
&= 3.1416 \times 100 \\
&= 314.16 \text{ sq ft}
\end{aligned}
$$

If the height "h" of the tank is 30 feet, what is the exterior wall area? It would be the circumference of the base multiplied by the height.

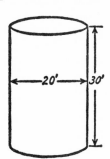

Wall surface area of a tank	$= \pi D\ h$
Circumference	$= \pi\ D$
Circumference	$= 3.1416 \times 20'$
Circumference	$= 62.832$ lin ft
Surface area	$= 62.832 \times 30'$
Surface area	$= 1,884.96$ sq ft

VOLUME MEASUREMENT

In the measurement of volume three dimensions are considered, namely length, width and height. Volume is expressed in terms of cubic units of inches, feet, and so forth.

Rectangular Solids

The rectangular solid is commonly encountered in estimating.

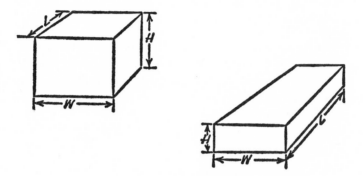

The formula for the volume of a rectangular solid is:
$$V = LWH$$

Illustrative Example

$$V = 20' \times 16' \times 10'$$
$$= 3,200 \text{ cu ft}$$

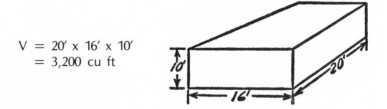

Cylinders

The cylinder is another common type of solid encountered in estimating, and generally its volume in terms of gallons capacity is required.

The cubic volume of a cylinder may be found by multiplying the area of its base by the height or length of the tank. The formula would be expressed: $V = \quad R^2 H$

Illustrative Example

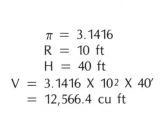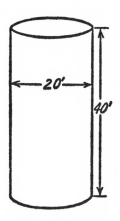

$$\pi = 3.1416$$
$$R = 10 \text{ ft}$$
$$H = 40 \text{ ft}$$
$$V = 3.1416 \times 10^2 \times 40'$$
$$= 12,566.4 \text{ cu ft}$$

To convert the contents in cubic feet to gallons multiply the number of cubic feet by 7.4805 the number of gallons in one cubic foot.

$$12,566.4 \times 7.4805 = 94,003 \text{ gallons}$$

Table 6-3
TABLE OF WEIGHTS AND MEASURES

Linear Measure

12 inches	= 1 foot
3 feet	= 1 yard
5 ½ yards	= 1 rod
320 rods	= 1 mile
5,280 feet	= 1 mile

Surface Measure

144 square inches	= 1 square foot
9 square feet	= 1 square yard
30 ¼ square yards	= 1 square rod
160 square rods	= 1 acre
43,560 square feet	= 1 acre
640 acres	= 1 square mile or one section

Volume Measure

1,728 cubic inches	= 1 cubic foot
27 cubic feet	= 1 cubic yard
128 cubic feet	= 1 cord
(A cord of wood	= 4' x 4' x 8')

Table 6-3
TABLE OF WEIGHTS AND MEASURES

Liquid Measure

4 gills	= 1 pint
2 pints	= 1 quart
4 quarts	= 1 gallon
231 cubic inches	= 1 gallon
31½ gallons	= 1 barrel
63 gallons	= 1 hogshead
7.4805 gallons	= 1 cubic foot
32 ounces	= 1 quart

Dry Measure

2 pints	= 1 quart
8 quarts	= 1 peck
4 pecks	= 1 bushel

Weight Measure

16 ounces	= 1 pound
100 pounds	= 1 hundred weight
2,000 pounds	= 1 ton
2,240 pounds	= 1 long ton

Metric System

10 millimeters (mm)	= 1 centimeter	(cm) =	0.3937	inches
10 centimeters	= 1 decimeter	(dm) =	3.9370	inches
10 decimeters)	= 1 meter	(m) =	39.37	inches
100 centimeters)		=	3.28	feet
10 meters	= 1 dekameter	(dkm) =	393.7	inches
10 dekameters	= 1 hectometer	(hm) =	328.08	feet
10 hectometers	= 1 kilometer	(km) =	0.62137	miles
10 kilometers	= 1 myriameter	(mym) =	6.2137	miles

METRIC CONVERSION FACTORS

To Convert From	To	Multiply By
acres	square feet	43,560.0
acres	square meters	4,047.0
acres	square yards	4,840.0
board feet	cubic inches	144.0
centimeters	feet	0.03281
centimeters	inches	0.3937
cubic feet	cubic centimeters	28,317.0
cubic feet	cubic meters	0.028317
cubic feet	cubic yards	0.03704
cubic feet	gallons (U.S.A.)	7.481
cubic feet	gallons (Imperial)	6.22905
cubic feet of water	pounds	62.37

To Convert From	To	Multiply By
cubic inches	cubic centimeters	16.38716
cubic yards	cubic meters	0.764559
fathoms	feet	6.0
feet	centimeters	30.48
feet	meters	0.304801
gallons (U.S.A.)	cubic centimeters	3,785.0
gallons (U.S.A.)	cubic feet	0.13368
gallons (U.S.A.)	Imperial gallons	0.832702
gallons (Imperial)	gallons (U.S.A.)	1.20091
gallons (U.S.A.)	cubic inches	231.0
gallons (U.S.A.)	liters	3.78543
gallons (U.S.A.)	ounces	128.0
horsepower (U.S.A.)	horsepower (metric)	1.01387
inches	centimeters	2.54001
inches	meters	0.0254001
inches	millimeters	25.4001
liters	cubic feet	0.03532
liters	gallons (U.S.A.)	0.26418
miles (statute)	kilometers	1.60935
miles (statute)	miles (nautical)	0.8684
miles (nautical)	feet	6,080.204
miles (nautical)	miles (statute)	1.1516
pounds (avoirdupois)	grams (metric)	453.592
pounds (avoirdupois)	kilograms	0.453592
square centimeters	square feet	0.0010764
square feet	square centimeters	929.0
square feet	square meters	0.0929034
square inches	square centimeters	6.452
square yards	square meters	0.83613
tons (metric)	tons (short)	1.1023
tons (long)	pounds	2,240.0
tons (short)	pounds	2,000.0
tons (metric)	pounds	2,204.6
yards	meters	0.91442

Chapter 7.

Concrete

Concrete is a mortar containing aggregate coarser than sand, usually gravel, stone or cinders. Plain concrete, as distinguished from reinforced, is used when the stresses are in compression as in footings and certain types of foundations. When tensile stresses are involved as in beams, girders and floor slabs, steel reinforcing is placed in the concrete near the tension area, in the form of rods, or mesh, or both, depending on the design. Concrete is made by mixing together cement, sand and a coarse aggregate, and allowing the mass to harden in forms. Other additives may be introduced to produce hardness, to prevent freezing, or to make the concrete waterproof.

The problem of estimating any complicated or extensive concrete work in connection with building losses is not frequently encountered. The most common cases that arise will be those which involve replacing or repairing damaged sidewalks, driveways, basement or garage floors, and occasionally repairs to foundations, columns, slabs, and so forth.

In preparing an estimate for concrete, four factors are to be considered in addition to any earth handling: (1) Forms, (2) Concrete, (3) Reinforcement, (4) Finishing.

FORMS

The unit of measurement of forms should be the actual square feet of concrete *in contact with the forms*. A foundation wall 20 feet long, 8 feet high and 12 inches thick would require 8' x 20' = 160 square feet of forms for each side or a total of 320 square feet excluding the ends. Wood forms are conventionally constructed of 2 by 4-inch studs and sheathing or plywood with necessary bracing. The studding is spaced from 12 inches to 24 inches on center depending on the height of the form. Where the wall height is 3 or 4 feet, spacing of 24 inches is adequate. For higher walls the studs are placed closer together to withstand the outward pressure exerted by the plastic mass of wet concrete.

ESTIMATING MATERIALS AND LABOR

For average estimating purposes the quantity of lumber (studding, sheathing and bracing) required *per square foot of concrete surface* is between 2 1/2 to 3 board feet. Using this basis, the board feet of lumber needed for the forms for a wall 8 feet high and 20 feet long, assuming 2 1/2 FBM per square foot of surface, would be:

$$
\begin{array}{rl}
2(8' \times 20') = 320 & \text{sq ft} \\
\times\ 2\frac{1}{2} & \text{FBM} \\
\hline
800 & \text{FBM}
\end{array}
$$

In pricing the lumber, unless quantities in excess of two or three thousand board feet are involved, an average price between the cost of sheathing boards and the studding may be used. If sheathing, for example is $110.00 per 1,000 FBM, and 2" x 4" studding is $130.00 per 1,000 FBM, an average price of $120.00 may be used.

Wood forms are usually built and erected on the job by carpenters. In some localities, carpenter helpers, or common laborers assist in the work. Laborers generally strip the forms. The number of hours required to build, erect, and strip concrete forms varies with the type of structure for which the form is being made.

Figures 7-1 and 7-2 show several common types of forms, and the method of estimating the board feet of lumber for each type, when an accurate quantity of form material is desired.

Table 7-1 shows the approximate board feet of lumber and the average hours of labor required to build, erect, and strip 100 square feet of forms of various types.

Forms for Slabs on Fill

Forms for concrete slabs that are placed directly on the surface of the ground such as sidewalks, garage or basement floors, or the floors of one-story dwellings without basements, will require a minimum of form work.

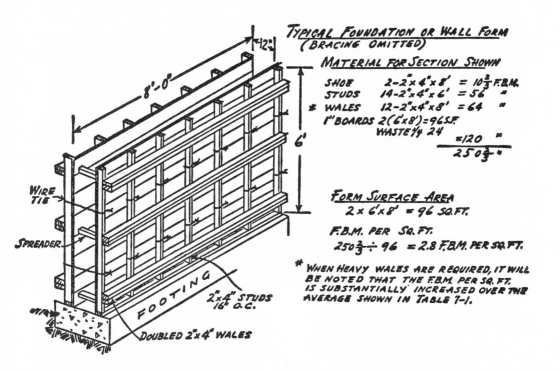

TYPICAL FOUNDATION OR WALL FORM
(BRACING OMITTED)

MATERIAL FOR SECTION SHOWN

SHOE   2-2"x 4"x 8' = 10⅔ F.B.M.
STUDS   14-2"x 4"x 6' = 56   "
* WALES 12-2"x 4"x 8' = 64   "
1" BOARDS 2(6'x8')=96 S.F.
              WASTE ¼ 24   =120 "
                           ─────────
                            250⅔ "

FORM SURFACE AREA
   2 x 6'x 8' = 96 SQ. FT.

F.B.M. PER SQ. FT.
   250⅔ ÷ 96 = 2.8 F.B.M. PER SQ. FT.

* WHEN HEAVY WALES ARE REQUIRED, IT WILL
BE NOTED THAT THE F.B.M. PER SQ. FT.
IS SUBSTANTIALLY INCREASED OVER THE
AVERAGE SHOWN IN TABLE 7-1.

8'-0"
12"
6'

WIRE TIE
SPREADER
FOOTING
2"x 4" STUDS 16" O.C.
DOUBLED 2"x 4" WALES

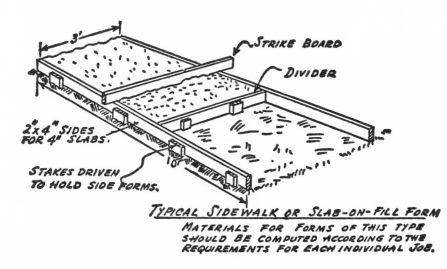

STRIKE BOARD
DIVIDER
3'
2"x 4" SIDES FOR 4" SLABS.
STAKES DRIVEN TO HOLD SIDE FORMS.

TYPICAL SIDEWALK OR SLAB-ON-FILL FORM
MATERIALS FOR FORMS OF THIS TYPE
SHOULD BE COMPUTED ACCORDING TO THE
REQUIREMENTS FOR EACH INDIVIDUAL JOB.

Figure 7-1
Concrete Forms

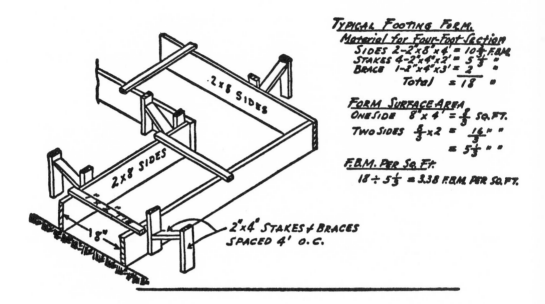

TYPICAL FOOTING FORM.
Material for Four-Foot Section
  SIDES 2-2"x8"x4' = 10⅔ F.B.M.
  STAKES 4-2"x4"x2' = 5⅓ "
  BRACE 1-2"x4"x3' = 2 "
            Total = 18 "

FORM SURFACE AREA
  ONE SIDE 8"x 4' = 8/3 SQ.FT.
  TWO SIDES 8/3 x2 = 16/3 "
                   = 5⅓ " "

F.B.M. PER SQ.FT.
  18 ÷ 5⅓ = 3.38 F.B.M. PER SQ.FT.

2"x4" STAKES & BRACES
SPACED 4' O.C.

2x8 SIDES

2x8 SIDES

8"

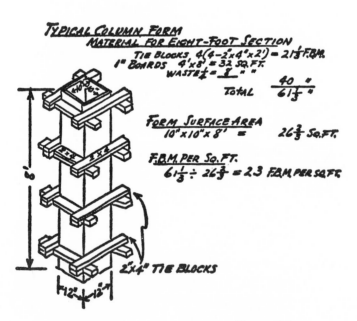

TYPICAL COLUMN FORM
  MATERIAL FOR EIGHT-FOOT SECTION
    TIE BLOCKS, 4(4-2"x4"x2') = 21⅓ F.B.M.
    1" BOARDS 4'x8' = 32 SQ.FT.
      WASTE ⅛ = 8 " "
              TOTAL   40 "
                      61⅓ "

FORM SURFACE AREA
  10"x10"x8' =   26⅔ SQ.FT.

F.B.M. PER SQ.FT.
  61⅓ ÷ 26⅔ = 2.3 F.B.M. PER SQ.FT.

2"x4" TIE BLOCKS

12" 12"

Figure 7-2
Concrete Forms (footing and column).

Table 7-1
Approximate Board Feet of Lumber and Hours of Labor Required to
Build, Erect, and Strip 100 Square Feet of Wooden Forms

Type of Form	FBM Lumber Per 100 Sq Ft Surface	Hours Labor Per 100 Sq Ft Surface		
		Build and Erect	Strip	Total Hours
Footings	350	6	2	8
Foundations and walls	250	8	2	10
Floor slabs above grade	250	10	4	14
Columns	300	10	4	14
Beams and girders	300	12	4	14
Stairs and steps	300	12	2	16

After the area has been prepared, stakes are driven and side pieces to
contain the concrete are put in place using braced boards, 2 x 4's, 2 x 6's or
other stock of suitable dimensions. These curbing pieces are usually placed
so that the upper edges are even with the top of the finished concrete slab.
In this way they serve as leveling guides. When estimating the amount of
material, and the labor to set forms of this kind, each job should be figured
according to its individual circumstances and requirements.

A rule of thumb for determining the number of board feet of form
lumber for slabs on fill is to use the same number of lineal feet of side
curbing as the distance around the slab and add 25 per cent for stakes and
braces.

Illustrative Example

Determine the board feet of lumber for a walk 4 inches thick, 3 feet wide by
30 feet long.

Perimeter = 2(30' + 3') = 66 lin ft
66 lin ft 2" x 4" = 44 FBM
Plus 25% Braces
and Stakes   11 FBM
      Total   55 FBM

A carpenter can place, level, drive the stakes and brace ground slab
forms at the average rate of 20 lineal feet per hour. The number of hours of
labor to install the forms in the above example would be

$$\frac{55}{20} = 2\frac{3}{4} \text{ hours}$$

Estimating Costs

Since the unit of measurement for concrete forms is a square foot of surface area in contact with the concrete, the unit cost of forms may be obtained using Table 7-1 for the number of board feet of lumber and hours of labor per 100 square feet of form.

Nails and wire are not included in the following examples as the quantity required per square foot is negligible when related to the total unit cost. If large quantities of form work are involved, an allowance of 30 pounds of nails per 1,000 board feet of lumber, and 40 feet No. 8 form wire per 100 square feet of form may be added.

CONCRETE
Illustrative Examples of Unit Costs For Forms
(Assumed Prices and Wages)

Foundation and Wall Forms

Table
(7-1) 250 FBM lumber (average price)	@ $160.00	=	$40.00
(7-1) 8 hours-carpenter, build and erect	@ 6.00	=	48.00
(7-1) 2 hours laborer to strip	@ 5.00	=	10.00

Unit cost for 100 sq ft = $98.00

$$\text{Unit cost per sq ft } \frac{\$98.00}{100} = \$.98$$

Floor Slabs Above Ground

Table
(7-1) 250 FBM lumber (average price)	@ $160.00	=	$ 40.00
(7-1) 10 hours-carpenter, build and erect	@ 6.00	=	60.00
(7-1) 4 hours laborer to strip	@ 5.00	=	20.00

Unit cost for 100 sq ft = $120.00

$$\text{Unit cost per sq ft } \frac{\$120.00}{100} = \$ \ 1.20$$

Columns

Table
(7-1) 300 FBM lumber (average price)	@ $160.00	=	$ 48.00
(7-1) 10 hours carpenter, build and erect	@ 6.00	=	60.00
(7-1) 4 hours laborer to strip	@ 5.00	=	20.00

Unit cost for 100 sq ft = $128.00

$$\text{Unit cost per sq ft } \frac{\$128.00}{100} = \$ \ 1.28$$

Beams and Girders

Table

(7-1) 300 FBM lumber (average price)	@ $160.00 =	$ 48.00
(7-1) 12 hours carpenter, build and erect	@ 6.00 =	72.00
(7-1) 4 hours laborer to strip	@ 5.00 =	20.00

Unit cost for 100 sq ft = $140.00

Unit cost per sq ft $\dfrac{\$140.00}{100} = \$\ 1.40$

Application of Unit Costs for Form Work

When the surface area of the concrete structure has been measured and computed in square feet, a unit cost is developed. The square foot area is then multiplied by the unit cost to obtain the total cost of building, erecting, and stripping the forms.

Using the unit costs shown in the foregoing example for *Foundations and Walls,* the cost of forms for a 12-inch wall 8 feet high and 20 feet long would be:

1 Side 8′ x 20′ =	160 sq ft
2 Sides	x 2
	320 sq ft
Unit cost per sq ft	$.98
Total cost	$313.60

As another example, using the unit cost shown, $1.28 per square foot, the total cost of the forms for 10 columns 12″ x 12″ x 10' high would be (perimeter of each column is 4 lin ft):

1 Column 4′ x 10′ =	40 sq ft
10 Columns	x 10
	400 sq ft
Unit cost per sq ft	$ 1.28
Total cost	$512.00

Salvage in Forms

Ordinarily when forms are built with new lumber specifically for footings, foundations, and so forth, on moderate-sized concrete jobs, the value of the lumber salvage is seldom considered in estimating unless there is a definite reuse for the lumber in the balance of reconstruction. In that event an allowance of approximately 2 to 3 hours labor per 100 square feet should be allowed for cleaning, removing nails, and for breakage, in addition to stripping time.

Some contractors keep on hand old form lumber that has been salvaged from previous concrete jobs and that can be used again. They may also have available prefabricated plywood panel forms, or patented metal forms that can be used many times, cleaned and stored for reuse. In estimating form work, consideration should be given to the availability of reusable forms as a means of reducing the cost. A reasonable charge should be included in the estimate for the rental of such forms, and also for cleaning and trucking them to and from the job.

CONCRETE

Mix

Concrete is a mixture of cement, sand and a coarse aggregate, usually gravel or crushed stone graded from 3/4 inch to 1 1/2 inches. The *mix* in concrete refers to the proportions of these three components and is stated in cubic feet of cement, cubic feet of sand, and cubic feet of stone. A mix of 1:3:5 means that the proportions are:

1 cu ft cement
3 cu ft sand
5 cu ft stone

In actual construction the proper proportions, the quality of the materials, and the amount of water used are important and controlled factors. The Portland Cement Association publishes considerable reference material on the subject of mixes for various kinds of work. It has been said that it costs very little more to make good concrete than it does to make concrete of a poor quality. For estimating purposes, materials may be figured on the basis of a mix of 1 part cement, 2 1/4 parts sand and 3 parts stone. The size of the stone should not exceed 1 inch. This proportion is generally considered satisfactory for good concrete to be used in footings, foundations, floors, walks, driveways, beams, girder, columns and swimming pools.

In terms of a cubic yard of concrete the quantity of materials required for a 1:2 1/4:3 mix would be: (Table 7-2, Mix No. 3)

6¼ sacks of cement
14 cu ft of sand (.52 cu yds)
19 cu ft of stone (.70 cu yds)

A sack of cement contains one cubic foot and weighs 94 pounds. A barrel of cement contains four sacks.

A yard of sand or stone contains 27 cubic feet, and weighs 2,600 to 2,800 pounds.

Where it is desired to use other proportions of cement, sand and stone, Table 7-2 shows the recommended quantity of each of these materials for one cubic yard of concrete.

Estimating Quantities

The unit of measurement for concrete is expressed in cubic yards, or sometimes in cubic feet (there are 27 cubic feet in a cubic yard). Coarse aggregate like gravel or stone contains voids between individual particles which are taken up with the sand. Similarly there are voids between the particles of sand which are taken up with the finer particles of cement. For this reason, in any prescribed *mix,* the combined cubic feet of cement, sand, and aggregate will exceed 27 cubic feet. When the materials are mixed, however, the voids are filled and the resultant bulk is 27 cubic feet, or one cubic yard. For example, a mix of 1:2 1/4:3 is composed of 6 1/4 cubic feet of cement, 14 cubic feet of sand and 19 cubic feet of coarse aggregate, a total of 39 1/4 cubic feet. After mixing, the mass will occupy 27 cubic feet.

Table 7-2
Quantities of Cement, Sand and Stone Required
for 1 Cubic Yard Concrete

| | | *Materials Per Cu Yd Concrete* | | | |
No. * Mix		Sacks of Cement	Cu Yds Sand	Cu Yds Stone	Max. Aggregate
1	1:1:1¾	10	.37	.63	3/8″
2	1:2:2¼	7¾	.56	.65	¾″
3	1:2¼:3	6¼	.52	.70	1″
4	1:3:4	5	.56	.74	1½″

*Mix No. 1—Concrete for heavy wearing surface.
Mix No. 2—Concrete exposed to weak acid or allkali soultions.
Mix No. 3—Concrete for watertight floors, foundations, swimming pools, septic tanks and for structural reinforced concrete, beams, columns, slabs, residence floors, etc.
Mix No. 4—Concrete for foundations and walls not subjected to weather or water pressure.

Typical concrete shapes are shown in cross section in Figure 7-3. To obtain the quantity of concrete, the *square foot* area of the cross section is computed and this is multiplied by the length to get the number of cubic feet.

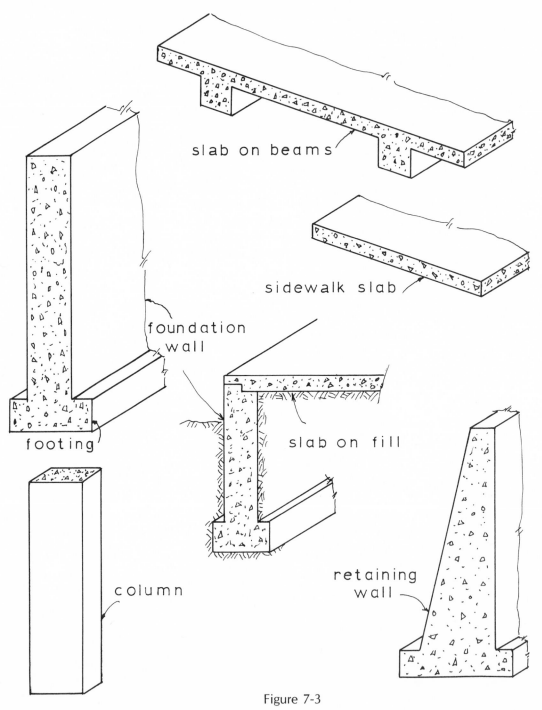

slab on beams

sidewalk slab

foundation wall

footing

slab on fill

column

retaining wall

Figure 7-3

Typical concrete shapes

Divide the cubic feet by 27 to obtain the number of cubic yards. Short-cuts are shown in the following section.

Shortcuts to Estimating Quantities

Rather than referring to lengthy Tables to determine the number of cubic yards of concrete in footings, walls, slabs and columns, there is an easier method. It is also much simpler than multiplying out the cubic feet and dividing by 27 to obtain the cubic yards.

Footings.

Multiply the cross-section in square inches by .000257 to get the cubic yards in 1-lineal foot of footing. Multiply that result by the length of the footing.

Illustrative Example

Find the number of cubic yards of concrete in a footing 24″ wide, 10″ deep and 40′ long.

24″ x 10″ x .000257 x 40′ = 2.47 yards

Wall.

Multiply the square foot surface area by the wall thickness in inches. Multiply that result by .0031.

Illustrative Example

Find the number of cubic yards of concrete in a foundation wall 8′ high, 50′ long and 12″ thick.

8′ x 50′ x 12″ x .0031 = 14.88 yards

Floor slabs and Concrete walks.

Multiply the floor or walk area in square feet times the thickness in inches times .0031.

Illustrative Example

Find the number of cubic yards of concrete in a garage floor 24′ wide, 25′ long and 4″ thick.

24′ x 25′ x 4″ x .0031 = 7.44 yards

Columns.

Compute the cubic yards in a column as though it were a footing. Multiply the cross-section in square inches by .000257. Multiply that result by the height of the column.

Illustrative Example

Find the number of cubic yards of concrete in a column 12″ by 12″ by 12′ high.

12″ x 12″ x .000257 x 12′ = .44 cu yds

Table For Slabs and Walks

Sometimes the estimator wants to know how many square feet of floor or slab of a given thickness will be covered by one cubic yard of concrete. Table 7-3 shows this for slabs 2 inches to 12 inches thick.

TABLE 7-3
SQUARE FOOT AREA THAT ONE CUBIC YARD WILL COVER

Thickness In Inches	Number Square Feet	Thickness In Inches	Number Square Feet
2	162	6	54
3	108	7	46
3½	93	8	40
4	81	9	36
5	65	10	32
5½	59	12	27

The number of square feet that a cubic yard of concrete will cover in floor slabs, driveways and walks, is easily determined. Simply divide the constant 324, by the thickness in inches.

Illustrative Example
How many square feet will one cubic yard of concrete cover in a floor slab 4.5″ thick

$$\frac{324}{4.5} = 72 \, \text{sq ft}$$

Estimating Material Costs

The cost of the materials for a cubic yard of concrete for a mix of 1:2 1/4:3 is readily obtained as follows, when the local cost of cement, sand and stone (or gravel) has been established:

Illustrative Example

6¼ sacks cement	@	$1.40 =	$ 8.75
.52 cu yds sand	@	3.75 =	1.95
.70 cu yds stone	@	3.50 =	2.45
		Total cost per cu yd =	$13.15

Estimating Labor Costs

Where small quantities of concrete are needed, or where either a mixing machine or ready-mix concrete is not available, it is mixed by hand. Most builders however are equipped with a concrete mixing machine, or more than one with different capacities. The machines may also be rented.

The rental cost varies with the capacity of the mixer and the length of time it is kept.

Ready-mix concrete is usually more economical than job-mixing, since it eliminates the use of a mixer and also the labor of handling materials to be mixed. In many situations ready-mix concrete reduces the labor cost of wheeling and pouring where the truck can back in close to the position where the concrete is to be placed. The concrete can be distributed to the proper spot with chutes attached to the rear of the truck.

Before an estimate of labor costs to mix and place concrete is made, the project should be carefully studied to determine the best procedure that will give proper consideration to all factors affecting the labor required. The following check list of major items that the estimator should give particular attention to will serve as a guide:

1. Kind of work.
> Footings or foundations
> Walls, beams, girders, columns
> Floor slabs, slabs on ground
2. Method of mixing concrete.
> By hand
> By machine on the job
> Ready-Mix delivered
3. How is concrete to be placed?
> Directly from Ready-Mix truck
> By wheelbarrow from site of mixing
> What distance must it be wheeled?
4. Are plank runways to be built?
5. Is hoisting necessary?
> By hand
> By power hoist
6. Will weather conditions affect labor?
> Winter or summer
> Protection against rain, snow, or frost

Mixing by hand is slower than by machine. Ready-mix concrete does away with the labor of mixing on the job, but the cost of the ready-mix should be checked against on-the-job mixing. The labor required to wheel concrete long distances or hoist it up or down to the forms must be estimated in hours per cubic yard over and above actual mixing.

The labor cost to set up and shift runways should be added as a separate item rather than be included in the unit cost per yard.

Mixing and placing concrete in the winter requires great care. The aggregate has to be heated or some type anti-freeze must be mixed with the batch. In some cases the poured concrete is protected by covering with straw, salt hay or tarpaulins to prevent frost action or actual freezing.

Table 7-4 shows the average number of hours of labor required for mixing and placing concrete for different kinds of work.

Table 7-4*
Approximate Hours of Labor Required for Mixing
and Placing Concrete

Kind of Work. (Forms 25 Feet or Less From Mixer)	Hours Per Cubic Yard
Mixing by hand (small jobs)	2
Mixing by machine	1
Placing in footings and foundations	2
Placing in columns and 12 inch walls	4
Placing in walls 12 to 24 inches thick	3
Placing in floors and walks	2
Placing in stairs	4
Add for heating aggregate	½
Add for each additional 25 feet of wheeling	½

Finishing Concrete	Hours Per 100 Sq ft
Wood float finish on slabs	2
Steel trowel finish on slabs	3
Machine finishing	1-2
Cement wash on walls	3

*The labor shown in Table 7-4 is the total man-hours. The estimator should base his costs on the wages of the labor available. Mixing and placing of concrete is principally done by common labor. Cement finishing is done by masons. Under certain conditions and in non-union areas it may be necessary to figure a mason's time for the entire job.

REINFORCEMENT

Reinforcing in concrete should only be figured when the size, quantity, and shape of bars are known to the person preparing the estimate; otherwise someone familiar with reinforced concrete design should be called upon to submit a sub-bid.

Reinforcing is estimated by the *pound* or by the *ton* for both material cost and the labor to place it. The material cost is computed by multiplying the *lineal feet* of reinforcing times the weight per foot times the price per

TABLE 7-5
WEIGHTS OF REINFORCING BARS

No.	Nominal Diameter Inches	Area in Square Inches	Weight in Pounds Per Lineal Foot
*2	.250	0.05	0.167
3	.375	0.11	0.376
4	.500	0.20	0.668
5	.625	0.31	1.043
6	.750	0.44	1.502
7	.875	0.60	2.044
8	1.000	0.79	2.670
9	1.128	1.00	3.400
10	1.270	1.27	4.303
11	1.410	1.56	5.313

* No. 2 bar is a plain round bar.

pound. Table 7-5 shows the weight per foot, diameter and area of various sizes of bars.

The labor to bend and place reinforcing depends on the weights of the bars and whether they must be shaped and wired in place. Where the bars are less than 3/4-inch, a man should be able to place them in position without tying at the rate of approximately 20 hours per ton. Where they are to be tied in place, the rate can be figured at about 24 hours per ton. These rates do not include unloading or hoisting to floors above grade. On jobs where bars over 3/4-inch are used, the rate for bending, placing and tying is approximately 18 hours per ton. Subcontractors substantially reduce these rates.

Wire Mesh Reinforcing

Wire mesh, or welded steel fabric reinforcing for slabs, driveways and walks comes in square or rectangular mesh and is sold by the roll usually 5 feet wide and 150 feet long. It is also available in flat sheets.

A popular size used in residential work is 6 x 6 - 10/10. This indicates the wires are 6 inches on center each way forming 6 inch squares and that the wire is No. 10 gauge.

When laid, the mesh is lapped a full square.

The labor putting wire mesh reinforcing in floors and driveways varies with the job. In straight slab work like garage floors a man should handle 200 square feet per hour, or a roll 5 feet by 150 feet in 4 hours.

FINISHING CONCRETE

After concrete slabs have had an initial set, the surface is given a float finish or a smooth troweled finish. The labor for doing this operation is added to the cost of mixing and placing the concrete.

When forms have been removed from concrete foundations, walls, ceilings, and so forth, the surface is sometimes given a cement wash to provide a more attractive finish. Table 7-4 shows the number of hours per 100 square feet of surface for finishing concrete. For cement plastering of surfaces, refer to Chapter 15.

TOPPING FOR FLOORS AND WALKS

Sometimes the top inch of a floor slab or walk consists of cement mortar which is laid over the concrete base. The mixture is generally sand and cement but occasionally a coarse aggregate not exceeding 3/8 inch is used. When a 1 inch topping is to be laid, the thickness of the concrete base is reduced 1 inch. On a slab, for example, 4 inches thick, the base would be 3 inches with a 1-inch topping.

The mixture of the mortar may be figured at 1 part cement to 2 parts sand. A cubic yard would require 9.5 sacks cement and 1.1 cu yds of sand.

To estimate the number of cubic yards required, the square-foot area is multiplied by 1/12 (1 inch is 1/12 of 1 foot) and the product is divided by 27 (1 cubic yard = 27 cubic feet).

Illustrative Example

To obtain the cubic yards of 1″ topping required on a slab that measures 20′ x 30′.

$$20' \times 30' = 600 \text{ sq ft}$$

$$\frac{600}{12''} = 50 \text{ cu ft}$$

$$\frac{50}{27} = 1.85 \text{ cu yds}$$

(See page 157 for shortcut estimating of slabs. Also see Table 7-3 for square feet one cubic yard covers.)

The labor to mix and place the topping may be estimated on the same basis as for concrete shown in Table 7-4. The labor for mixing is figured separately from the placing. Finishing by hand or machine should be added by using Table 7-4 as a guide.

ESTIMATING COSTS

The cost of material and labor for a specific concrete job may be figured either at a unit cost per cubic yard or the total material and total labor may be estimated. For example, a foundation wall 12" thick 9' high and 60' long would contain (1' x 9' x 60') 540 cubic feet or 20 cubic yards of concrete. The cost, then, of the material may be estimated by obtaining the cost of 1 cubic yard and multiplying by 20. The cost of the labor to mix and pour the concrete may be obtained from Table 7-4.

Illustrative Example
Cost of 1 Cubic Yard of Concrete, 1:2 ¼:3 Mix
(Assumed Prices and Wages)

6¼ sacks cement	@	$1.40 =	$ 8.75
.52 cu yds sand	@	3.75 =	1.95
.70 cu yd stone	@	3.50 =	2.45
		Total =	$13.15

Cost of 20 cu yds = $13.15 x 20 = $263.00

Labor to Mix and Pour
2 cu yds per hour (see Table 7-4)
Labor for 20 cu yds = 2 x 20 = 40 hours
40 hours laborer's time @ $5.00 = $200.00

Total Cost for Job

Material	$263.00
Labor	200.00
Total	$463.00

The same job estimated on a unit cost per cubic yard basis would be:

Unit Cost Per Cu Yd

Material for 1 cu yd	= $ 13.15
(Table 7-4) 2 hours laborer @$5.00	= 10.00
Unit cost for 1 cu yd	= $ 23.15
Total cost for 20 cu yds = 20 x $23.15	= $463.00

Unit Costs

The following basic formula may be used to obtain unit costs per cubic yard of concrete by inserting local labor wages and material prices.

The mix of 1:2 1/4:3 is recommended in estimating for all but concrete that requires a special mixture. It does not include additives for quick setting, hardening, antifreeze or coloring.

Basic Formula

6¼ sacks cement	@	$	= $
.52 cu yds sand	@		=
.70 cu yds stone	@		= _____
Cost of material for 1 cu yd			= $
Hours labor to mix concrete	@		=
Hours labor to place concrete	@		= _____
Unit cost for 1 cu yd			= $

The number of hours necessary to mix and place the concrete should be selected from Table 7-4, depending upon the method of mixing, the kind of work, and on the distance from the site of mixing to the forms. The cost of building runways, hoisting or other operations should be added to the job cost as a separate item.

Illustrative Examples

Assuming the following material costs and wages, and using a mix of 1:2 ¼:3, compute a unit cost per cubic yard for machine mixed concrete to be placed in a foundation form.

Cement $1.40 per sack
Sand   3.75 per cu yd
Stone   3.50 per cu yd
Common labor wages $5.00 per hour.

6¼ sacks cement	@	$1.40 =	$ 8.75
.52 cu yds sand	@	4.00 =	1.95
.70 cu yds stone	@	3.50 =	2.45
		Material cost per cu yd =	$13.15
1 hour labor mixing by machine	@	5.00 =	5.00
2 hours labor placing	@	5.00 =	10.00
		Unit cost per cu yd =	$28.15

Application of Unit Costs

The following illustrative examples are shown for applying unit costs to obtain the total cost of concrete work. Material prices and labor wage rates are the same as those assumed in the foregoing illustrations of unit costs.

Illustrative Example

A concrete walk 4 inches thick, 3 feet wide and 40 feet long is to be laid using a mix of 1:2 ¼:3. The concrete is machine mixed, and a float finish is given the top by a mason.

Quantity of concrete $\frac{4}{12}$ x 3′ x 40′ = 40 cu ft

$$\frac{40}{27} = 1.48 \text{ cu yds}$$

Unit cost for 1 cu yd = $28.15
   Cost for 1.48 cu yds = $41.66
Surface to be finished 3′ x 40′ = 120 sq ft
(Table 7-4) 2.4 hours mason's time @ $6.00 = $\underline{14.40}$
                                              $56.06

The cost per square foot may be obtained by dividing $56.06 by 120 sq ft = $.476

Illustrative Example

To find the total cost of concrete to be machine mixed and placed in a retaining wall 12 inches thick, 6 feet high and 20 feet long, using a mix of 1:2¼:3. Develop a unit cost (page 164), and the total cost

1′ x 6′ x 20′ = 120 cu ft

$$\frac{120}{27} = 4.44 \text{ cu yds}$$

Cost for wall = 4.44 x $28.15 = $124.99

DAMAGEABILITY OF CONCRETE

Concrete has demonstrated its fire-resistive qualities in innumerable extreme fires. Its high value as a protection for steel lies in the similarity of its temperature coefficient to that of steel. Generally 2 inches of concrete protection is considered adequate against temperatures in a fire which would cause steel to buckle and warp.

The rate of heat conductivity of concrete is relatively slow, due in part to its porosity. As the surface of the concrete is dehydrated, the porosity increases, improving the insulating qualities with the result that, except in long-burning, hot fires, the main damage is to the outer surface. Spalling of the surface (popping off of thin shale-like fragments) occurs frequently in hot fires, and this condition may be aggravated by alternate heating by the fire and cooling by water. The degree of spalling of concrete varies with the kind of aggregate used. Quartz, gravel, and chert tend to spall quicker than aggregates composed of trap rock or limestone.

Fire damage to concrete floor slabs such as garage and basement floors, is seldom of a serious nature. There are a number of reasons for this. The burning of the structure or its contents is above the level of the floor and the heat rises. Water from fire hoses or sprinkler systems tends to keep the floor cooled down. Also contents that have not burned and falling debris from above frequently will cover portions of the floor surface affording a protection.

Soilage, caused by smoke, heat, tar or asphalt drippings from melted roofing, or stains from various contents materials during a fire, is not uncommon.

Cracks in concrete due to shrinkage, may develop during a fire. They may be merely surface cracks, or may penetrate completely through a wall. Concrete walls and floors may also develop serious cracking as a result of settlement, earthquake, or from hydrostatic pressure during flooding conditions which take place during excessive rains and overflow of rivers and streams.

Methods of Repair

Concrete, either plain or reinforced, in most cases can be readily repaired. Foundations, floor slabs, columns, beams, girders and walls which have been seriously injured can be repaired by removing the affected portions, placing new reinforcement where required, erecting forms and placing new concrete to be bonded to the existing structure. Surfaces of floor slabs that have been damaged can be chipped off, cleaned, and a new 1 to 1 1/2-inch topping applied over the old floor.

Each situation must be treated according to the particular conditions that are encountered. When preparing the details of a repair estimate, the estimator should consider whether or not it will be necessary to:

1. Cut away the damaged concrete.
2. Build, erect and strip necessary forms.
3. Place new reinforcing bars or mesh.
4. Prepare old concrete surface to be bonded to the new concrete.
5. Mix and place the concrete.
6. Finish concrete surface as required.

(1) All building or contents debris in the working area should be removed, and any undamaged machinery or fixtures that are in the way of the operation will have to be moved or shored up. The concrete that is to be replaced should be cut away with power chisels, or by hand. Reinforcing bars that have been affected by heat should be cut off, allowing enough of the old steel to protrude in order to tie or weld it to the new.

(2) The building, erecting, and stripping of forms have already been discussed in this chapter.

(3) For estimating reinforcing materials and labor refer to that section in this chapter.

(4) The preparation of the old surface for bonding to the new concrete is probably the most important requirement for successful repair work. After the surface has been hacked and chipped, it must be thoroughly cleaned with stiff wire brushes and washed with a caustic solution to

remove all dirt and grease. In some instances steam hosing is necessary, or a solution of muriatic acid is applied followed by washing with clean water.

There are several reliable bonding products available which either can be mixed in the concrete batch, or can be made into a grout by mixing with cement and water and brushing on to the old surface.

(5) The method of estimating materials and labor for mixing and placing concrete has been previously discussed in this chapter.

(6) Where concrete structures have been repaired, if appearance is important, it may be necessary to chip or machine grind the joints where the new meets the old. In addition to smoothing the joining surfaces, a cement wash coat may have to be brushed over the entire old and new concrete to provide a uniform color.

Repairs to Concrete Floors

Floor surfaces that are spalled, chipped, or otherwise injured will have to be cut away an inch or so deep to expose the old aggregate and provide a rough bonding surface for the new concrete.

Estimating the labor required to cut away old concrete, or to scarify and chip the surface for bonding, is so variable that each job must be studied and judged according to the hardness of the old concrete, whether power hammers are available, or hand tools will have to be used.

The surface must be carefully cleaned and prepared for bonding as outlined under "Methods of Repair." A new topping of cement mortar 1 to 1 1/2 inches thick is then placed and finished (see Topping for Floors and Walks).

Concrete floors that are broken through or heaved will have to be taken up and replaced with a new section bonded to the adjacent old section.

Repairs to Cracks

Cracks in structural concrete can be repaired by chiseling out the cracks, removing all dirt and loose material, and then filling them with cement mortar of a 1:3 mixture. Repairs should be estimated on the basis of time and materials.

Repairing Spalled Walls

Spalled walls, not injured structurally, can be repaired by applying a coat of cement plaster to the surface. All loose material should be chipped away. After cleaning the surface a wash coat is usually applied; it consists of cement, a small amount of sand, and one of the patented bonding admixtures. The cement plaster or stucco is then put on. (For estimating, see Chapter 15.)

Repairs to Soiled Walls

Soilage of concrete from smoke, roof tar drippings, and stains from various contents, can be removed by one of the following methods:

(1) Wash with caustic solution.
(2) Steam clean.
(3) Sand blast.
(4) Cement wash.
(5) Paint.

The estimator must determine from actual conditions what steps are necessary, and what method is most likely to restore the appearance of the surface.

Chapter 8.

Masonry

Masonry is work done by a mason who lays up either stone found in nature or units which have been manufactured such as brick, tile or cement block. Mortar is used between the joints to bind the masonry units together. It may be either cement mortar or lime mortar, or a combination of both.

There are many different types of masonry units used in building construction, particularly those that are manufactured. There is a variety of material used in making bricks, and they come in a number of sizes, shapes and finishes. Hollow tile and concrete, cinder or gypsum blocks are made in different sizes and shapes.

This chapter deals with the more commonly encountered kinds of masonry. The principles outlined, however, are applicable to all of them. With careful attention given to the cost of materials, and with proper allowances for labor variations on any specific type of masonry, little difficulty should be found in estimating costs.

MORTAR

The desirable physical properties of mortar are that it shall be easily worked by the mason, will not lose its water too rapidly to the masonry units, will be sufficiently strong and durable for the purpose used, and that it will have good bonding qualities. These properties are dependent on the proportions of cement, sand, water, and lime (if used), and also on the quality of those materials.

MATERIALS

Structurally the proper proportioning of the ingredients of mortar is very important. In estimating the *cost* of masonry, it is relatively unimportant. The cost of the materials for mortar when compared to the over-all cost of the masonry, is small. Any difference between the cost of mortar materials as a result of varying the proportions will make an insignificant difference in the total estimate.

The Portland Cement Association recommends for different types of mortar service the proportions shown in Table 8-1.

Table 8-1
Recommended Mortar Mixtures—Proportions by Volume

Type of Service	Cement	Hydrated lime or lime putty	Mortar sand in damp loose condition.
For ordinary service	1 masonry cement*	. . . . . .	2¼ to 3
	—— or ——		
	1 portland cement	½ to 1¼	4½ to 6
Subject to extremely heavy loads, violent winds, earthquakes or severe frost action.	1 masonry cement* plus 1 portland cement	. . . . . .	4½ to 6
	—— or ——		
Isolated piers	1 portland cement	0 to ¼	2¼ to 3

*ASTM Specifications C91, Type II.

The addition of hydrated lime to mortar makes it easier for the mason to handle; otherwise it would be stiff and difficult. Furthermore the addition of lime in quantities up to 10 or 15 per cent makes the cement adhere better to the sand. Quantities in excess of that amount reduce both the tensile strength of the mortar and its adhesive, or bonding qualities.

The *volumetric* proportions shown in Table 8-1 can be converted to *weight* proportions by multiplying the unit volumes shown by the weight per cubic foot of the materials, which may be assumed to be as follows:

Masonry cement . . . . Weight printed on bag
Portland cement . . . . 94 lbs
Hydrated line . . . . 40 lbs
Mortar sand, damp and
    loose . . . . 85 lbs (approximately)

Illustrative Example

Using a volume mix of 1 portland cement, ¼ hydrated lime, and 3 of sand, the proportions by weight would be:

1 cement	@	94 lbs =	94 lbs
¼ lime	@	40 lbs =	10 lbs
3 sand	@	85 lbs =	225 lbs

Using 1 sack of cement, the mixture would require 10 pounds of lime and 255 pounds of sand.

To convert this to a yardage basis, assuming that 1 yard of sand weighs approximately 2,550 pounds, it would require 1 cubic yard of sand, 10 sacks of cement and 2½ sacks of lime (100 pounds).

For convenient and practical estimating purposes, since the amount of lime is subject to judgment, the following quantities of materials are recommended for estimating the cost of *1 cubic yard* of mortar.

Cement	10 sacks
Sand	1 cubic yard
Hydrated lime	2 sacks

Any admixtures that are to be combined with the mortar to produce hardness or color, or to prevent freezing, should be included in the cost of the materials.

MIXING

Mortar is usually mixed by laborers, or masons' helpers, while they are tending masons by carrying materials to be placed within their reach, and shifting scaffold planks or horses. On smaller jobs the mixing is done by hand; otherwise a portable mixer is used. The labor to mix the mortar is included in the over-all labor allowance made for masons' helpers. No special or separate computation is necessary.

ESTIMATING MORTAR COST

By using the previously recommended quantities of materials for a cubic yard of mortar, and by applying local prices, a cost per cubic yard of mortar can be obtained.

Illustrative Example
(Assume Local Material Prices As Shown)

10 sacks cement	@	$1.40 =	$14.00
1 cu yd sand	@	3.75 =	3.75
2 sacks hyd. lime	@	.85 =	1.70
		Cost per cu yd =	$19.45

Mortar Required for Masonry

The amount of mortar needed to lay up masonry units varies with the type and size of unit involved and also with the thickness of the joints between the units. The larger the unit and the narrower the joint between, the less mortar will be required.

When estimating losses, very little attention is given to such refinements as the joint thickness unless an unusually large quantity of masonry is being considered. As a general rule for estimating, using a 3/8 inch joint, Table 8-2 shows the quantities of mortar to be used for the types of masonry shown. An allowance of 10 per cent waste is included.

Where some of the cells of the concrete block, or the hollow tile are to be filled, an allowance for more mortar must be made accordingly.

Table 8-2
Approximate Quantities of Mortar Required for Various
Types of Masonry Units

Type of Masonry 3/8" Joints	Nominal size of Unit in Inches	Cubic Yards of Mortar		Cubit Feet of Mortar	
		Per 1,000 Units	Per 100 Units	Per 1,000 Units	Per 100 Units
4" brick masonry	2¼ x 3¾ x 8	.33	.033	9.0	.9
8" brick masonry	2¼ x 3¾ x 8	.50	.05	13.5	1.35
12" brick masonry	2¼ x 3¾ x 8	.60	.06	16.2	1.62
Hollow terra cotta wall tile*	6 x 12 x 12	1.0	.10	27	2.7
(Laid cells horizontal)	8 x 12 x 12	1.3	.13	35	3.5
(Laid cells horizontal)	10 x 12 x 12	1.6	.16	42	4.2
(Laid cells horizontal)	12 x 12 x 12	2.0	.20	54	5.4
8" Concrete block masonry	8 x 8 x 16	1.5	.15	40	4.0
12" Concrete block masonry	12 x 8 x 16	1.7	.17	46	4.6
Gypsum partition tile	3 x 12 x 30	. . .	. . .	. . .	5.0
Gypsum partition tile	4 x 12 x 30	. . .	. . .	. . .	6.0
Gypsum partition tile	6 x 12 x 30	. . .	. . .	. . .	9.0

*For hollow terra cotta tile laid on end, add 20 per cent more mortar to figures shown

CONCRETE BLOCK MASONRY

Concrete block is manufactured in two weights. One is made of lightweight aggregate such as cinders and is frequently referred to as cinder block. A unit actual size 7 5/8" x 7 5/8" x 15 5/8" but called 8" x 8" x 16" weighs about 30 pounds. Its light weight reduces the cost of handling. It has better insulating qualities, and also provides a good nailing base for furring strips. The heavier block is made with aggregates of pebbles, crushed stone or slag. A nominal unit 8" x 8" x 16" weighs about 45 pounds. Both units may be used interchangeably in construction.

Concrete block is an economical type of construction. It is used in dwellings, and may be finished by painting or by application of brick veneer, cement plaster or stucco. The lightweight cinder block provides an excellent bonding surface for plaster both inside and outside. Because of its fire-resistive qualities it is used frequently in the construction of farm buildings, garages, and one-story mercantile and manufacturing buildings. It is also used in retaining walls and fences. Figures 8-1 and 8-2 show the common shapes and sizes of concrete block.

Bond Beams

In some jurisdictions concrete block walls must, by regulation, be stabilized at the top with a continuous concrete beam or cap to which is anchored the timber wall plate. The width of the bond beam conforms with that of the wall, and the depth may range from 8 to 10 inches. It is reinforced with rods. Side forms are placed on top of the wall and concrete is placed in the form with proper reinforcing.

Reinforced Concrete Block

In areas subject to frequent high winds, or earthquake shocks, concrete block walls must be reinforced both vertically and laterally and tied in with the footing. If the type of reinforcing used cannot be determined by inspection, it is recommended that the local building code be consulted for proper specifications.

Waterproofing

The earth side of the wall should have two coats of portland cement plaster (1 cement to 2 1/2 parts of fine sand). Each coat should be about 1/4 inch thick. The first coat is applied over dampened block and scratched to make a bond for the finish coat which is applied 24 hours later.

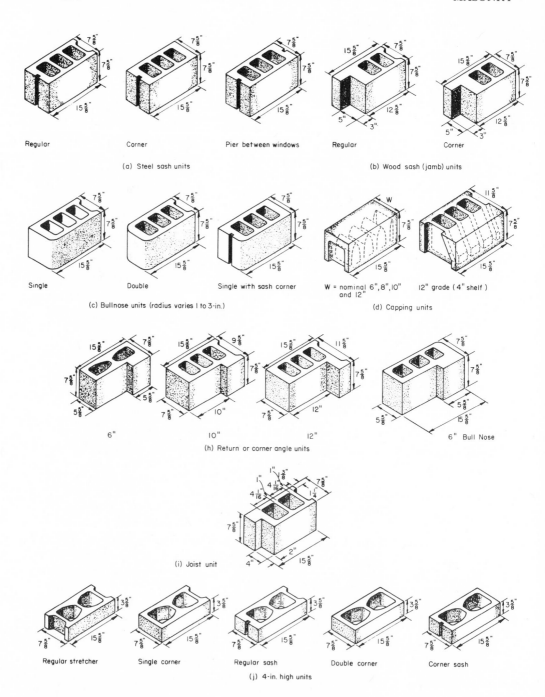

Figure 8-1

Common shapes and sizes of concrete block. Courtesy Portland Cement Association, Skokie, Ill. 60076

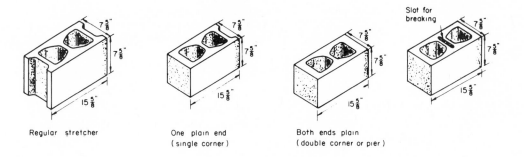

Regular stretcher

One plain end
(single corner)

Both ends plain
(double corner or pier)

Slot for
breaking

(a) Two-core 8x8x16-in. units

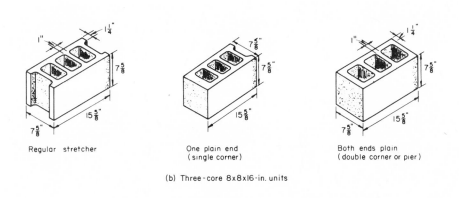

Regular stretcher

One plain end
(single corner)

Both ends plain
(double corner or pier)

(b) Three-core 8x8x16-in. units

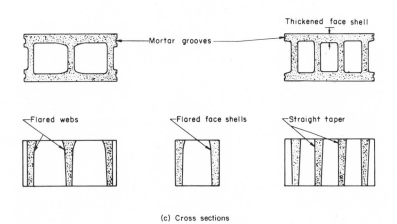

Mortar grooves

Thickened face shell

Flared webs

Flared face shells

Straight taper

(c) Cross sections

Figure 8-2

Usually a coat of hot asphalt or tar is applied over the plaster to seal the wall against moisture. The entire waterproofing operation should extend down and over the edge of the footing.

ESTIMATING QUANTITIES

Concrete blocks are made in a variety of sizes, although the most popular is the 8" x 8" x 16". Its actual measurements are 7 5/8" x 7 5/8" x 15 5/8", the difference being an allowance made in manufacturing for a 3/8" joint. Units of special shape are available for door jambs, and for corners.

Illustrative Example

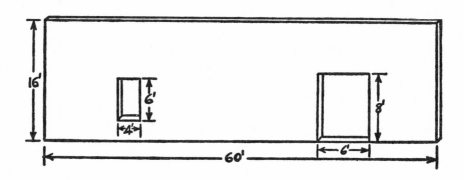

Gross wall area . . . . 60' x 16'		= 960 sq ft
Deduct openings: window . 4' x 6' = 24 sq ft		
Door . . . . 6' x 8' = 48 sq ft		= -72 sq ft
Wall area less openings		888 sq ft
Plus 1/8		111 sq ft
Number of units required		999

A unit 8" x 8" x 16" contains (8" x 16") 128 square inches of surface area which is 128/144 or 8/9ths of a square foot. To estimate the number of 8" x 8" x 16" blocks needed in a wall, compute the square foot area of the wall, and *deduct all openings in full*. Add 1/8 to the number of square feet obtained to get the required number of concrete blocks, or multiply the square foot area by 1.125.

Corners should be deducted, making certain that it is done only once. A building 20 feet by 30 feet has a perimeter of 100 lineal feet. In computing the number of 8-inch block, two of the walls should be shortened by 16 inches. The lineal feet of wall to be used in estimating are:

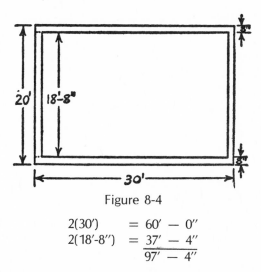

Figure 8-4

$$2(30') \quad = 60' - 0''$$
$$2(18'-8'') = \underline{37' - 4''}$$
$$97' - 4''$$

If the height of the wall is 12 feet, the area is 12 feet by 97 1/3 feet = 1,168 sq ft. To obtain the number of 8" x 8" x 16" concrete block, assuming there are no openings to be deducted, one-eighth is added.

Wall surface area = 1,168 sq ft
Plus 1/8   146
Number of block   1,314

Gable ends of walls are sometimes constructed by carrying the concrete blocks up to the underside of the roof boards. The blocks are cut to the roof slope. A few extra blocks should be allowed for cutting waste in such cases, but no additional labor need be made as the laying rate is not sufficiently affected.

ESTIMATING LABOR

A mason and a laborer work together in laying concrete block. The laborer mixes and carries mortar to the mason and keeps him supplied with blocks. He also shifts, or assists in shifting, scaffold plant and wood horses when the mason moves to a new area of operation.

The unit of measurement used in estimating concrete block masonry is 100 blocks. The number of hours required for a mason and a laborer to lay 100 blocks varies with the kind of work and the size and weight of block being handled. The rate of laying block decreases where the work involves pilastered walls, numerous window or door openings, piers, or breaks and corners. The lightweight cinder blocks lay up faster than the heavyweight units. The 12-inch-thick blocks lay up slower than those 8 inches or less in thickness.

Table 8-3 shows the approximate number of hours required for a mason and a laborer to lay 100 units of lightweight concrete block under average conditions. The cost of structural scaffolding to be erected by carpenters and the cost of hoisting should be added as a separate item whenever either is required.

ESTIMATING COSTS

The cost of concrete block masonry is the combined cost of the block, the mortar required to lay them up, and the cost of labor. If 1 1/2 yards of mortar will lay 1,000 8-inch blocks, it will take .15 yards for each 100 blocks. The cost of a yard of mortar should be computed and multiplied by .15 to obtain the mortar cost for each 100 blocks.

The cost of labor is determined by multiplying the rate of wages for a mason and a laborer times the number of hours shown in Table 8-3.

Illustrative Example

10 sacks cement	@ $1.40 =	$14.00
1 cu yd sand	@ 3.75 =	3.75
2 sacks hydrated lime	@ .85 =	1.70
	Cost of 1 cu yd mortar =	$19.75
	(Table 8-2) Mortar for 100 concrete block =	.15
		$ 2.92
100 (8″ x 8″ x 16″) concrete block	@$ .29 =	29.00
(Table 8-3) 5 Hours-mason	@ 6.00 =	30.00
(Table 8-3) 5 Hours-laborer	@ 5.00 =	25.00
	Unit Cost For 100 Block =	$86.92
	Unit Cost Per Block $86.92 = $ .87	
	100	
	Unit Cost Per Sq Ft = $.87 + 1/8 = $ .98	

Unit Costs

The basic formula for the unit cost of concrete block may be further simplified as follows:

	100 concrete block @	$	= $
(Table 8-2)	Cu yds mortar @		=
(Table 8-3)	Hours—mason @		=
(Table 8-3)	Hours—laborer @		= ____
	Unit Cost For 100 Block		= $
	Unit Cost Per Block	$___	= $
		100	

Table 8-3
Approximate Hours of Labor Required to Lay 100 Light-Weight
Concrete Masonry Units of Various Sizes

Nominal Size of Unit in Inches	Kind of Work	Mason	Laborer	Blocks per Hour
8 x 8 x 16	Foundations and piers	5.5	5.5	18
8 x 12 x 16	Foundations and piers	6.5	6.5	15
4 x 8 x 12	Superstructures	3.5	3.5	29
4 x 8 x 16	Superstructures	4	4	25
8 x 8 x 12	Superstructures	4.5	4.5	22
8 x 8 x 16	Superstructures	5	5	20
8 x 12 x 16	Superstructures	6	6	17

For heavyweight concrete blocks add 10 per cent to the hours shown for lightweight units. These hours include pointing and cleaning.

CONCRETE BLOCK CHIMNEYS

Manufacturers of concrete block make a light-weight unit 16 by 16 by 8 inches (in height). A single 8 by 8-inch tile flue lining can be set inside the unit. The blocks laid one on top of the other with mortar joints form a chimney.

Mortar is filled in between the flue lining and the concrete block. A cubic yard of mortar will lay approximately 100 lineal feet of chimney. To find the cost of mortar required, multiply the cost of 1 cubic yard by the length of the chimney divided by 100.

Illustrative Example

If the cost of 1 cubic yard of mortar is computed at $20.00 and the chimney is 30 feet high, the cost of the mortar required would be:

$$\frac{30}{100} \times \$20.00 = \$6.00$$

Since the blocks are 8 inches thick, the number required is obtained by multiplying the height of the chimney by 1.5. A chimney 30 feet high would require 1.5 x 30 = 45 blocks.

A mason and a laborer working together can lay on the average 4 feet per hour including setting of the flue lining.

Flue lining 8" x 8" comes in 2-foot lengths.

The total cost of building a 30-foot concrete block chimney can be obtained in the following manner, assuming that the flue lining begins 5 feet from the base. Prices and wages are assumed.

45 concrete blocks 16″ x 16″ x 8″   @ $1.50 ea. = $67.50

$(\frac{3}{2} \times 30' = 45)$

13 lengths 8″ x 8″ TC flue lining   @   .60 =   7.80
   (30′ − 5′ = 25′, 26 lin ft needed)

Mortar − $\frac{30'}{100}$ x $20.00   =   6.00

   Mason labor $\frac{30'}{4}$ = 7½ hours   @ 6.00 =   45.00

Laborer (same as mason) 7½ hours   @ 5.00 =   37.50
                            Total cost $163.80

DAMAGEABILITY OF CONCRETE BLOCKS

In general the damageability of concrete as discussed in Chapter 7 applies to concrete blocks. Whether composed of aggregate of slag, gravel, or cinders, the 8-inch and thicker concrete blocks provide excellent insulation against heat up to temperatures similar to those common in chimneys, 600° to 1000° F. In moderate fires they have been found to stand up well. There is usually no spalling such as is characteristic of concrete, but the blocks will tend to crumble after long exposure to extreme heat. Because of the thinness of the face shell, which is approximately 1 1/2 inches thick, the small amount of water used in their manufacture, and the air cells, concrete blocks are less effective as an insulation against the heat of long burning fires than a solid concrete wall of equal thickness.

Cracks may develop in concrete block walls from intense heat, subsidence, vibration, blasting, earthquake, impact or pressure from an external force such as earthslide, and hydrostatic pressure. If the forces are of sufficient magnitude, the walls will bulge and finally collapse. Numerous cases are on record of *unbraced* concrete block walls of buildings under construction having been blown over during strong windstorms.

Concrete block masonry is subject to soilage by smoke, tar or asphalt roof drippings, and also from contact with contents materials. Concrete block walls, unless reinforced by steel or pilasters, offer somewhat less resistance to lateral pressures caused by explosion, impact, or by expanding water-soaked contents than similar walls of solid concrete or brick masonry.

METHODS OF REPAIR

Sections of concrete block walls that have been damaged are readily cut out and new sections can be installed. The blocks are removed at mortar joints stepping back the sides. Old mortar is removed where the new blocks are to be set in place. When the patching-in of new sections is being estimated, special consideration should be given to the labor necessary to lay the stepped-back blocks where the new wall joins the old. Unless a very large area of wall is involved, it is better to prepare the estimate on a time and material basis rather than trying to approximate a unit cost. The cost of demolishing the damaged wall in preparation for laying the new block should be treated as a *separate labor item*. In some cases the walls adjacent to the damaged section will require propping or bracing to keep them in line and plumb during the operation. If the area being replaced extends very far vertically down the surface of the wall, leaving the sides unsupported, they should be checked for plumbness and held in place by timber supports until the new wall has thoroughly set. Where a roof has been destroyed, has collapsed into the interior, or is so badly damaged that its replacement is necessary, the side walls generally need stabilizing if any extensive repairs are contemplated. Each situation should be carefully examined and analyzed to determine the extent and method of repairs that will be undertaken.

Soilage of concrete block masonry is estimated in the same manner as soilage of concrete masonry, which is discussed in Chapter 7. The basis of estimating cleaning depends on the kind of stain or soilage. Scraping, steam cleaning, and sandblasting may have to be resorted to to remove such deposits as asphalt tar dripping that comes from roofing. In certain instances it will be necessary to provide a uniform color to the surface by whitewashing, painting, cement washing, and in extreme cases by cement plastering.

Concrete block walls sometimes escape structural damage but the surface that was exposed to heat may have chipped, pitted, crumbled, or been otherwise scarred in various ways. Rather than replacing an otherwise sound wall, it may be satisfactory to apply portland cement plaster to the entire surface. The specifications for such restoration are a matter of negotiation and agreement between the parties concerned.

Fire surface cracks in concrete block may be pointed with mortar and if necessary a coating of cement wash or paint can be applied. Larger cracks caused by heat, settlement, blasting or other perils can be pointed or grouted with cement mortar as long as no structural damage is evident that would require replacing the units of masonry.

BRICK MASONRY

Brick masonry is a term applied to masonry employing solid units made of burned clay, the "Standard Brick" size being 2 1/4 x 3 3/4 x 8 inches long. There are many sizes, kinds and qualities of brick depending upon the type of clay, the molding process, and the manner of firing.

Figure 8-5 shows the sizes of the various Modular and Non-Modular brick.

Common brick are machine molded of ordinary clay. The position in the kiln affects the shape and hardness of the brick, and to some extent the finished color. Overburned brick may be darker in color and some of them are irregular in shape. In certain localities, brick are sold *kiln-run,* while in others, those of uniform color and shape are selected and sold at a slightly higher price.

Face brick are usually wire-cut, and have a uniform color and finish. They are used as face or veneer brick. Pressed brick are also known as face brick.

Glazed brick has one surface that is glazed during firing by being coated with mineral salts. They are mainly used where sanitary wall surfaces are specified.

Firebrick, made from fire clay to stand high temperatures, are used in lining furnace fire boxes, and also fire places.

The use of brick for structural purposes is almost unlimited. Although cement block has supplanted them in foundations, brick are used extensively in exterior and interior walls of mercantile and manufacturing buildings, dwelling walls, porches and steps, chimneys, and columns. They are also used in sidewalks, patio floors, and garden walls.

Usually when brick is used as a veneer in a solid brick wall, the veneer is bonded to the main wall by making every sixth course a header course. When brick is used as a veneer on the exterior of a frame building, or one of concrete block or hollow tile, metal clips are used, preferably at every second or third course. In the case of a frame building the metal ties are nailed to the sheathing and imbedded in the mortar joints of the veneer masonry. When brick is being used as veneering against concrete block or hollow tile construction, the clips also may be nailed or imbedded in the backing joints.

ESTIMATING QUANTITIES

Walls

The unit of measurement for brick masonry is 1,000 brick. A "Standard Brick" is approximately 3 3/4 inches thick. A wall one brick thick is nominally referred to as a 4-inch wall; one 2 bricks thick is an 8-inch wall; one 3 bricks thick a 12-inch wall, and so on, in multiples of 4 inches. The actual measured thickness of the wall in inches may be 3 3/4 or 4 inches, 8 or 9 inches, or 12 or 13 inches. A cubic foot of brick wall with joints of about 1/4 to 3/8 inches contains approximately 21 brick. For

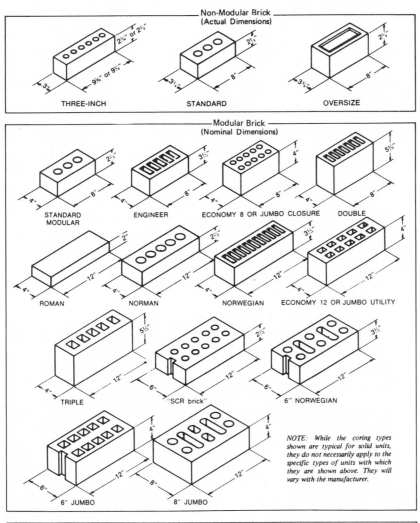

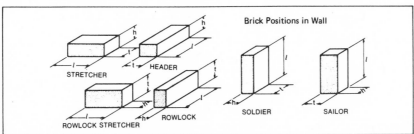

Figure 8-5

Sizes of Modular and Non-Modular Brick, Courtesy Brick Institute of America, McLean, Va. 22101

convenience in estimating, the following figures are applicable and include allowance for breakage.

Thickness of Wall (Inches)	Number of Brick Per Square Foot
4	7
8	14
12	21
16	28
20	35
24	42

To obtain the number of brick in a wall, compute the surface area, and *deduct all window and door openings in full.* Multiply this figure by the proper number of brick per square foot for the particular wall thickness.

Illustrative Example

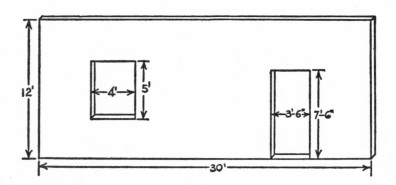

Gross wall area 12″ x 30″	360 sq ft
Deduct openings:	
Window 4′ x 5′ = 20 sq ft	
Door 7′6′ x 3′ x 6′ = 26 sq ft	—46
Wall area less openings	314 sq ft
(21 brick for a 12″ wall)	x21
Number of brick required	6,594

(If this wall were 8 inches thick, the number of brick would be 314 x 14 = 4,396.)

Pilasters

Where pilasters are part of a wall, the total area of the wall is first obtained, and the projecting pilasters ignored. The area of the pilasters alone is then computed and the number of brick is determined by the thickness of the pilaster.

Illustrative Example

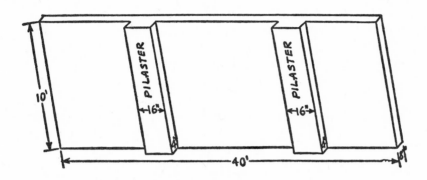

Wall area excluding pilasters 10 x 40 ft   400 sq ft
    Times 14 brick for 8 inch wall   X14
      Number of bricks per square foot in the 8 inch wall   5,600
Area of pilasters 2 x 1⅓' x 10'   26⅔ sq ft
    Times 7 brick for a 4 inch wall   X7
      Number of brick in pilasters   187
        Total number of brick   5,787

Piers

The number of brick in piers and columns can be figured either by taking the surface area of one side and multiplying by the number of brick according to thickness, or simply by determining the number of cubic feet in the pier or column and multiplying by 21 brick per cubic foot.

Illustrative Example

Method 1 by Surface Area
(16 in = 1⅓ ft)   1⅓' x 9'   = 12 sq ft
Times 28 brick for a 16-inch wall   X28
   Total number of brick   336

Method 2 by Cubic Feet
1⅓' x 1⅓' x 9'   = 16 cu ft
Times 21 brick per cubic ft   X21
   Total number of brick   336

Brick Chimneys

Brick chimneys are built with different sizes and numbers of flues. Nominal flue sizes are 8" x 8" x 12" and 12" x 12" x 12". Interior chimney walls may be 4 inches thick when lined with terra cotta flue, but exterior lined chimney walls should be 8 inches thick. When the exterior wall of a building forms one side of a chimney, care should be taken not to include it in computing the number of brick in the chimney.

To figure the number of brick in a chimney, draw a sketch (as in Figure 8-9) showing the number of brick in each course. Multiply the number of brick *per course* by the number of courses in a lineal foot of chimney. The height of the chimney in feet multiplied by the number of brick per lineal foot will give the total brick required.

Illustrative Example
An interior chimney with 4-inch walls has an 8" x 8" and an 8" x 12" flue. Using Figure 8-9, we find there are 11 brick per course or 50 brick per lineal foot of height. If the chimney is 30 feet high, it will require 1,500 brick.

Mortar for chimney brick may be obtained from Table 8-2. A mason and a laborer work as a team building a chimney. Table 8-4 shows the approximate hours of labor required per 1,000 brick. Flue lining is set in place by the mason as he erects the chimney and is included in the hours of labor shown.

Corners, Sills, and Waste

In figuring the quantity of brick in building walls, it should be remembered that corners are to be deducted.

Waste in brick masonry is a small factor for consideration and it is not customarily added in estimating losses unless large quantities of brick are involved. Generally between 2 and 5 per cent is adequate for normal breakage in handling and cutting if it is desired to add waste.

Window sills and window or door lintels are treated in the same manner discussed under concrete block masonry. Where *brick* windown sills are laid the number of brick may be obtained by dividing by 2.5 the length of the sill in inches.

ESTIMATING LABOR

In laying brick, as in concrete block masonry, a mason and a laborer work together. It is customary, therefore, in most brick masonry to figure the labor on the basis of an equal number of hours for both a mason and a

NUMBER OF BRICKS PER LINEAL FOOT FOR CHIMNEYS
WITH FLUE SIZES SHOWN

	WALL THICKNESS	BRICK PER COURSE	BRICK PER LINEAL FOOT
	4"	6	28
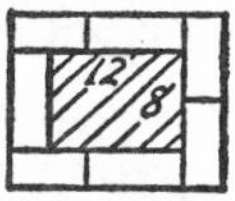	4"	7	32
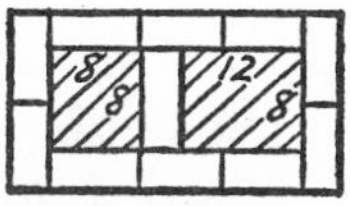	4"	11	50
	8"	16	73
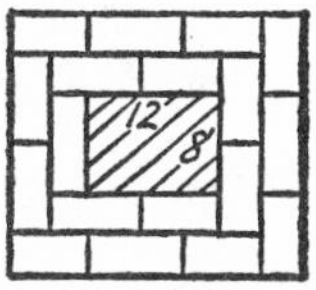	8"	18	82
	8"	20	91
	8"	29½	139

(BASED ON 3/8" MORTAR JOINTS.)

Figure 8-9

laborer. The mason is attended by the laborer who mixes the mortar, arranges the plank and horse scaffolding, and keeps him supplied with materials.

The unit of measurement in brick masonry is 1,000 brick. The number of hours required to lay 1,000 brick depends on the kind of work. Straight walls without openings take less time per thousand than walls with many openings, or pilasters, where proper plumbing of jambs and corners requires extra time.

Table 8-4 shows the approximate number of hours required for a mason and a laborer to lay 1,000 brick under normal working conditions for the kind of work indicated. The cost of scaffolding to be erected by carpenters and any necessary hoisting should be added as a separate item to the estimate.

The number of hours it will take for a mason to lay 1,000 brick varies generally with those factors that affect all labor production as outlined in Chapter 2. The time also varies specifically with the type of structure involved, that is, whether it is a solid wall, a chimney, piers, 4-inch veneer, a fireplace, and so forth. Table 8-4 covers the more common types of work.

Table 8-4
Approximate Hours of Labor Required to Lay 1,000 Brick

Kind of Work	Mason	Laborer
8″ or 12″ foundation walls	10	10
8″ common brick walls, normal openings	12	12
8″ common brick walls, large openings, pilasters	14	14
12″ common brick walls, normal openings	10	10
12″ common brick walls, large openings, pilasters	12	12
4″ veneer on frame or masonry—ordinary grade	16	16
4″ veneer on frame or masonry—high grade	20	20
Masonry piers and columns	14	14
Common brick chimneys	15	15
Ordinary brick fireplaces	20	20
Brick veneering fireplaces	24	24
Washing down brick masonry (per 100 sq ft)	1	1

ESTIMATING COSTS

The factors to be included in estimating the cost of brick masonry are the brick, the mortar, and the labor.

Table 8-2 shows the approximate amount of mortar required for 1,000 brick in masonry walls of various thicknesses. Multiply the cost of 1 cubic yard of mortar by the percentage of a yard indicated.

The cost of labor for laying 1,000 brick may be obtained by multiplying the prevailing rate of wages for a mason and a laborer by the number of hours shown in Table 8-4 for the kind of work being done.

Illustrative Example

Assume that it has been determined that 28,000 common brick are required in the 8-inch exterior walls of a ranch type dwelling. Prices and wages are for illustration only.

Cost of 1 cu yd of mortar

10 sacks cement	@ $1.40 =	$14.00
1 cu yd sand	@ 3.75 =	3.75
2 sacks hyd lime	@ .85 =	1.70
	Total =	$19.45

Cost of mortar for 1,000 brick (Table 8-2): .5 x $19.45 = $9.73

Labor to lay 1,000 brick (Table 8-4)

12 mason's hours	@ $6.00 =	$ 72.00
12 laborer's hours	@ 5.00 =	$ 60.00
	Total labor cost for 1,000 brick =	$132.00

Application of material and labor cost per 1,000 brick

28,000 common brick	$ 45.00 per M =	$1,260.00
Mortar (28,000)	9.73 per M =	272.44
Labor (28,000)	132.00 per M =	3,696.00
	Total =	$5,228.44

Cost per 1,000 brick: $\frac{\$5,228}{28} = \$186.73$

UNIT COST

Using the same material and labor wages shown in the previous illustrative example, the *unit cost* for laying 1,000 common brick would be:

1,000 common brick	@ $45.00 =	$ 45.00
½ cubic yard mortar	@ 19.45 =	9.73
12 hours mason labor	@ 6.00 =	72.00
12 hours laborer	@ 5.00 =	60.00
	Unit cost per 1,000 brick =	$186.73

Having obtained a unit cost per 1,000 brick, the cost of 28,000 brick would be:

28 x $186.73 = $5,228.44

The unit cost of veneer brick, glazed brick, and other kinds may be obtained in the same way. The *local price* of the materials, and the *prevailing wage rates* should be used in each instance. The approximate hours of labor per 1,000 brick are shown in Table 8-4.

Illustrative Examples Using Assumed Prices and Wages

Illustrative Example No. 1

Unit Cost for Ordinary Grade 4-inch Veneer on Frame Wall.

	1,000 common brick	@$45.00 = $ 45.00
(Table 8-2)	.33 cu yds mortar	@ 19.45 = 6.48
(Table 8-4)	16 hours mason	@ 6.00 = 96.00
(Table 8-4)	16 hours laborer	@ 5.00 = 80.00
	Unit cost for 1,000 brick = $227.48	

Illustrative Example No. 2

Unit Cost for 12-inch Brick in Wall Having Large Window and Door Openings and Pilasters.

	1,000 common brick	@$45.00 = $ 45.00
(Table 8-2)	.60 cu yds mortar	@ 19.45 = 11.67
(Table 8-4)	12 hours mason	@ 6.00 = 72.00
(Table 8-4)	12 hours laborer	@ 5.00 = 60.00
	Unit cost for 1,000 brick = $188.67	

Illustrative Example No. 3

Unit Cost for Brick Piers and Columns 12 inches by 12 inches

	1,000 common brick	@$45.00 = $ 45.00
(Table 8-2)	.60 cu yds mortar	@ 19.45 = 11.67
(Table 8-4)	14 hours mason	@ 6.00 = 84.00
(Table 8-4)	14 hours laborer	@ 5.00 = 70.00
	Unit cost for 1,000 brick = $210.67	

DAMAGEABILITY OF BRICK MASONRY

Damage by Fire

Bricks are made principally of clay products, and during the manufacturing process they are fired at temperatures ranging as high as 2,000°F. It is therefore understandable that good quality brick masonry is able to withstand exposure to very hot fires without suffering serious structural damage. As is true with other kinds of masonry, the first injury is soilage or discoloration by smoke. As the fire progresses and the heat intensifies, the exposed face of the brick may crack or spall. Alternate heating and then cooling with water from the fire hoses tends to aggravate this condition. Soft, underburned brick will show the effect of fires more rapidly than hardburned brick under similar circumstances. Thin masonry walls suffer damage more readily than thicker walls. Old brick masonry, especially when laid with sand-lime mortar, reacts to severe fires less favorably than new brickwork laid with good portland cement mortar. The reason is mainly due to the effect of extreme heat and the water on the bond which has already been weakened by old age. It is not uncommon to have walls of this kind fall during a fire, particularly under pressure of dropping timbers or debris, or from side thrusts of collapsing floor and roof structures. Masonry of this type may also present a difficult problem when local authorities, for public safety, condemn entire walls even though on the surface they show only a slight damage resulting from the fire.

Brickwork usually does not disintegrate and crumble in extremely hot fires of long duration as does concrete block masonry, or certain types of inferior concrete. There are, however, instances of brick actually melting in temperatures in excess of 2,500°F. when exposed over a long period of time.

Good brick masonry has a modulus of elasticity approximating that of good concrete, but it does not have steel reinforcing imbedded in it to care for the tensile stresses induced during expansion and contraction. There is, therefore, a greater danger of cracks developing in brick masonry. They may run the full height of a wall or show up at the corners, near the window sills, or at the window or door heads. The cracks may follow the joints in the masonry or pass completely through the brick itself. Thin walls are more inclined to develop cracks than thick walls.

Damage by Water

Good quality brick masonry is not readily damaged by water, and it can withstand submersion over long periods of time. Poor quality mortar may be washed out of the joints by the strong action of a fire hose. It may also be washed out by the action of successive heavy rains or during submersion, as in the case of flooding or wave action. The brick itself will absorb water up to approximately 10 to 18 per cent by weight when submerged for 24 hours. Although the absorption of water slightly reduces the compressive strength of brick, they dry out readily and show no permanent damage. Freezing of water absorbed in the brick very seldom has any ill effects because the pores are only partially filled, leaving room for expansion during the frost action.

Soft brick weathers more rapidly than a hard-burned good quality brick. It is not uncommon to see actual erosion of soft brick in old buildings. This is especially noticeable at corbeled cornices, belt courses, window sills, and other parts of a structure where the brick projects out beyond the face.

Efflorescence, a white deposit on brickwork, is the crystallization of salts on the surface of the brick. It appears on new brickwork frequently, and is attributed to the presence of certain salts within the brick itself or sometimes in the mortar used in laying them up. It is inherent in the masonry. Linseed oil brushed over the surface removes the deposit but the treatment has to be repeated every few years as the oil is washed off.

Damage by Cracking

Cracks in brick masonry may be caused by heat, explosion, or by disturbance of the foundation supporting it. Blasting operations, earthquake, and, in rare cases, vibration caused by trucks or trains passing in the immediate vicinity will cause cracking in brickwork. Settlement due to soil subsidence, and the action of frost may also be a factor. Poorly constructed foundations and footings and inferior materials or workmanship in the masonry increase the possibilities of cracking. Brick walls that have been struck by vehicles, or subjected to either impact or pressure such as may occur during a collapse of a portion of the structure or during violent tornado-like windstorms, are subject to cracking in addition to other damage.

One of the tasks confronting the person making the survey and estimate of cracked masonry is to distinguish the cracks that may be directly attributed to the immediate casualty from the old unrelated cracks.

METHODS OF REPAIRING BRICK MASONRY

Brickwork that has been damaged can be stripped down to where it is undamaged, and rebuilt. The upper sections of multistoried brick buildings involved in a fire may have several courses of brick removed to reach a sound base for rebuilding or, when indicated, whole stories may be demolished and new walls erected. Most times it is preferable to engage a structural or consulting engineer to advise on the extent of repairs that are necessary. When faced with condemnation by local authorities it is good procedure to consult a well-qualified and reputable engineer as to whether any doubt exists concerning the justification of condemning the masonry.

Sectional Repairs

When a portion of the top of a brick wall must be removed and rebuilt, or a hole in a wall must be repaired, the primary factors to be considered are:

1. The labor to remove the damaged brick
2. The scaffolding necessary
3. Hoisting equipment
4. Shoring and sidewalk canopy
5. Rebuilding the new section

Laborers, as a general rule, demolish the damaged brickwork under the supervision of a mason or the contractor. The work on upper walls is usually done from the inside so that outside scaffolding is not required. If an exterior scaffold is needed, the cost must be added. The means of getting new materials up and debris down from upper stories should be considered according to the circumstances of each individual job. A building freight elevator may be available or it might be necessary to erect a hand- or machine-operated hoist. The site of the job, the height of the work above ground, and amount of material to be hoisted will have to be given consideration.

Shoring of beams, girders, rafters, or other structural supports bearing on the brick masonry should be added to the cost. When the need of a sidewalk canopy to protect pedestrians is indicated, or when required by local ordinance, the cost should be figured and included.

Rebuilding the new section should be figured as any other new job except when estimating the labor, adequate time should be allowed for the masons to tooth-in or step-in the new brick with the old where they join.

Unit costs may be used where there is a sufficient quantity of brick to be laid to approximate a rate of hours per thousand. Otherwise the repairs can be more accurately estimated on a material and time basis.

Repairing Spalled Brickwork

When the exterior face of a brick wall has been chipped and spalled with no structural injury, the damaged layer of brick can be cut out with power drills and hammers or chiseled out by hand. New matching brick can be laid. This method is recommended for veneer brick. When a large area of wall has been spalled, it is usually possible, with the consent of the owner, to clean it off and apply portland cement stucco over the entire surface.

On secondary structures, if the spalling is not serious, cement washing or oil painting may be satisfactory to all parties. Much depends on the individual situation.

Spalled interior walls that are to be furred and lined with wall board or plaster, will usually require little more than a coat of cement mortar, cement wash, or whitewash to remove smoke odor. In some instances nothing is done inasmuch as they are permanently concealed by the wall treatment. If the walls are exposed, cement mortar plaster, cement washing, or painting are generally acceptable as methods of repair.

Repairs to Cracks

The repairing of cracks in brick masonry can usually be done by a tuck-pointer, or mason. Where the crack is not large, and it follows the mortar joint, grouting and filling the joint is generally satisfactory. The old mortar is chiseled out as deeply as possible and, when filled, the new joint is struck.

Large cracks sometimes present a more serious problem, depending on the importance of appearances after the repairs are made. If grouting and filling are not acceptable as a method of repair, it may be necessary to cut out the face brick and replace them with new ones. In difficult situations, because skillful repairs to brick masonry are highly specialized, an expert mason should be consulted and asked to make the estimate.

Repairs to Soiled Walls

Brick masonry that has been soiled by smoke, roof tar drippings, or stains from contact with contents can usually be cleaned. (See Repairs to Soiled Walls, Chapter 7 on "Concrete.") There are, in most cities, firms that make a specialty of cleaning exteriors of brick, stone, or concrete structures. The method used will depend on the type of stain. Most stains can be removed by washing down with caustic solutions or brushing and steam cleaning. On secondary buildings and some older buildings it is often satisfactory to paint over old brick work, thereby covering up the stain.

HOLLOW TERRA COTTA TILE MASONRY

Hollow terra cotta tile is used extensively in building construction because it is relatively light in weight (an 8" x 12" x 12" block weighs about 25 pounds) is sufficiently strong, and has excellent fire-retardant qualities. The name *terra cotta* comes from the Latin words meaning "cooked earth," which aptly names the process by which it is made. The hard-burned units are dense and strong. The semi-porous units are the more commonly used. They are exposed to less firing in the kiln but are strong enough for all but heavy load-bearing walls. The outer wall of a tile block is called the "shell." The inner "cells" are divided by thin partitions called "webs."

Hollow tile is used for the walls of dwellings and for small manufacturing and mercantile structures. It is also used for curtain and nonbearing walls in steel frame buildings, for fire-protecting steel columns, beams and girders, and for flat-arch type floors. The units are made in a variety of shapes and sizes depending on their use in the structure. They are made for corners, end blocks, and window and door jambs or lintels. The standard wall units are 12" x 12" giving them a surface area of 1 square foot; the thickness ranges from about 2 to 12 inches. The most commonly used are the 8-, 10- and 12-inch thick unit.

ESTIMATING QUANTITIES

The unit of measurement of hollow terra cotta tile is usually 100 blocks. Since a standard wall tile is 12" x 12", each unit covers 1 square foot. To obtain the number of units required, compute the surface area in square feet, being certain *to deduct all window and door openings in full*. The number of hollow tile will be equal to the number of square feet. Generally waste is not added unless large quantities are involved, or unless an unusual amount of cutting is to be done. When waste is added, 2 to 5 per cent is considered adequate.

Corners should be deducted, but one must be certain not to deduct them twice. Lintels and sills should be estimated as outlined under concrete block masonry.

The materials and quantity of mortar required are discussed under "Mortar" in this chapter.

ESTIMATING LABOR

A mason and a laborer work together in laying hollow tile block as with concrete block and brick masonry. The rate at which the units can be laid depends on the thickness of the tile blocks and on the kind of work being done. Table 8-5 shows the approximate number of hours required to lay 100 hollow tile.

Table 8-5
Approximate Hours of Labor Required to Lay 12" x 12"
Hollow Wall Tile

Size of Unit	Mason	Laborer
6" x 12" x 12"	5	5
8" x 12" x 12"	6	6
10" x 12" x 12"	7	7
12" x 12" x 12"	8	8

ESTIMATING COSTS

The cost of laying hollow terra cotta wall tile is obtained by determining the cost of the tile, the mortar, and the labor required to lay them.

From Table 8-2, the approximate quantity of mortar for 100 units may be found for the thickness of the wall being constructed. Multiply the percentage indicated by the cost of 1 cubic yard. Table 8-5 shows the approximate number of hours required to lay 100 wall tile of different thicknesses. Multiply the number of mason and laborer hours indicated by the prevailing rate of wages.

Illustrative Example

Cost of 1 Cubic Yard Mortar (page 190)		$19.45
Mortar required for 100 tile (Table 8-2)		x .13 cu yds
		$ 2.53
100 T.C. tile 8" x 12" x 12"	@$ .30 =	30.00
(Table 8-5) 6 hours mason	@ 6.00 =	36.00
(Table 8-5) 6 hours laborer	@ 5.00 =	30.00
	Unit cost per 100 tile =	$98.53
	Unit cost per tile =	.99

LINTELS

In concrete block construction a reinforced concrete lintel is usually provided for each door and window opening. The size of the lintel and the size and the number of reinforcing rods used is determined by the load above the arch. In some buildings a steel angle lintel is used. The type of construction employed can be observed by inspection. Where a reinforced concrete lintel exists, the cross section generally conforms with the cross section of the block, 8 by 8 inches or 8 by 12 inches, and the lintel extends approximately 8 inches beyond each side of the opening. Estimating the cost of cast-on-the-job lintels consists in figuring the wood forms, the concrete and reinforcing, plus the labor of mixing and placing the materials (see Chapter 7 on Concrete).

In brick veneer or solid brick wall construction, the lintel for windows and doors consists of a steel angle iron for veneer and two, back to back, for solid walls. The size and weight of the angle irons depend on the span and weight carried above. The exact size and shape can be determined in the field. (See Structural Steel, Chapter 20.)

Stone lintels were frequently used in older buildings and reinforced concrete lintels are still used in some solid masonry wall construction.

SILLS

Windows and doors in masonry walls are provided with sills. In commercial structures, particularly, precast concrete sills are used or they are cast in place on the job. Their cost should be estimated the same as for lintels.

Limestone sills are frequently used in better residential construction, and in offices and public buildings. Measurements can usually be obtained in the field and prices should be obtained from local supply sources. The price is governed by the shape, dimensions, number of sills required and shipping costs.

COPING

Masonry walls that project above the roof are customarily finished by capping with stone, concrete, or terra cotta tile coping. The main purpose of coping is to seal the top of the wall against the eroding effects of snow, water, and freezing. A secondary purpose is to provide an architectural finish to the wall.

Glazed terra cotta wall coping is available in widths of 9, 13, and 18 inches. Each section is 2 feet long. Also available are starter, end, angle, and corner coping.

Cut flagstone coping is occasionally used. Lengths vary from 18 to 30 inches and widths are adequate to overlap each side of the wall an inch or two.

Dressed limestone coping is cut according to specifications as to shape, width, thickness and length.

When estimating, the proper sizes of coping units may be obtained at the place of the job. Price should be determined locally.

All coping is set in regular mortar and laid by a mason and a laborer. A mason and a laborer can lay approximately 100 lineal feet of terra cotta or flagstone coping a day. Cut stone coping takes longer and should be estimated in accordance with the circumstances of each job.

The cost of hoisting, or otherwise getting the coping up to where it is to be laid must be added to the estimate as a separate item.

Chapter 9.

Rough Carpentry

Carpentry is work done by a carpenter, who builds with wood and with certain composition materials. About 80 per cent of all building construction involves a substantial amount of carpentry work. A knowledge of this trade is therefore essential for anyone who either makes or checks building loss estimates. It not only is a major factor in the construction of dwellings, barns and other farm buildings, but also plays a large part in commercial structures which have masonry walls, wood floors, partitions and roofs. Some of these are public garages, apartment houses, warehouses, motels, and older office and loft buildings, mills, schools, and hotels. There are very few buildings that do not contain some form of carpentry. Carpenters also build and erect concrete forms, scaffolding, fences, and the various types of wooden yard fixtures. The carpentry trade is very broadly divided into two classifications: *rough carpentry* and *finish carpentry* or *millwork*.

TYPES OF ROUGH CARPENTRY

While the line of separation between rough and finish carpentry cannot always be sharply defined, the following work and materials are generally considered as applying to rough carpentry.

Sleepers.
Forms.
Framing.
  Sills, plates, ridge pieces,
  Studding, fire stops, etc.
  Joists, bridging and bracing,
  Rafters, wood trusses, and purlins.
Sheathing, roof boards and rough floors.
Furring and grounds.
Window and door frames and bucks.
Temporary and rough cellar stairways.
Fences, trellises, arbors and other yard fixtures.
Rough hardware.
Cutting for other trades.
Scaffolding.
Debris chutes, canopies, etc.

TYPES OF BUILDING CONSTRUCTION EMPLOYING CARPENTRY

There are four common classifications of building construction which require a major or a substantial amount of carpentry. They are customarily referred to as *frame* (including frame and stucco), *brick veneer, ordinary brick,* and *mill or slow-burning.*

Frame Construction

Buildings of frame construction, except for the foundations, interior wall treatment, or stucco finish when it exists, are constructed almost entirely of wood. The principle for erecting the basic framework or skeleton is the same in all frame buildings. A lateral bottom member called a *sill* is laid on the ground (*a mudsill*), on piers, or on top of a masonry type foundation. *Studding* (called *scantlings* in some sections of the country) are fastened vertically to the sill and capped with a member similar to the sill but called a *plate*. This assembly provides the general framework for exterior walls. *Rafters,* which are members that support the roof, are set on top of the upper plates. In gable roofs the rafters are either butted together at the ridge or joined to a *ridge pole*. Floor *joists,* frequently called floor *beams,* are laid horizontally across the building and supported on the exterior plates, and on girders carrying partitions. Solid or cross-*bridging* is placed between the joists at intervals to stiffen and strengthen them.

Interior frame partitions, erected to separate the area into rooms, closets, and other service quarters, are constructed in a manner similar to the exterior walls. The partition studs may be secured at the bottom directly to doubled joists or to a *shoe* (interior sill) which is laid across the joists, or on the *subflooring*.

Wood or composition *sheathing* is applied to the exterior walls. Roof *sheathing* (called *roof decking* in some areas) is laid over the rafters, and rough or *subflooring* is put down over the floor joists.

A weather surface of roofing material is laid over the roof decking, and the exterior is appropriately finished with wood, metal, or composition *siding*. In a *frame* and *stucco* building the exterior finish is composed of wire lath and portland cement stucco. *Windows* and *exterior doors* are installed. *Cornices* and other exterior *wood trim* are put into place.

Interior finishes are then applied. These may be *lath and plaster, wood paneling,* or *composition board* such as gypsum board, Celotex, or Masonite. *Finish floors* are laid, *interior doors, trim, cupboards, stairways,* and other *millwork* is installed.

During the construction, and before covering the walls and ceilings, the *electrical wiring, plumbing* and *heating* facilities are roughed in. The final operations are generally the installation of *fixtures* for the service equipment, and the *decorating*.

There are many modifications of all types of construction depending upon the purpose for which the building is to be used, its climatic location, architecture, local custom and practice, and the amount of money to be spent.

Brick Veneer

A building which in most respects is constructed as that described for frame construction but which has a masonry veneer (usually stone or brick) is classified as *brick veneer* construction.

Ordinary Brick

A building which is constructed similar to a frame building, but which has solid 8-inch masonry exterior walls, is known as *ordinary brick* construction. While the exterior walls are usually brick, they may also be cement block, or some other type of masonry. In multiple story structures such as apartment houses, mercantile or manufacturing buildings, the exterior walls are thicker varying with the height and local building code requirements.

Mill, or Slow-Burning

This type of construction employs heavy timber framing. Exterior walls are masonry, being either brick, stone or concrete. Wood *girders* of large dimensions are set on iron, steel, or wood *columns* or *posts,* the ends bearing in or on the walls. *Beams* not less than 10 inches deep and spaced 8 to 10 feet on center are placed between the girders on *beam hangers.* The floors are usually 3- to 4-inch tongue and groove or *splined* plank flooring spiked to the beams. The roof construction is relatively the same as the floor construction but somewhat lighter. It is surfaced with roll type or built-up asphalt roofing.

ESTIMATING MATERIALS

The material used most in carpentry is lumber. Building lumber comes from a dozen or more commercially useful trees. The *conifers* or *needle-leaved* trees comprise the group that supplies the majority of woods used for structural purposes. They are the so-called *evergreens* or soft woods such as fir, spruce, hemlock, pine and cedar. The *broad-leaved* trees generally supply wood for interior floors, trim and cabinet work. These are the hardwoods such as oak, maple, birch, ash, walnut, and white wood (basswood). In some sections of the country, depending on choice and local supply, the different types of lumber are used interchangeably.

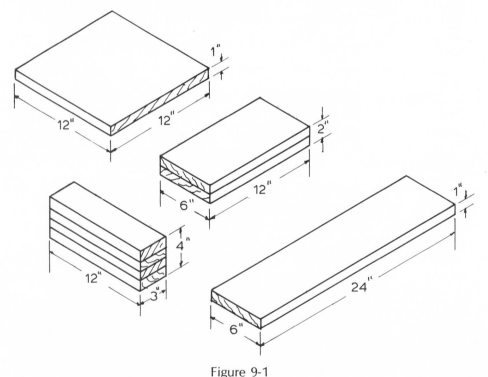

Figure 9-1

Each a Board Foot

BOARD MEASURE

Nearly all lumber, from the time it is surveyed in the forest until it is sawed from the logs into timbers and boards, is measured and sold by the unit of the *board foot.* Special millwork like moldings, window and door trim, lattice, handrail, balusters, closet pole, shelf cleats, some types of baseboard, molded exterior trim, wood gutters, and fence pickets are sold by the lineal foot. Plywood, particle board, flakeboard etc., as distinguished from lumber is sold by the square foot.

A board foot is one square foot of wood one inch thick. The symbol for board feet is FBM (feet board measure). See Figure 9-1.

The number of board feet in a piece of wood is obtained by multiplying the width times the thickness in inches, times the length in feet, and dividing by 12.

Table 9-1 shows the number of board feet in various sizes of lumber of different lengths. Table 9-2 gives the factor to multiply a piece of lumber by to determine the number of board feet it contains.

Table 9-1
NUMBER OF BOARD FEET IN VARIOUS SIZES OF LUMBER
Length of Piece

Size Inches	8′	10′	12′	14′	16′	18′	20′	22′
2 x 3	4	5	6	7	8	9	10	11
2 x 4	5⅓	6⅔	8	9⅓	10⅔	12	13⅓	14⅔
2 x 6	8	10	12	14	16	18	20	22
2 x 8	10⅔	13⅓	16	18⅔	21⅓	24	26⅔	29⅓
2 x 12	16	20	24	28	32	36	40	44
3 x 4	8	10	12	14	16	18	20	22
3 x 6	12	15	18	21	24	27	30	33
3 x 8	16	20	24	28	32	36	40	44
3 x 10	20	25	30	35	40	45	50	55
3 x 12	24	30	36	42	48	54	60	66
4 x 4	10⅔	13⅓	16	18⅔	21⅓	24	26⅔	29⅓
4 x 6	16	20	24	28	32	36	40	44
4 x 8	21⅓	26⅔	32	37⅓	42⅔	48	53⅓	58⅔
6 x 6	24	30	36	42	48	54	60	66
6 x 8	32	40	48	56	64	72	80	88
6 x 10	40	50	60	70	80	90	100	110
8 x 8	42⅔	53⅓	64	74⅔	85⅓	96	106⅔	117⅓
8 x 10	53⅓	66⅔	80	93⅓	106⅔	120	133⅓	146⅔
8 x 12	64	80	96	112	128	144	160	176
10 x 10	66⅔	83⅓	100	116⅔	133⅓	150	166⅔	183⅓
10 x 12	80	100	120	140	160	180	200	220
12 x 14	112	140	168	196	224	252	280	308

$$\frac{\text{Width in inches X Thickness in inches X Length in feet}}{12}$$

Illustrative Example

A 2″ x 6″ x 10′ contains 10 FBM

$$\frac{2'' \ X \ 6'' \ X \ 10'}{12} \ = \ 10 \ FBM$$

The total number of FBM in a list of framing lumber would be obtained as follows:

$$6 \text{ pieces } 2'' \text{ x } 4'' \text{ x } 8' = \frac{\cancel{6} \ X \ \cancel{2} \ X \ 4 \ X \ 8}{\cancel{12}} \ = \ 32 \ FBM$$

$$10 \text{ pieces } 2'' \text{ x } 6'' \text{ x } 12' = \frac{10 \ X \ 2 \ X \ 6 \ X \ \cancel{12}^{\ 1}}{\cancel{12}} \ = \ 120 \ FBM$$

$$4 \text{ pieces } 2'' \text{ x } 10'' \text{ x } 10' = \frac{\cancel{4} \ X \ 2 \ X \ 10 \ X \ 10}{\cancel{12}} \ = \ 66\tfrac{2}{3} \ FBM$$

$$8 \text{ pieces } 4'' \text{ x } 4'' \text{ x } 18' = \frac{8 \ X \ 4 \ X \ 4 \ x \ \cancel{18}^{\ 6}}{\cancel{12}_{\ 3}} \ = \ \underline{192} \ FBM$$

Total = $410\tfrac{2}{3}$ FBM

TABLE 9-2

MULTIPLIERS TO COMPUTE THE NUMBER OF BOARD FEET
IN ANY LENGTH OF DIMENSION LUMBER

Nominal Size In Inches	Multiply Length By	Nominal Size In Inches	Multiply Length By
2 x 2	0.333	4 x 4	1.333
2 x 3	0.500	4 x 6	2.000
2 x 4	0.667	4 x 8	2.667
2 x 6	1.000	4 x 10	3.333
2 x 8	1.333	4 x 12	4.000
2 x 10	1.667		
2 x 12	2.000	6 x 6	3.000
		6 x 8	4.000
3 x 3	0.750	6 x 10	5.000
3 x 4	1.000	6 x 12	6.000
3 x 6	1.500		
3 x 8	2.000	8 x 8	5.333
3 x 10	2.500	8 x 10	6.667
3 x 12	3.000	8 x 12	8.000

Example: Find the number of board feet in
6 - 2″ x 10″ joists 14′ long
6 x 1.667 x 14 = 140 FBM

LUMBER INDUSTRY ABBREVIATIONS
(American Softwood Lumber Standard)

These abbreviations are commonly used for softwood lumber, although all of them are not necessarily applicable to all species. Additional abbreviations which are applicable to a particular region or species may be included in approved grading rules.

Abbreviations are commonly used in the forms indicated, but variations such as the use of upper- and lower-case type, and the use or omission of periods and other forms of punctuation are optional.

AD	Air-dried
ADF	After deducting freight
ALS	American Lumber Standards
AV or AVG	Average
Bd	Board
Bd. ft.	Board foot or feet
Bdl	Bundle
Bev	Beveled
B/L	Bill of lading
BM	Board Measure
Btr	Better
B&B or B& Btr	B and better
B&S	Beams and stringers
CB1S	Center bead one side
CB2S	Center bead two sides
CF	Cost and freight
CG2E	Center groove two edges
CIF	Cost, insurance, and freight
CIFE	Cost, insurance, freight, and exchange
Clg	Ceiling
Clr	Clear
CM	Center matched
Com	Common
CS	Caulking seam
Csg	Casing
Cu. Ft.	Cubic foot or feet
CV1S	Center Vee one side
CV2S	Center Vee two sides

D&H . . . . Dressed and headed
D&M . . . . Dressed and matched
DB. Clg. . . . . Double-beaded ceiling (E&CB2S)
DB. Part . . . . Double-beaded partition (E&CB2S)
DET . . . . Double end trimmed
Dim . . . . Dimension
Dkg . . . . Decking
D/S or D/Sdg . . . . Dropped siding
EBIS . . . . Edge bead one side
EB2S . . . . Edge bead two sides
E&CBIS . . . . Edge and center bead one side
E&CB2S . . . . Edge and center bead two sides
E&CVIS . . . . Edge and center Vee one side
E&CV2S . . . . Edge and center Vee two sides
EE . . . . Eased edges
EG . . . . Edge (vertical) grain
EM . . . . End matched
EVIS . . . . Edge Vee one side
EV2S . . . . Edge Vee two sides
Fac . . . . Factory
FAS . . . . Free alongside (named vessel)
FBM . . . . Foot or feet board measure
FG . . . . Flat (slash) grain
Flg . . . . Flooring
FOB . . . . Free on board (named point)
FOHC . . . . Free of heart center or centers
FOK . . . . Free of knots
Frt . . . . Freight
Ft . . . . Foot or feet
GM . . . . Grade marked
G/R or G/Rfg . . . . Grooved roofing
HB . . . . Hollow back
H&M . . . . Hit-and-miss
H or M . . . . Hit-or-miss
Hrt . . . . Heart
Hrt CC . . . . Heart cubical content
Hrt FA . . . . Heart facial area
Hrt G . . . . Heart girth
IN. . . . . Inch or inches
J&P . . . . Joists and planks
KD . . . . Kiln-dried
LCL . . . . Less than carload
LFT or Lin. Ft. . . . . Linear foot or feet
Lgr . . . . Longer

Lgth . . . . Length
Lin . . . . Linear
Lng . . . . Lining
M . . . . Thousand
MBM . . . . Thousand (feet) board measure
MC . . . . Moisture content
Merch . . . . Merchantable
Mldg . . . . Moulding
No . . . . Number
NIE . . . . Nosed one edge
N2E . . . . Nosed two edges
Og . . . . Ogee
Ord . . . . Order
Par . . . . Paragraph
Part . . . . Partition
Pat . . . . Pattern
Pc . . . . Piece
Pcs . . . . Pieces
PE . . . . Plain end
PO . . . . Purchase order
P&T . . . . Post and timbers
Reg . . . . Regular
Res . . . . Resawed or resawn
Rfg . . . . Roofing
Rgh . . . . Rough
R/L . . . . Random lengths
R/W . . . . Random widths
R/W&L . . . . Random widths and lengths
Sdg . . . . Siding
Sel . . . . Select
S&E . . . . Side and Edge (surfaced on)
SE Sdg . . . . Square edge siding
SE & S . . . . Square edge and sound
S/L or S/LAP . . . . Shiplap
SL&C . . . . Shipper's load and count
SM. or Std. M . . . . Standard Matched
Spec . . . . Specifications
Std . . . . Standard
Stpg . . . . Stepping
Str. or Struc . . . . Structural
S1E . . . . Surfaced one edge
S1S . . . . Surfaced one side
S1S1E . . . . Surfaced one side and one edge
S1S2E . . . . Surfaced one side and two edges

S2E . . . . Surfaced two edges
S2S . . . . Surfaced two sides
S2S1E . . . . Surfaced two sides and one edge
S2S&CM . . . . Surfaced two sides and center matched
S2S&SM . . . . Surfaced two sides and standard matched
S4S . . . . Surfaced four sides
S4S&CS . . . . Surfaced four sides and caulking seam
T&G . . . . Tongued and grooved
VG . . . . Vertical grain
Wdr . . . . Wider
Wt . . . . Weight

STANDARD SIZES OF LUMBER
(American Lumber Standard 1970)

Type of Lumber	Nominal Size in Inches		Actual Size in Inches	
	Thickness	Width	Thickness	Width
	2	4	1.5	3.5
	2	6	1.5	5.5
Dimension	2	8	1.5	7.25
Plank	2	10	1.5	9.25
and	2	12	1.5	11.25
Joist	4	4	3.5	3.5
	4	6	3.5	5.5
S 4 S	4	8	3.5	7.25
	4	10	3.5	9.25
	6	6	5.5	5.5
Timbers	6	8	5.5	7.5
	6	10	5.5	9.5
S 4 S	8	8	7.5	7.5
	8	10	7.5	9.5
	1	4	¾	3.5
		6		5.5
Common		8		7.25
Boards		10		9.25
S 4 S		12		11.25

			Width	
			Overall	Face
	1	6	5½	5 1/8
		8	7¼	6 7/8
Shiplap		10	9¼	8 7/8
Boards		12	11¼	10 7/8
	1	4	3 3/8	3 1/8
Tongue		6	5 3/8	5 1/8
and		8	7 1/8	6 7/8
Grooved		10	9 1/8	8 7/8
Boards		12	11 1/8	10 7/8

MILLING WASTE

Matched boards are those which have the edges milled in a *tongue and groove,* or a *lap joint* (shiplap), to join them tightly together. Stock of this kind is used as surface lumber, as distinguished from framing lumber, and consists of such types as rough and finish flooring, exterior sheathing, siding, roof boards, paneling, and ceiling boards.

Most sawed lumber is planed (dressed) at the lumber mill before shipping. The result is that the nominal size, for example a 2" x 4" stud, turns out to measure 1 1/2" x 3 1/2". Prior to September 1, 1970, when the new Lumber Standards were accepted by the government, the same 2" x 4" stud measured 1 5/8" x 3 5/8". And before that the same 2" x 4" actually measured 1 3/4" x 3 3/4". The difference (in the case of the 2" x 4") was termed *milling waste.* That term is no longer accurate because the present 1 1/2" x 3 1/2" stud is *not* cut from a piece 2" x 4" but rather something less in size.

Milling waste is not significant in estimating where planks, timbers, joists, and similar dimension lumber are involved. It does become important, however, when estimating quantities of surface lumber such as common boards, shiplap, tongue and grooved boards, and bevel siding. The supplier will compute his charges on the basis of the *nominal size* and if a factor is not added for the so-called "milling waste," when placing the order, the quantity delivered will fall short of what is needed.

Table 9-3 (page 210) shows the factor by which the area to be covered is multiplied to obtain the exact amount needed. This factor is obtained for *any piece of surface lumber* by dividing the *nominal* size by the *actual* face size. For example, the factor for a 1" x 6" square edge board (actual face size 5.5") is 6 divided by 5.5 or 1.09.

This formula, $\dfrac{\text{nominal size}}{\text{actual face size}}$ = factor, is useful when there is no Table to refer to. Normal cutting and fitting waste is to be added.

Use of Table 9-3 to Compute Material Needed

To obtain from Table 9-3 the exact quantity of material to cover an area, multiply the measured square foot area by the factor in the Table for the particular type of material to be used.

Illustrative Examples

1. The measured area in which 1" x 8" shiplap is to be applied to the exterior walls of a building is 100' x 9' or 900 sq ft. The Table gives a factor of 1.16 for 1" x 8" shiplap.

900 sq ft x 1.16 = 1,044 FBM needed to cover the area.

2. Tongue and grooved rough flooring, 1″ x 6″, is to be layed in a room 15′ x 30′ or 450 sq ft. The Table gives a factor of 1.17 for 1′ x 6″ T&G boards.

450 sq ft x 1.17 = 527 FBM needed to cover the area.

3. The area of the exterior walls of a building to be covered with 1″ x 8″ bevel siding is 1,200 sq ft. The Table gives a factor of 1.28 for 1″ x 8″ bevel siding.

1,200 sq ft x 1.28 = 1,536 FBM needed to cover the area.

TABLE 9-3
FACTOR BY WHICH AREA TO BE COVERED IS MULTIPLIED TO
DETERMINE EXACT AMOUNT OF SURFACE MATERIAL NEEDED.
(From Western Wood Products Association)

Item	Nominal Size	Width Overall	Face	Area Factor
	1″ x 6″	5½″	5 1/8″	1.17
	1 x 8	7¼	6 7/8	1.16
Shiplap	1 x 10	9¼	8 7/8	1.13
	1 x 12	11¼	10 7/8	1.10
	1 x 4	3 3/8	3 1/8	1.28
Tongue	1 x 6	5 3/8	5 1/8	1.17
and	1 x 8	7 1/8	6 7/8	1.16
Grooved	1 x 10	9 1/8	8 7/8	1.13
	1 x 12	11 1/8	10 7/8	1.10
	1 x 4	3½	3½	1.14
	1 x 6	5½	5½	1.09
S 4 S	1 x 8	7¼	7¼	1.10
	1 x 10	9¼	9¼	1.08
	1 x 12	11¼	11¼	1.07
	1 x 6	5 7/16	5 7/16	1.19
Solid	1 x 8	7 1/8	6¾	1.19
Paneling	1 x 10	9 1/8	8¾	1.14
	1 x 12	11 1/8	10¾	1.12
	1 x 4	3½	3½	1.60
*Bevel	1 x 6	5½	5½	1.33
Siding	1 x 8	7¼	7¼	1.28
	1 x 10	9¼	9¼	1.21
	1 x 12	11¼	11¼	1.17

Note: This area factor is strictly so-called milling waste. The cutting and fitting waste must be added.
* 1-inch lap

CUTTING WASTE

In addition to the so-called *milling waste*, an allowance should be made for material wasted in cutting and fitting it into place. This will average 3 to 5 per cent in framing lumber, which includes a certain amount used on jobs for such things as saw-horses or minor scaffolding. On surface materials such as rough floor and roof boards, also board sheathing, paneling etc., the waste will average 5 to 10 per cent depending on whether the material is applied horizontal or diagonal.

ESTIMATING FRAMING MATERIALS

When estimating quantities of framing lumber, it should be remembered that the lengths needed for plates, sills, joists, and rafters will have to be cut from the nearest length sold at the lumber yard. A rafter or joist measuring 11 feet 6 inches must be cut from a 12-foot length.

Where studs can be obtained *pre-cut* at the lumber yard, that may be the most economical way to buy them rather than cutting them on the job and wasting part. This will be governed by the length of studs needed.

SILLS AND PLATES

Sills and plates are taken off by the lineal foot. The sill in a shed 10 feet wide and 20 feet long would have 60 lineal feet which is the equivalent of the perimeter of the building. If the plate on top of the studs carries around the entire building, as it would if the roof were flat or slightly pitched, the number of lineal feet of plate would, for all practical purposes be the same as the lineal feet of sill. Assuming that the sill was 2 by 6 inch stock, and that the plate was 4 by 4 inch stock, then the material would be listed in this manner.

Sill—60 lin ft 2″ x 6″   = 60 FBM
Plate—60 lin ft 4″ x 4″ = 80 FBM

EXTERIOR STUDDING

Exterior studding is estimated by determining the number of studs required, their size, and their length. In house framing, studding is usually spaced 16 inches center to center. Many rural and secondary buildings, and some dwelling type structures located outside of the jurisdiction of building code requirements, will be found to have studding on centers of 18, 20 and sometimes 24 inches. Except in barn structures, ice houses, refrigerated warehouses or buildings requiring unusual wall thickness and strength, studding is customarily 2" x 4" stock. Non-bearing curtain partitions frequently have 2" x 3" studs.

Some estimators prefer to take off the exact number of studs, and window and door framing, and add a percentage for cutting waste. This may be done by counting and measuring the members if they have not been destroyed. When they cannot be physically counted, a sketch should be drawn to be used for the take-off.

Where the exact number of studs is desired, the lineal feet of wall is divided by the spacing in feet. One stud is added for the end. See also Tables 9-4 and 9-5.

Illustrative Example
A partition is 40 feet long and the studding is 16 inches on center.

$$40 \div \frac{16}{12} = 30 + 1 \text{ for the end} = 31 \text{ studs}$$

TABLE 9-4
NUMBER OF STUDS, JOISTS AND RAFTERS
REQUIRED IN WALLS, FLOORS OR ROOFS

Length of Wall, Floor or Roof	Spacing 12"	on 16"	Center 20"	24"
8'	9	7	6	5
9	10	8	6	6
10	11	9	7	6
11	12	9	8	7
12	13	10	8	7
13	14	11	9	8
14	15	12	9	8
15	16	12	10	9
16	17	13	11	9
17	18	14	11	10
18	19	15	12	10
19	20	15	12	11
20	21	16	13	11
21	22	17	14	12
22	23	18	14	12
23	24	18	15	13
24	25	19	15	13
25	26	20	16	14
26	27	21	17	14
27	28	21	17	15
28	29	22	18	15
29	30	23	18	16
30	31	24	19	16
32	33	25	20	17
34	35	27	22	18
36	37	28	23	19

To the above there should be added members for doubling studs at windows, doors, corners; doubling joists under partitions, at stairwells and around chimneys; doubling rafters at roof openings around chimneys, etc.

TABLE 9-5
MULTIPLIERS FOR OBTAINING THE NUMBER OF STUDS,
JOISTS AND RAFTERS IN THE LENGTH
OF FLOORS, WALLS OR ROOFS

Spacing In Inches	Factor to Multiply the Width of Room, Length of Wall or Roof	
12	1.00	
16	.75	
18	.67	Add one for
20	.60	end in each
24	.50	case

Examples

A room 16′ wide requires joists 16″ o.c.

Ans. 16′ x .75 + 1 = 13 joists.

A wall 30′ long requires studs 18″ o.c.

Ans. 30′ x .67 + 1 = 21 studs.

A gable roof 40′ long requires rafters 24″ o.c.

Ans. 40′ x .50 + 1 = 21 rafters.

Illustrative Example

The walls of a ranch house are 24 feet by 36 feet with a 6 foot rise at the gable end. The 2″ x 4″ x 8′ studs are 16 inches on center.

Perimeter 2(24′ + 36′) = 120 feet

Studs required 120-2″ x 4″ x 8′ = 640 FBM

2 Gable Ends 24 feet

Studs required   24-2″ x 4″ x 6′ = 96 ″

Total = 736 FBM

An acceptable rule-of-thumb for determining the number of studs that are spaced 16 inches on center, is to figure one for each lineal foot of wall. This takes care of the normal amount of window and door framing, and also cutting waste. The number of studs required for a gable would be equivalent to the width of the gable. The average length of the gable studs would be one half of the distance from the eave-line to the ridge.

PARTITION STUDDING

Studs for partitions are determined in the same way as exterior studs. Members may be counted individually taking into consideration extra studs at door openings and corners such as at closets, or where joining other partitions or exterior walls.

Using one stud for each lineal foot of partition, as explained under *Exterior Studding,* is also acceptable.

The use of either Table 9-4 or Table 9-5 is a quick method for determining the number required.

PRE-CUT STUDS

Where pre-cut studs are available in the size desired, rather than cutting them on the job, the number required may be taken from Table 9-4 (page 212), or computed using Table 9-5 (page 213).

Pre-cut studs come in lengths 92 5/8", 93", 94 1/2" and 96".

FLOOR JOISTS

Joists in a dwelling are generally 2-inch stock and their depth ranges from 6 to 12 inches. In other types of buildings the thickness may be 2, 3, or even 4 inches depending on the floor load and the span. The depth of the large timbers may be as much as 16 inches. Table 9-7 shows the maximum safe span for joists under various live loads with either plastered or un-plastered ceilings below.

Where double or sometimes triple joists are needed under load bearing partitions, or at stairwells, they must be added in the material list.

The number of joists may be counted directly at the premises if still visible, or may be computed from Table 9-4 or Table 9-5.

GIRDERS, DOOR AND WINDOW HEADERS

A *girder* is a horizontal beam of iron or wood, or a combination, used to support vertical loads. Where an I-beam is used as a girder in frame construction, a 2" x 4" or 2" x 6" is bolted to the top and the joists are nailed to that. There is less shrinkage and less sagging or deflection in this arrangement than in girders composed solely of doubled or tripled joists. A recommended type of basement girder is made up of 2-2" x 12" members with a 12" steel plate in between and the three members bolted firmly together. Sometimes, in place of a steel plate, a 1/2" piece of plywood is glued and nailed between members. This works well as it brings the header flush with the 2" x 4" studding. (Illustrations of girders in Figure 9-2)

Window and door headers can have 2-2" x 4"s *on edge* up to spans of 2 feet. Larger spans require stronger framing. As in girders, where 2" x 8" to 2" x 12" members are used a piece of 1/4" to 1/2" plywood is glued and nailed between two members. Table 9-6 shows header sizes and maximum spans recommended.

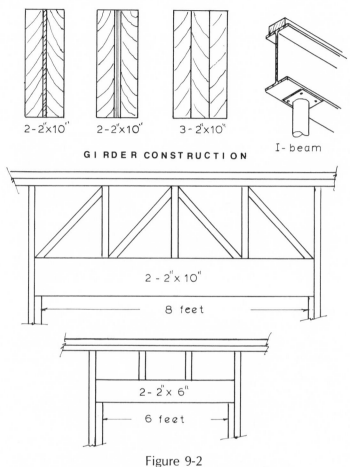

2 - 2"x10"   2 - 2"x10"   3 - 2"x10"

GI RDER CONSTRUCTI ON

I- beam

2 - 2" x 10"

8 feet

2- 2" x 6"

6 feet

Figure 9-2

Girder Construction and Framed Openings

TABLE 9-6
WINDOW AND DOOR OPENING MAXIMUM
SPANS FOR VARIOUS HEADERS

Header Sizes	Maximum Span
2 - 2" x 4"	2'
2 - 2 x 6	4'
2 - 2 x 8	5'
2 - 2 x 10	8'
2 - 2 x 12	10'

Note: The use of plywood interlay is recommended. (See Illustrations in Figure 9-2)

Table 9-7*

Span Calculations provide for carrying the live loads shown and the additional weight of the joists and double flooring.

SIZE	SPACING	20# L.L. Plaster Clg.	30# LIVE LOAD Plaster Clg.	30# LIVE LOAD No Plaster	40# LIVE LOAD Plaster Clg.	40# LIVE LOAD No Plaster	50# LIVE LOAD Plaster Clg.	50# LIVE LOAD No Plaster	60# LIVE LOAD Plaster Clg.	60# LIVE LOAD No Plaster
2 x 4	12"	8'-8"								
	16"	7'-11"								
	24"	6'-11"								
2 x 6	12"	13'-3"	11'-6"	14'-10"	10'-8"	13'-2"	10'-0"	12'-0"	9'-6"	11'-1"
	16"	12'-1"	10'-6"	12'-11"	9'-8"	11'-6"	9'-1"	10'-5"	8'-7"	9'-8"
	24"	10'-8"	9'-3"	10'-8"	8'-6"	9'-6"	8'-0"	8'-7"	7'-7"	7'-10"
2 x 8	12"	17'-6"	15'-3"	19'-7"	14'-1"	17'-5"	13'-3"	15'-10"	12'-7"	14'-8"
	16"	16'-0"	13'-11"	17'-1"	12'-11"	15'-3"	12'-1"	13'-10"	11'-5"	12'-9"
	24"	14'-2"	12'-3"	14'-2"	11'-4"	12'-6"	10'-7"	11'-4"	10'-1"	10'-6"
2 x 10	12"	21'-11"	19'-2"	24'-6"	17'-9"	21'-10"	16'-8"	19'-11"	15'-10"	18'-5"
	16"	20'-2"	17'-6"	21'-6"	16'-3"	19'-2"	15'-3"	17'-5"	14'-6"	16'-1"
	24"	17'-10"	15'-6"	17'-10"	14'-3"	15'-10"	13'-5"	14'-4"	12'-8"	13'-3"
2 x 12	12"	26'-3"	23'-0"	29'-4"	21'-4"	26'-3"	20'-1"	24'-0"	19'-1"	22'-2"
	16"	24'-3"	21'-1"	25'-10"	19'-7"	23'-0"	18'-5"	21'-0"	17'-5"	19'-5"
	24"	21'-6"	18'-8"	21'-5"	17'-3"	19'-1"	16'-2"	17'-4"	15'-4"	16'-0"
3 x 8	12"	20'-0"	17'-7"	24'-3"	16'-4"	21'-8"	15'-4"	19'-10"	14'-7"	18'-4"
	16"	18'-6"	16'-1"	21'-4"	14'-11"	19'-1"	14'-1"	17'-4"	13'-4"	16'-0"
	24"	16'-5"	14'-3"	17'-9"	13'-2"	15'-9"	12'-4"	14'-4"	11'-9"	13'-3"
3 x 10	12"	25'-0"	22'-0"	30'-2"	20'-6"	27'-1"	19'-3"	24'-10"	18'-4"	23'-0"
	16"	23'-2"	20'-3"	26'-8"	18'-10"	23'-10"	17'-8"	21'-9"	16'-10"	20'-2"
	24"	20'-7"	17'-11"	22'-3"	16'-7"	19'-10"	15'-7"	18'-1"	14'-10"	16'-8"

SPAN OF JOISTS

* Weyerhaeuser Company, Lumber and Plywood Division, St. Paul, Minn.

Table 9-8*

SPAN OF RAFTERS

Span Calculations provide for carrying the live loads shown and the additional weight of the rafters, sheathing and wood shingles

SIZE	SPACING	15# L.L.		20# L.L.		30# L.L.		40# L.L.	
		No Plaster	Plaster	No Plaster	Plaster	No Plaster	Plaster	No Plaster	Plaster
2 x 4	12"	12'-7"	8'-11"	11'-4"	8'-4"	9'-8"	7'-6"	8'-7"	6'-11"
	16"	11'-0"	8'-2"	9'-11"	7'-7"	8'-5"	6'-10"	7'-5"	6'-3"
	24"	9'-1"	7'-2"	8'-2"	6'-8"	6'-11"	6'-0"	6'-1"	5'-6"
2 x 6	12"	19'-2"	13'-8"	17'-4"	12'-9"	14'-10"	11'-6"	13'-2"	10'-8"
	16"	16'-10"	12'-6"	15'-2"	11'-8"	12'-11"	10'-6"	11'-6"	9'-8"
	24"	13'-11"	11'-0"	13'-0"	10'-3"	10'-8"	9'-3"	9'-6"	8'-6"
2 x 8	12"	25'-1"	17'-11"	22'-9"	16'-10"	19'-7"	15'-3"	17'-5"	14'-1"
	16"	22'-1"	16'-6"	20'-0"	15'-5"	17'-1"	13'-11"	15'-3"	12'-11"
	24"	18'-5"	14'-7"	16'-7"	13'-8"	14'-2"	12'-3"	12'-6"	11'-4"

* Weyerhaeuser Company, Lumber and Plywood Division, St. Paul, Minn.

Illustrative Example

The local building code specifies a live load for a dwelling of 40 pounds per square foot of floor area. With a plastered ceiling below, and a span of 16 feet, 2" x 10" joists placed 16 inches on center would be adequate.

RAFTERS

In dwellings of more recent construction the rafters are usually of 2-inch stock. A rafter of 2" x 6" is very common in house construction, although on bungalows, porches, and many secondary buildings 2" x 4" rafters are used. In many of the earlier built homes, especially those with large steeply pitched roofs, rafters are heavier stock. In many commercial type buildings with flat roofs the size of the rafters may be 3 or 4 inches thick and 8 to 12 inches deep. The dimensions of rafters depends to a large degree on the span, the weight of the roofing material, and the expected wind and snow load as shown in Table 9-8.

Illustrative Example

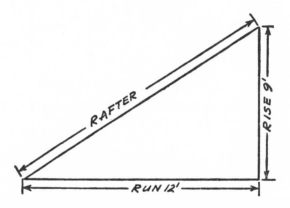

A live load of wind and snow is determined to be approximately 20 pounds on a wood shingle roof surface. With an open attic (no plaster), and a span of 15 feet from plate to ridge, 2" x 6" rafters 16 inches on center will be adequate.

The length of rafters may be measured by making a scale sketch of the rise and horizontal run. They may also be determined from the rise and run by application of the Rule of Pythagoras. (See Chapter 6.)

$$
\begin{aligned}
\text{Rafter} &= \sqrt{\text{Run}^2 + \text{Rise}^2} \\
&= \sqrt{144 + 81} \\
&= \sqrt{225} \\
&= 15 \text{ Feet}
\end{aligned}
$$

Table 9-9
TABLE OF RAFTER LENGTHS

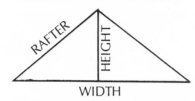

Height (Feet & Inches)

Width of Building in Feet

	18		20		22		24		26		28		30		32		34		36	
	Length of Rafters in Feet and Inches																			
4 6	10	2	10	11	11	9	12	10	13	9	14	9	15	8	16	8	17	7	18	7
5 0	10	4	11	3	12	1	13	0	13	11	14	11	15	10	16	9	17	9	18	8
5 6	10	6	11	5	12	4	13	3	14	2	15	1	16	0	16	11	17	11	18	10
6 0	10	9	11	8	12	6	13	5	14	4	15	3	16	2	17	1	18	1	19	0
6 6	11	0	11	11	12	9	13	8	14	6	15	5	16	4	17	3	18	3	19	2
7 0	11	4	12	3	13	0	13	11	14	8	15	8	16	7	17	5	18	5	19	4
7 6	11	8	12	6	13	4	14	2	15	0	15	11	16	9	17	8	18	7	19	6
8 0	12	0	12	10	13	8	14	5	15	3	16	2	17	0	17	10	18	9	19	8
8 6	12	4	13	2	13	11	14	8	15	6	16	5	17	3	18	2	18	11	19	11
9 0	12	9	13	5	14	3	15	0	15	10	16	8	17	6	18	5	19	3	20	2
9 6	13	2	13	10	14	6	15	4	16	1	16	11	17	9	18	8	19	6	20	5
10 0	13	6	14	2	14	10	15	8	16	5	17	3	18	1	18	11	19	9	20	7
10 6	13	10	14	6	15	3	15	11	16	9	17	6	18	4	19	3	20	0	20	10
11 0	14	3	14	10	15	7	16	4	17	0	17	10	18	8	19	6	20	3	21	1
11 6	14	7	15	3	15	11	16	8	17	4	18	2	18	11	19	9	20	6	21	4
12 0	15	0	15	8	16	4	17	0	17	9	18	5	19	3	20	0	20	10	21	8
12 6	15	5	16	0	16	8	17	4	18	0	18	9	19	7	20	4	21	1	21	11
13 0	15	10	16	5	17	0	17	8	18	5	19	2	19	10	20	8	21	5	22	2
13 6	16	3	16	10	17	5	18	0	18	9	19	6	20	3	20	11	21	9	22	6
14 0	16	8	17	3	17	10	18	5	19	2	19	10	20	7	21	3	22	0	22	10
14 6	17	1	17	8	18	3	18	10	19	6	20	2	20	10	21	7	22	4	23	2
15 0	17	6	18	0	18	8	19	3	19	11	20	7	21	3	21	11	22	8	23	6
15 6	17	11	18	5	19	0	19	8	20	3	20	11	21	7	22	4	23	0	23	10
16 0	18	4	18	10	19	5	20	0	20	8	21	4	21	11	22	8	23	4	24	1
16 6	18	10	19	4	19	10	20	5	21	0	21	8	22	4	23	0	23	8	24	5
17 0	19	3	19	9	20	3	20	11	21	5	22	1	22	8	23	4	24	1	24	9
17 6	19	9	20	2	20	8	21	3	21	10	22	5	23	0	23	8	24	5	25	1
18 0	20	1	20	7	21	0	21	8	22	3	22	10	23	5	24	1	24	9	25	6
18 6	20	7	21	0	21	6	22	0	22	8	23	3	23	10	24	5	25	2	25	10
19 0	21	0	21	5	22	0	22	6	23	1	23	8	24	3	24	10	25	6	26	2
19 6	21	5	21	11	22	5	22	11	23	5	24	0	24	8	25	3	25	10	26	7
20 0	21	11	22	4	22	10	23	4	23	10	24	5	25	0	25	8	26	3	26	11
20 6	22	5	22	10	23	3	23	9	24	4	24	10	25	5	26	0	26	8	27	4

The overhang should be added to the computed length, and the next even foot length of rafter is used. If the overhang is 18 inches then the length of the rafter is to be used would be 18 feet. The computed length is $15 + 1.5 = 16.5$ (use 18-foot rafter).

The number of rafters are obtained in the same way as the number of joists. The length of the roof in feet is divided by the rafter spacing in feet, and one rafter is added for the end.

Table 9-9 shows the rafter length when the height and width are known. For example, if the height is 8' and the width is 30' the rafter length is 17'.

A *ridge pole* (ridge piece) is the horizontal framing member at the ridge of the roof to which the rafters are framed and attached. It is listed in the framing schedule by length and size, which should be nominally 2 inches deeper than the rafter. In other words where 2" x 6" rafters are used, the ridge pole should be 2" x 8".

BRIDGING

Floor joists need to be reinforced so that the floor load is distributed over many joists and any deflection of the joists is uniform. To accomplish this bridging is placed between the joists. There are three types in use: *wood cross bridging, wood solid bridging,* and *metal cross bridging.* (See Figure 9-3)

Wood cross bridging can be purchased at most lumber yards in 50 piece bundles or it can be cut on the job. The material used can be 1" x 3", 1" x 4", or 2" x 2". The length of the pieces (struts) will depend on the depth and spacing of the joists as will the angle of the cut, since the piece is nailed even with the upper edge of one joist and even with the lower edge of the next joist. Two pieces of bridging are placed between each joist, crossing one another. In small rooms one line of bridging is adequate but in larger rooms the lines of bridging should be spaced 6' to 8' feet apart. The bridging is placed before the floor or subfloor is layed.

Wood solid bridging as its name implies is composed of a line of solid pieces usually of the same stock as the joists. Obviously solid bridging has to be staggered in line in order to nail it between joists.

Metal cross bridging are factory-made 18 ga. steel braces placed in the same manner as wood cross bridging. They are available in cartons of 200.

Bridging costs

The labor to install bridging is shown in Table 9-15 and varies with the type used. The following examples illustrate both redicut wood cross bridging, also cut on the job; and metal bridging.

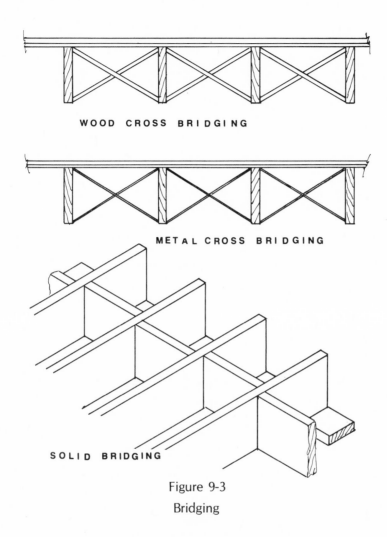

WOOD CROSS BRIDGING

METAL CROSS BRIDGING

SOLID BRIDGING

Figure 9-3

Bridging

Illustrative Example

A room with 2″ x 10″ x 12′ joists is 20 feet long. They are on 16″ inch centers so there are 15 bridging spaces.

Wood cross bridging cut on job:

Table 19-10 shows 2.15 lin ft of bridging is needed for each foot of row. The row of bridging is 20 ft. Therefore (20 x 2.15) 43 lin ft is required plus 10 per cent waste or 47 lin ft.

Using 1″ x 3″ at 6¢ per foot and labor at $6 per hour this bridging will cost

47 lin ft 1″ x 3′ @ 6¢ = $ 2.82

(Table 9-15) 8 sets per hr = 1.875 hrs@$6 = $ 11.25

Cost for 15 sets = $14.07

Cost per set = $ .94

Wood cross bridging ready-cut:
By purchasing ready-cut wood 1″ x 3″ bridging the cost would be: (50 pc bundle $5.25)

30 pcs or, 15 sets bridging =	$ 3.15
(Table 9-15) Labor 15 sets per hour =	6.00
15 Sets Total =	$ 9.15
Cost per set =	$ .61

Metal cross bridging:
Metal bridging for the same job would cost, installed: (material cost .05 ea)

30 pcs (15 sets)@ 5¢ ea =	$1.50
(Table 9-15) Labor 15 sets per hr =	6.00
15 sets Total =	$7.50
Cost per set =	$ .50

TABLE 9-10
MATERIAL REQUIRED FOR WOOD BRIDGING CUT ON THE JOB

Size Joist In Inches	Spacing In Inches	Lineal Feet Per Set (2)	Lineal Feet Per-Foot-of-Row
2 x 6	16	2.57	1.92
2 x 8	16	2.70	2.02
2 x 10	16	2.87	2.15
2 x 12	16	3.06	2.30
2 x 8	20	3.31	2.00
2 x 10	20	3.45	2.07
2 x 12	20	3.61	2.17
2 x 8	24	3.94	2.00
2 x 10	24	4.05	1.97
2 x 12	24	4.19	2.10

Note: Add to the total lineal feet developed from the Table at least 10% cutting waste.

Example: A room 20 feet wide has two rows of bridging. The 2″ x 12″ joists are 16″ on center
2 x 20 lin ft x 2.30 ft per-foot of-row = 92.0 lin ft
Add 10 % waste 9.2 ″ ″
Total lin ft = 101.2 ″ ″
Round out to 101 lin ft
(Per set method would be, 30 sets x 306 = 91.8 to which must be added 10% waste.)

ROOF—TRUSS CONSTRUCTION

Trusses are pre-cut and assembled on the ground, at the job site or at a prefabricating plant that is set up for mass production. Trusses are triangular frames for gable roofs and in residences can be designed to span

40 feet. Special trusses are available for hip roofs. It is estimated that the roofs in 60 to 70 per cent of all new residential construction are or will be truss-construction. Figure 9-4 shows the construction of a typical truss.

The size of the members, that is the bottom cord, rafter-cord and braces, depend on the span and load requirements. On residential roofs, where the spans are generally less than 32 feet, the entire truss can be fabricated from 2" x 4" 's. Local building codes and FHA requirements govern the use of these light framing materials, particularly since nominal 2" x 4" timbers are now actually 1 1/2" x 3 1/2". Much also depends on the wind and snow loads in the area. An acceptable and practical design employs 2" x 6" top cord with all other units 2" x 4" material.

There are numerous methods of connecting the members of a truss. Earlier designs used 1/2" plywood "gussets" which were applied to both sides and nailed and glued. More recently truss components are attached with one of the numerous types of metal connector plates. These range from plain zinc coated steel, 22 to 14 gauge, with pre-drilled holes; to crimped, barbed, corrugated, and flared types. In truss assembly plants these are applied with presses or rollers, or pneumatic hammers.

Trusses in residential construction are usually placed 24" on center. The lower cord serves as the ceiling joist and the roof boards or plywood sheathing is nailed to the rafter-cord.

Where trusses can be purchased from a local fabricator, the cost is considerably less than attempting to make them on the job with inexperienced labor. The following is an example of the material and estimated labor to build one truss on the job.

Illustrative Example

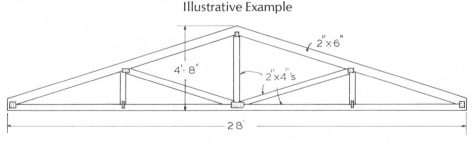

Figure 9-4

Metal-Plate Type Truss

To build a truss with 2" x 6" rafters, 2" x 4" cord and braces, using metal connectors. The span is 28 feet and the pitch is 1/6, 4" rise per-foot-of-run. Assume first truss fabricated and templates in place.

Using the truss in Figure 9-4 the length of the members can be scaled directly. (A line drawing or sketch to scale is recommended.)

Top cord	2 - 2″ x 6″ x 18′ - 0″ =	36.0 FBM
Bottom cord	2 - 2″ x 4″ x 16′ - 0″ =	21.3 ″
Center post	1 - 2″ x 4″ x 5′ - 0″ =	3.3 ″
Vertical braces	2 - 2″ x 4″ x 2′ - 6″ =	3.3 ″
Diagonal braces	2 - 2″ x 4″ x 7′ - 0″ =	9.3 ″
		73.2 FBM

Assume cost @ 20¢	$ .20
	$14.64
16 Metal truss plates @ 25¢	4.00
	$18.64
2 hours labor @ $6	12.00
Toal cost, labor and material	$30.64

Note: Laying out and fabricating the first truss takes about 5 or 6 hours. With templates in place 2 hours is adequate.

Table 9-11 shows how to compute the board feet of material in trusses with spans from 20 feet to 30 feet, and pitches or 1/6, 5/24 and 1/4. For practical purposes the members in Table 9-11 may be used to figure either King Post trusses or W-Trusses. Add any overhang to the top cord. Also add connectors, whether metal, or wood gussets glued and nailed.

TABLE 9-11
BOARD FEET OF MATERIAL FOR TRUSSES WITH 2″ x 6″ TOP CORD,
OTHER MEMBERS 2″ x 4″. NO OVERHANG CONTEMPLATED

	SPAN IN FEET				
	20	24	26	28	30
	Top Cord (2″ x 6″)				
Pitch					
1/6	2 - 12′	2 - 14′	2 - 16′	2 - 16′	2 - 18′
5/24	2 - 12′	2 - 14′	2 - 16′	2 - 18′	2 - 18′
¼	2 - 14′	2 - 16′	2 - 16′	2 - 18′	2 - 18′
	Bottom Cord (2″ x 4″)				
Pitch					
1/6	2 - 12′	2 - 14′	2 - 16′	2 - 16′	2 - 18′
5/24	2 - 12′	2 - 14′	2 - 16′	2 - 16′	2 - 18′
¼	2 - 12′	2 - 14′	2 - 16′	2 - 16′	2 - 18′
	Diagonals (2″ x 4″)				
Pitch					
1/6	2 - 8′	2 - 10′	2 - 10′	2 - 12′	2 - 12′
5/24	2 - 10′	2 - 12′	2 - 12′	2 - 12′	2 - 14′
¼	2 - 10′	2 - 12′	2 - 14′	2 - 14′	2 - 16′

(1/6 pitch = 4 inch per-ft-run, 5/24 = 5 inch per-ft-run, ¼ = 6 inch per-ft-run).
Note: Add 2 or 3 ft of 1″ x 4″ to join 2″ x 4″ bottom cord. Add metal connecting members, nails etc.

RAPID ESTIMATING OF FRAMING BY AREA OF WALLS, FLOORS, OR ROOF

Sometimes it is necessary to rapidly estimate the framing in a given area of floor, wall or roof, knowing the size of the framing members and the center to center spacing. From Table 9-12 the framing in any part of a structure can be quickly computed, or the framing in an entire building can be determined.

For example, assume without even seeing the building, the question arises as to the total amount of framing it has. The building submitted is 24' x 40' long and has a roof rise of 6'. The exterior walls average 10' high and there are 160' of partitioning in it. Total framing is determined from Table 9-12, page 226, as follows:

Exterior Walls - 2" x 4" studs 16" o.c.
    Area   2(24' + 40') x 10' = 1280 sq ft x .50 = 640 FBM
Partitions - 2" x 4" studs 16" o.c.
  Area 160' x 10'   = 1600 sq ft x .50 = 800"
Floor Joists - 2" x 10" - 16" o.c.
  Area 24' x 40'   = 960 sq ft x 1.25 = 1,200"
Ceiling Beams - 2" x 6" - 16" o.c.
  Area 24' x 40'   = 960 sq ft x .75 = 720"
Rafters - 2" x 6" -   16" o.c.
  (Rafter length from Table 9-9 is 13.5 ft)
  2 x 13.5' = 27' x 40'   = 1080 sq ft x .75 810"
                                                       4,170 FBM
                                          Plus 10% 417"
                                           Total 4,587 FBM

To the total framing developed from Table 9-12 a percentage should be added to cover miscellaneous framing such as bridging, door and window heads, doubling of beams and rafters where necessary and doubling studs at walls, openings, etc. In the example shown 10 per cent is considered adequate.

NAILS AND NAILING

Figure 9-5 shows 14 common nails with their gauge, length and "penny" (d) designation. Most of these sizes are also available in finishing nails that have a small head which can be set below the surface in finish carpentry. Up until a few decades past these nails, along with flooring cut nails, lath nails, and roofing nails were, with some exceptions, the only types available and those commonly used in house construction.

TABLE 9-12
BOARD FEET OF FRAMING PER SQUARE FOOT OF
WALLS, FLOORS AND ROOFS

Size of Framing In Inches	Center to Center In Inches		
	12	16	20
1 x 2	.167	.125	.10
2 x 2	.333	.250	.20
2 x 3	.500	.375	.30
2 x 4	.667	.500	.40
2 x 6	1.00	.750	.60
2 x 8	1.33	1.00	.80
2 x 10	1.67	1.25	1.00
2 x 12	2.00	1.50	1.20
3 x 6	1.50	1.13	.90
3 x 8	2.00	1.50	1.20
3 x 10	2.50	1.88	1.50
3 x 12	3.00	2.25	1.80

Note: 1. When spacing is 24 inches on center, double the figure shown for 12 inches on center.

2. Always add a percentage judgment factor for miscellaneous framing as bridging, doubling of studs, joists or rafters, catting, and the normal cutting waste. The *total* of these will average from 10 to 15 per cent.

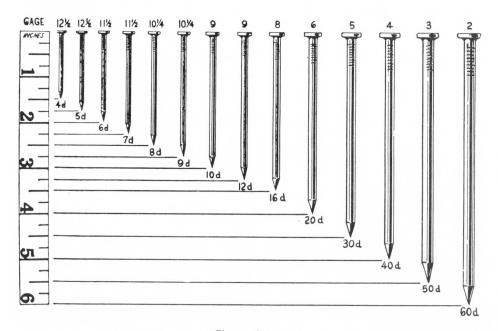

Figure 9-5

Common Nail Sizes

Since then the development of new types of nails designed for specific purposes has been spectacular. There is a nail for almost every type of construction material, some with special heads, points or size. Some have screw threads, spiral, annular and knurled threads. There are nails for framing, roofing, drywall, underlayment, flooring, hardboard, trim, cedar shakes, asbestos siding, wood siding, and gypsum lath to name a few of the special purpose nails. Some are coated with rosin or cement for better holding power although the threaded nail, which can be driven, holds best.

Table 9-13 lists the penny system of nails, both common and finishing, and shows the length, gauge and approximate number to the pound.

TABLE 9-13
Penny Nail System

| Size | Length in Inches | Gauge Number | Approximate Number to Pound | | |
			Common	Finishing	Casing
2d	1	15	850		
3d	1¼	14	550	640	
4d	1½	12½	350	456	
5d	1¾	12½	230	328	
6d	2	11½	180	273	228
7d	2¼	11	140	170	178
8d	2½	10¼	100	151	133
9d	2¾	9½	80	125	100
10d	3	9	65	107	96
12d	3¼	9	50		60
16d	3½	8	40		50
20d	4	6	31		
30d	4½	5	22		
40d	5	4	18		
50d	5½	3	14		
60d	6	2	12		

Special purpose nails

There are a number of special purpose nails manufactured with particularly designed heads, threads and points, depending upon their use. These three features are shown in Figure 9-6. Some of the more common special purpose nails are shown and described in Figures 9-7 and 9-8 by courtesy of Independent Nail, Inc., Bridgewater, Mass. 02324.

Nails required

Builders are not always in agreement on the pounds of nails needed for specific items of rough carpentry. In fact not all agree on the size or type to use which has an important bearing on the amount used. Table 9-14

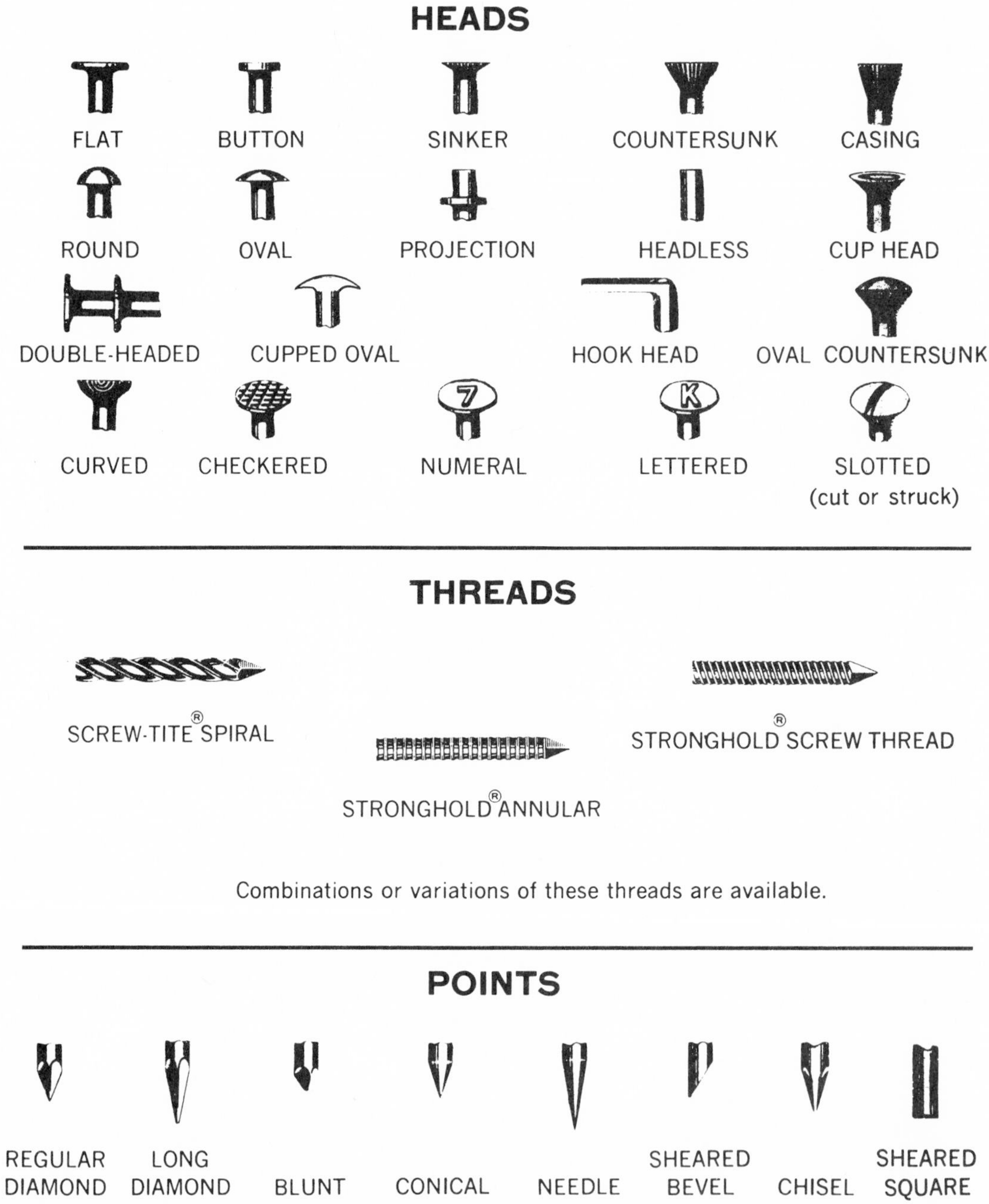

HEADS

FLAT   BUTTON   SINKER   COUNTERSUNK   CASING

ROUND   OVAL   PROJECTION   HEADLESS   CUP HEAD

DOUBLE-HEADED   CUPPED OVAL   HOOK HEAD   OVAL COUNTERSUNK

CURVED   CHECKERED   NUMERAL   LETTERED   SLOTTED
                                                          (cut or struck)

THREADS

SCREW-TITE® SPIRAL   STRONGHOLD® SCREW THREAD

STRONGHOLD® ANNULAR

Combinations or variations of these threads are available.

POINTS

REGULAR   LONG   SHEARED   SHEARED
DIAMOND DIAMOND   BLUNT   CONICAL   NEEDLE   BEVEL   CHISEL   SQUARE

Figure 9-6

Nail Heads, Threads and Points, Courtesy Independent Nail, Inc., Bridgewater, Mass. 02324

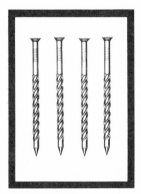

INTERIOR HARDBOARD NAILS

- **STRONGHOLD**® Annular Thread for tight, smooth decorative, pre-finished plywood and interior hardboard paneling applications.
- Small, button head for flush driving helps make head inconspicuous.
- Color-matching eliminates need for nail-setting and puttying.
- Hardened Steel — heat treated, quenched, and tempered steel to make this slender shank nail easily driveable.

EXTERIOR HARDBOARD SIDING NAILS

- **SCREWTITE**® Spiral Thread for weathertight siding applications and ease of driving into hardboard siding.
- Stiff Stock Steel — high carbon, high tensile steel aids driving.
- Pilot point to help start the nail straight and ease a path for the thread.
- Flat, countersunk head for flush driving and best apearance of installed hardboard siding.
- Hot-dipped Galvanized for maximum corrosion resistance for a steel nail.

BLUNT-POINT WOOD SIDING NAILS

- **STRONGHOLD**® Annular Thread for weathertight siding applications.
- Stiff Stock Steel — high carbon, high tensile steel aids driving.
- Slender shank minimizes splitting.
- Large, slightly countersunk head (checkered) for ease of flush driving.
- Blunt diamond point crushes wood fibers, effectively minimizing splitting in wood siding patters, or close to end or edge of wood siding patterns.
- Hot-dipped Galvanized for maximum corrosion resistance for a steel nail.

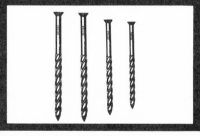

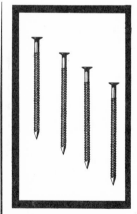

FLOORING NAILS

- **SCREWTITE**® Spiral Thread effectively minimizes "squeaks" in floor systems by pulling flooring tight. Turning action provided by thread minimizes splitting of hardwood flooring.
- Flat, countersunk head permits flush driving.
- Stiff Stock Steel — high carbon, high tensile steel for driving into less dense flooring.
- Hardened Steel — heat treated, quenched, tempered steel for ease of driving in dense hardwood flooring.

COPPER SLATING NAILS

- **STRONGHOLD**® Annular Thread for maximum holding power in wood or plywood.
- **SCREWTITE**® Spiral Thread for maximum shear, racking, lateral load resistance and high holding power.

ASBESTOS SIDING FACE NAILS

- **STRONGHOLD**® Annular Thre for weathertight siding appli tion.

Figure 9-7

Special Purpose Nails, Courtesy Independent Nail, Inc., Bridgewater, Mass. 02324

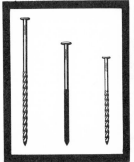

UNDERLAYMENT NAILS

TRUSSED RAFTER POLE BARN NAILS

- **STRONGHOLD®** Annular Thread for maximum withdrawal resistance in engineered wood construction.
- **SCREWTITE®** Spiral Thread for maximum shear resistance in engineered wood construction.
- The generally smaller wire diameters than those of comparable common nails minimize splitting of drier, thinner lumber, framing, and decking.
- Stiff Stock Steel — high carbon, high tensile steel for driving into medium density wood species.
- Hardened Steel — heat treated, quenched, and tempered steel for driving into high density, dry wood species and for maximum rigidity.

ROOFING NAILS

- **STRONGHOLD®** Annular Thread for best holding power when fastening asphalt shingles and roll roofing over wood or plywood decking or sheathing.
- Large, flat head for maximum bearing area.
- Hot-dipped Galvanized for maximum corrosion resistance for a steel nail.

- **STRONGHOLD®** Annular Thread effectively minimizes "squeaks" in floor systems by minimizing nail-popping, which can also mar floor coverings.
- Flat, slightly countersunk head permits flush driving for smooth underlayment surface.
- Regular steel for driving into less dense underlayment and subflooring.
- Hardened Steel — heat treated, quenched and tempered steel for ease of driving into denser underlayment and subflooring.

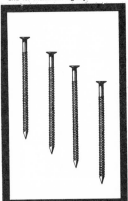

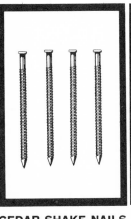

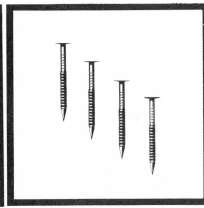

ASBESTOS SIDING FACE NAILS

- **STRONGHOLD®** Annular Thread for weathertight siding application.
- Hot-dipped Galvanized for maximum corrosion resistance for a steel nail.

CEDAR SHAKE NAILS

- **STRONGHOLD®** Annular Thread for weathertight cedar shingle and shake applications.
- Slender shank minimizes shake and shingle splitting.
- Small, checkered, button head blends with "textured" shingle and shake surfaces.
- Hot-dipped Galvanized for maximum corrosion resistance for a steel nail.

DRYWALL NAILS (type 500)

- **STRONGHOLD®** Annular Thread effectively minimizes nail-popping that can mar finished drywall appearance.
- Long diamond point for the "thumb" start.
- Flat, slightly countersunk, thin-rimmed head aids flush driving with minimum disruption of drywall surfaces.

Figure 9-8

More Special Purpose Nails, Courtesy Independent Nail, Inc., Bridgewater, Mass. 02324

gives the approximate number of pounds and size of nail for various types of rough carpentry using the "penny" system. For special purpose nails the manufacturer or supplier should be consulted.

TABLE 9-14
COMMON NAILS REQUIRED FOR ROUGH CARPENTRY
(Based on Framing Members 16″ o.c.)

Type Material	Kind of Nail	Pounds
Studding 16″ o.c.	10d, 16d	10 per 1,000 FBM
Joists and Rafters		
2″ x 6″	16d	9 per 1,000 ″
2″ x 8″	″	8 per 1,000 ″
2″ x 10″	″	7 per 1,000 ″
2″ x 12″	″	6 per 1,000 ″
Average house framing	8d, 10d, 16d, 20d	10-15 per 1,000 ″
Wood bridging 1″ x 3″	8d	1 per 12 sets
Shiplap and boards for walls, floors and roof		
1″ x 6″	8d	25-30 per 1,000 FBM
1″ x 8″	″	20-25 per 1,000 ″
1″ x 10″	″	15-20 per 1,000 ″
Plywood (Plyscord) sheathing and subfloor		
5/16″, 3/8″	6d	10 per 1,000 Sq ft
5/8″, ¾″	8d	20 per 1,000 ″ ″
Insulating board (Composition)	2d galv.	10 per 1,000 ″ ″
Furring on masonry	8d masonry	½ per 100 lin ft
Furring on studs	8d common	½ per 100 ″ ″

LISTING FRAMING IN AN ESTIMATE

A systematic, orderly listing in an estimate of the rough carpentry materials or operations, is most essential for accurate results. The location of the work should be clearly identified as to floor, room or portion of the structure involved. The number of framing pieces, their dimensions, and size, should be properly noted. The kind of work should be indicated either immediately before, or after, each item listed, showing that it is *sheathing, rough flooring, roof boards, studding, rafters, joists, plates, blocking, girders, bridging,* and so forth. In this way an estimator can easily recheck his take-off, and the board feet may be readily extended and totaled. Each type of rough carpentry in the listing can also be examined to establish an adequate labor charge. The illustrative detailed estimate shown on page 190 offers an excellent example of this recommended system of recording rough carpentry.

TABLE 9-15
AVERAGE HOURS TO INSTALL ROUGH CARPENTRY

Kind of Work	Hours Per 1,000 FBM	FBM Per Hour
Sills and plates, bolted and grouted	20	50
Studding, 2″ x 4″, 16″ o.c.	25	40
Joists, 2″ x 6″ and 2″ x 8″	22	45
2″ x 10″ and 2″ x 12″	20	50
Girders, built-up from 2″ stock	20	50
Rafters, 2″ x 6″ and 2″ x 8″, plain gable roof	30	34
″ 2″ x 6″ and 2″ x 8″, plain hip roof	33	31
″ 2″ x 6″ and 2″ x 8″, flat roof	26	39
Average for completely framing house	20 - 30	34 - 50
Exterior wall sheathing 1″ x 6″, and 1″ x 8″	16	63
Subflooring 1″ x 4″	18	56
″ 1″ x 6″, and 1″ x 8″	15	67
Roof boards, flat roof 1″ x 6″, and 1″ x 8″	15	67
″ gable roof 1″ x 6″, and 1″ x 8″	20	50
″ hip and cut up 1″ x 6″, and 1″ x 8″	22	45

	Per 1,000 Sq ft	Sq Ft Per Hour
Plywood subfloor . . . . 5/8″	12	83
″ subfloor . . . . ¾″	16	63
″ sheathing . . . . 5/16″	14	71
″ sheathing . . . . 3/8″	16	63
″ roof decking (flat roof) . . . . ½″	12	83
″ roof decking (gable roof) . . . . 5/8″	16	63
″ roof decking (cut-up) . . . . 5/8″	18	56
Insulating (composition) sheathing . . . . ½″	12-20	50-83
″ ″ roof decking . . . . 1½″	20-30	33-50

Bridging - wood 1″ x 3″, cut on job	1-hour per 8 sets
″ wood 1″ x 3″, redi-cut	1-hour per 15 sets
″ metal	1 hour per 15 sets
″ wood - solid	1-hour per 10 pcs
Furring on masonry incl shimming	4 hours per lin ft
″ on studding and shimming	3 hours per lin ft

Note: Adjust hourly rates to unusual difficult work conditions and where boards are laid diagonally. Matched boards take a little longer than square edged.

ESTIMATING FRAMING COSTS

Table 9-15 shows the average number of hours required for carpenters to install various kinds of framing under normal working conditions. These rates may be used for developing unit costs for material and labor, or they may be used when it is desired to show the labor cost separately from the material cost. There may also be occasions when an hourly rate per 1,000 FBM is not applicable to a specific job operation. In that case the cost of the material should be estimated, and the number of hours of labor may be determined by judgment. This judgment rate is based on the estimator's best approximation of the length of time it will take one, two, or more carpenters to perform the particular work. Care should be exercised not to use this method on any extensive amount of carpentry that includes numerous kinds of framing, as it then becomes an over-all guess rather than a thoughtful analysis.

Unit cost for framing

Unit cost for framing are developed from a unit of 1,000 FBM of lumber. The total cost is the sum of the cost of the lumber, the nails, and the labor to install the material. The material cost is obtained in the locality where the loss occurred. The number of hours of labor may be taken from Table 9-15 and multiplied by local wage rates. The following basic formula is used to develop a unit cost for framing.

Basic Formula

1,000 FBM lumber	@ $ = $	
Pounds of nails (Table 9-14)	@ =	
Hours carpenter labor (Table 9-15)	@ = _____	

$$\text{Unit cost per 1,000 FBM} = \$$$
$$\text{Unit cost per board foot} = \frac{\$\_\_\_\_}{1,000} = \$$$

The following illustrative examples show the method of developing various unit costs by using *assumed material prices and wage rate* for carpenters. The unit costs have been carried out to the nearest whole cent per board foot.

Illustrative Examples

Sills and Plates

1,000 FBM lumber	@ $160.00 =	$160.00
8 lb nails	@ .20 =	1.60
20 hours carpenter labor	@ 6.00 =	120.00
	Unit cost per 1,000 FBM =	$281.60
	Unit cost per board foot =	$.2816 or $.28

Studding

1,000 FBM lumber	@	$160.00 =	$160.00
10 lbs nails	@	.20 =	2.00
25 hours carpenter labor	@	6.00 =	150.00
	Unit cost per 1,000 FBM	=	$312.00
	Unit cost per board foot	=	$.312 or $.31

(2″ x 10″) Floor Joists

1,000 FBM lumber	@	$160.00 =	$160.00
6 lbs nails	@	.20 =	.20
20 hours carpenter labor	@	6.00 =	120.00
	Unit cost per 1,000 FBM	=	$281.20
	Unit cost per board foot	=	$.2812 or $.28

(2″ x 6″) Rafters-Gable Roof

1,000 FBM lumber	@	$160.00 =	$160.00
8 lbs nails	@	.20 =	1.60
30 hours carpenter labor	@	6.00 =	180.00
	Unit cost per 1,000 FBM	=	$341.60
	Unit cost per board foot	=	$.3416 or $.34

(2″ x 10″) Rafters-Flat Roof

1,000 FBM lumber	@	$160.00 =	$160.00
6 lbs nails	@	.20 =	1.20
24 hours carpenter labor	@	6.00 =	144.00
	Unit cost per 1,000 FBM	=	$305.20
	Unit cost per board foot	=	$3052 or $.30

Mill-Type Beams and Girders

1,000 FBM lumber	@	$160.00 =	$160.00
5 lbs spikes	@	.20 =	1.00
30 hours carpenter labor	@	6.00 =	180.00
	Unit cost per 1,000 FBM	=	$341.00
	Unit cost per board foot	=	$.341 or $.34

Bridging

1,000 FBM lumber	@	$160.00 =	$160.00
15 lbs nails	@	.20 =	3.00
80 hours carpenter labor	@	6.00 =	480.00
	Unit cost per 1,000 FBM	=	$643.00
	Unit cost per board board	=	$.643 or $.64

Furring on Masonry

100 lin ft 1″ x 2″ furring	@	$ .03 =	$ 3.00
1 lb nails	@	.20 =	.20
4 hours carpenter labor	@	6.00 =	24.00
	Unit cost per 100 lin ft	=	$27.20
	Unit cost per lin ft	=	$.272 or $.27

Average Framing Entire House

1,000 FBM lumber	@	$160.00	=	$160.00
10 lbs nails	@	.20	=	2.00
25 hours carpenter labor	@	6.00	=	150.00
	Unit cost per 1,000 FBM		=	$312.00
	Unit cost per board foot		=	$.312 or $.31

Application of Framing Unit Costs

The following framing lumber list was taken from a small, inexpensive ranch-type residence measuring 24' x 30' x 7' to the plate, with a 6-foot rise in a gable roof.

Member	Quantity	FBM	Unit Cost	Total Cost
Sill	108 lin ft 2″ x 4″	72	.28	$ 20.16
Plate	108 lin ft 4″ x 4″	144	.28	40.32
Exterior studs	108—2″ x 4″ x 7′	504	.31	156.24
Gable studs	24—2″ x 4″ x 6′	96	.31	29.76
Rafters	48—2″ x 6″ x 14′	672	.34	228.48
Ridge	30 lin ft 2″ x 8″	40	.34	13.60
Ceiling joists	22—2″ x 6″ x 24′	528	.28	147.84
Partition shoe	84 lin ft 2″ x 4″	56	.28	15.68
Partition plate	84 lin ft 2″ x 4″	56	.28	15.68
Partition studs	84—2″ x 4″ x 7″	392	.31	121.52
		2,560		$789.28

By applying the foregoing unit costs to each individual kind of framing, the total cost of framing 2,560 FBM is $789.28. If the unit cost of $.31, as developed for the average rate for framing of an entire house, is applied to 2,560 FBM, the cost is $793.60, or a difference of $4.32. This is an insignificant difference. Where the framing that is being estimated will fit into a category of "average" house framing, it is recommended that an average unit cost be used.

SPECIAL CONSIDERATIONS IN ESTIMATING ROUGH CARPENTRY LABOR

The principal factors that affect labor in general are discussed under "Estimating Labor" in Chapter 2. Each of these should be reviewed and carefully considered when estimating carpentry labor specifically.

In most instances it is not good practice to include in carpentry items, the labor cost of tearing out and removing the damaged materials. This work, sometimes done by carpenters, sometimes by *common labor,* should be treated as a separate cost for the entire job under a "Wrecking and

Debris" item, or similar classification. Any attempt to intermingle the time to tear out and cart debris with the installing of new materials, particularly when done by labor of a different wage rate, is apt to result in inaccuracies. There are exceptions where a carpenter who is repairing a particular part of a structure will have to tear out damaged materials as he progresses in repairing or replacing the damaged section. Such situations should be recognized when encountered. The estimator should not be influenced or confused by the existence of burned or otherwise damaged materials when he is determining the time required to install *new* material. He should visualize the area surrounding the operation as being completely cleared of all damaged material, and consider that the workmen generally will be unhampered by it.

For this reason, carpenters can cut, fit, or install materials at approximately the same hourly rate when repairing damaged buildings as they can when working on new construction, provided the working conditions are *reasonably* comparable. This is especially true where a large quantity of framing, sheathing, siding, and rough flooring is being replaced. When new work joins the old, as in the case of applying new siding, laying new flooring, or roof boards that have to be notched or stepped back to fit in with the undamaged part, more time will have to be given for that kind of an operation. The estimator should anticipate job conditions that will slow down the production rate of a carpenter. Working areas that restrict free movement, or require awkward handling of materials, affect the labor factor accordingly. Difficult patch jobs that need care and study on the part of a carpenter while doing the actual work, should be thought out intelligently before the probable hours of labor are decided upon.

The unit of material for lumber is 1,000 FBM. The rate of production for a carpenter, working with lumber, is based on the average number of hours it will take to erect, install, or apply 1,000 FBM. The rate varies with the kind of work being done, and is affected by conditions on the job which require more or less handling. Small dimension framing members generally require more hours per 1,000 FBM than larger timbers. For example, it takes about 33 hours per 1,000 FBM to frame 2" x 4" rafters whereas 2" x 6" rafters can be framed in the same roof at an average of 30 hours per 1,000 FBM. This is readily understandable because the cutting, and fitting of a 2" x 6" does not take much longer than a 2" x 4" rafter, but the former has 50 per cent more board feet per rafter; therefore, the rate per 1,000 FBM is less.

When it is necessary to hoist material mechanically or by hand to upper floors to get it to the place in a building where it is to be installed, the labor and the cost of using the equipment must be included. It may be added to the hours per 1,000 FBM if done by hand, or if it is a major operation, it can be treated as a specific item.

It is not uncommon for carpenters, who do very little estimating, to figure their rough carpentry labor as a percentage of the cost of the material. An old and completely unreliable rule of thumb was to figure the cost of the material and double it to take care of labor. Any method of this kind is dangerous because of the variations in labor rates, in labor production, and their relation to material costs.

Figures 9-9 and 9-10 show carpenters framing in a building using automatic nailing machines. Where a sizeable amount of work is to be done these machines save considerable time.

ESTIMATING ROUGH SURFACE MATERIALS (SHEATHING, SUBFLOORING, ROOF DECKING)

There is a wide difference of opinion among builders, contractors, and others who estimate, as to whether openings should be deducted when measuring areas for sheathing, subflooring and roof boards. Some are of the opinion that openings should be deducted in full, while others contend that the surfaces should be figured as solids. Yet there are those who figure that a percentage (usually one half) of each opening or that openings with an area in excess of 10 square feet should be deducted. The principal argument in favor of including all or part of the openings in the area measurement is that its inclusion takes care of the additional labor required to cut and fit around them. However, this does not dispose of the question of what becomes of the excess material for the area of the opening. If no waste for cutting and fitting is added, then the practice of including openings has some justification, but usually the waste is added.

As estimate is an approximation of the material and the labor required to install that material, it attempts to anticipate the exact quantity of material needed, and also the hours of labor to put it in place. Where unit costs are used, as against the method of estimating material and labor separately, the inclusion of openings adds to the estimate of both the labor and material for the area of the openings. If the openings are large or numerous, an accumulated error in cost may result.

A more accurate method is to deduct the areas of all openings, add the normal waste for cutting and fitting, and make allowances of extra labor for the openings encountered in each instance.

Board sheathing, subflooring and roof decking

These items of rough carpentry are nominal 1-inch boards that are applied to the surface area of walls, floors, and roofs. Except where rough

A.

B

Figure 9-9

Pneumatic Nailing Tool being used in framing a building, Courtesy Senco Products, Inc., Cincinnati, Ohio 45244

A.

B.

Figure 9-10

Pneumatic Nailing Tools being used to install bridging, Courtesy Senco Products, Inc., Cincinnati, Ohio 45244

sawed lumber is used, the materials are dressed, and may be either square-edged or matched lumber. The boards are available in widths that range from 4 inches to 12 inches, but the 4-, 6- and 8-inch are most commonly used.

Matched boards have a tongue and groove or ship-lap edge. Page 208 shows the *nominal* size of various boards and also the *actual* size. The lumber is sold by the board foot on the basis of the *nominal* face dimension. It is delivered in the *actual* size and, therefore, an order of 1,000 FBM of matched boards will not cover 1,000 square feet of area due to milling waste. The percentage to be added to the square foot area to obtain an adequate number of FBM is shown in Table 9-16. This percentage includes an allowance of between 5 and 10 per cent for end-cutting waste. The Table also shows the proper waste for dressed square-edge stock as well as matched stock.

TABLE 9-16
APPROXIMATE MILLING AND CUTTING WASTE*
In 1-INCH BOARDS, TO BE ADDED TO AREA.

Item	Nominal Size, Inches	Laid Horizontal	Laid Diagonal
Shiplap	6	22%	27%
	8	21	26
	10	18	24
	12	15	21
Tongue & Groove	4	33	38
	6	22	27
	8	20	26
	10	18	23
	12	15	20
Square edge	4	19	24
	6	14	20
	8	15	20
	10	13	18
	12	12	17
Matched Solid Paneling Patterns	6	24	29
	8	24	29
	10	19	24
	12	17	22

* This includes 5 percent cutting and fitting waste which should be adjusted to the particular operation being considered.

Composition sheathing

The most commonly used composition board sheathing is the type known as insulating sheathing. It is made of ground cane or wood fibers impregnated with water-proofing compounds. The outside, and sometimes the inside, is coated with asphalt. The material is manufactured in 1/2-inch and 25/32-inch thicknesses. The 1/2-inch is available in widths of 4 feet and lengths of 8, 9, and 12 feet. The 25/32-inch sheathing comes in standard sizes of 8, 9 and 12 feet. The edges are matched with tongue and groove, or lap joint. Larger sizes are also available.

Quantity measurements are obtained by determining the area to be covered and adding about 5 per cent for cutting waste. For estimating purposes about 1 pound of 3d nails is required per 100 square feet of 1/2-inch insulating board, and 1 1/2 pounds of 4d nails for 25/32-inch insulating board.

Plywood Sheathing, Subflooring and Roof Decking

Plywood is made in two types—Exterior and Interior. The former is moisture-resistant being made with a waterproof glue, and panels of "C" or better grade.

Interior plywood is made with a water-resistant glue and sometimes with a waterproof glue. The panel backs and interior plies are of lower grade than used in exterior types.

Plywood sheathing, called *Plyscord* when the panels are scored every 16 inches for ease of nailing, comes in sheets 3 and 4 feet wide and 6 to 10 feet long in multiples of 1 foot. Thicknesses range from 5/16, 3/8, 1/2, 5/8 to 3/4 inches.

Recommended thicknesses are as follows:

Exterior Walls	Minimum Thicknesses
2″ x 4″ Studs 16″ o.c.	5/16″
2″ x 4″ Studs 24″ o.c.	3/8″
Subflooring	
2″ x 10″ Joists 16″ o.c.	5/8″
2″ x 10″ Joists 24″ o.c.	¾ ″
Roof Decking	
2″ x 6″ Rafters 16″ o.c.	½″
2″ x 6″ Rafters 24″ o.c.	5/8″

Nails used for 5/16″ to 1/2″ are 6d, and for 5/8″ to 3/4″ are 8d. Sheets are nailed about 6 to 8 inches along the edges and 10 to 12 inches otherwise. Table 9-14 shows the quantity needed. Page 233 outlines how to estimate quantities, which is the same as for composition boards. In other words, computing the area and adding about 5% for cutting and fitting depending on the circumstances.

Labor to apply plywood sheathing, subflooring and roof decking is shown in Table 9-15.

LISTING SHEATHING, SUBFLOORING AND ROOF DECKING

A systematic listing, as discussed under framing, should be used for items of sheathing, subflooring, and roof decking. Most estimators list these items along with those of framing rather than putting them in a separate schedule. In this way they appear in their proper sequence with respect to actual repair operations. Roof decking follows rafter framing, sheathing follows exterior studs, and subflooring follows joist framing.

The measurements of the area to be covered, the extension of square feet and the addition of waste should be shown.

Illustrative Example

North Side of Roof
Rafters	15 pcs	2″ x 6″ x 16′	240 FBM
Ridge	20 lin ft	2″ x 8″	28 FBM

Roof boards (1″ x 6″) 16′ x 23′ =
         368 sq ft
Waste 20%   74
        442 FBM

North Exterior Wall
Wall studs 20 pcs		2″ x 4″ x 10′	134 FBM
Repair sill 12 lin ft		2″ x 6″	12 FBM
Repair plate 12 lin ft		4″ x 4″	16 FBM

Sheathing (1″ x 6″) 14′ x 20′ = 280 sq ft
    Waste 20%   56 FBM
                      336 FBN

Second Floor Bedroom
Floor joists 3 pcs		2″ x 10″ x 12′	60 FBM
Bridging 18 lin ft		1″ x 3″	5 FBM

Subflooring (1″ x 6″) 6′ x 12′ = 72 sq ft
    Waste 20%   15
                   87 FBM

ESTIMATING LABOR—SHEATHING, SUBFLOORING, ROOF DECKING

The cost to apply sheathing, subflooring, and roof boards is estimated by the board foot or square foot for plywood, particle board and composition boards. The labor required varies with the kind of work, the physical conditions under which the carpenters are working and the amount of cutting and fitting. When carpenters are working from ladders or scaffolds, or on a steeply pitched roof, the number of hours is increased. Patching small areas is time consuming and, in most cases, the labor is best determined on a job basis instead of a rate of hours per 1,000 FBM, or 1,000 square feet.

Except for small patch jobs (when the carpenters simply cut and nail boards over clear areas, however small), they generally can apply materials at an average rate. Table 9-15 shows the average number of hours of carpentry labor required for different kinds of work under normal working conditions.

Portable and Mechanical Nailers

Figures 9-11 and 9-12 show workmen using the portable and the mechanical nailers. Both of these tools save considerable time in the laying of subflooring and roof decking and in applying sheathing.

Figure 9-11

Pneumatic Nailing Tool being used to apply plywood subflooring, Courtesy Senco Products, Inc., Cincinnati, Ohio 45244

A.

B.

Figure 9-12

Porta-Nailer being used to nail subflooring and sheathing, Courtesy Rockwell International, Power Tool Division, Pittsburgh, Pa. 15208

Unit costs for sheathing, subflooring and roof boards

A unit cost for these materials is developed from a unit of 1,000 FBM. It includes the material and labor to install or apply it. The prices for material and the rate of wages should be obtained locally. The number of hours for the particular kind of work may be taken from Table 9-15.

The following basic formula is used to develop a unit cost for 1 inch sheathing, subflooring, and roof boards:

Basic Formula

1,000 FBM lumber	@ $ = $	
Lbs of nails	@ $ =	
Hours carpenter labor	@ $ =	
Unit cost per 1,000 FBM	= $_____	
Unit cost per board foot	= $_____ = $	
	1,000	

Illustrative Examples
(Prices and Wages Assumed)

Sheathing, applied horizontally

1,000 FBM 1″ x 6″ matched boards	@	$150.00	= $150.00
30 lbs nails	@	.20	= 6.00
16 hours carpenter labor	@	6.00	= 96.00
Unit cost per 1,000 FBM			= $252.00
Unit cost per board foot			= $.252 or $.25

Subflooring laid diagonally

1,000 FBM 1″ x 6″ matched boards	@	$150.00	= $150.00
30 lbs nails	@	.20	= 6.00
17 hours carpenter labor	@	6.00	= 102.00
Unit cost per 1,000 FBM			= $258.00
Unit cost per board foot			= $.258 or $.26

Roof boards (hip roof)

1,000 FBM 1″ x 6″ matched boards	@	$150.00	= $150.00
30 lbs nails	@	.20	= 6.00
22 hours carpenter labor	@	6.00	= 132.00
Unit cost per 1,000 FBM			= $288.00
Unit cost per board foot			= $.288 or $.29

Roof boards (plain gable)

1,000 FBM 1″ x 6″ matched boards	@	$150.00	= $150.00
30 lbs nails	@	.20	= 6.00
20 hours carpenter labor	@	6.00	= 120.00
Unit cost per 1,000 FBM			= $276.00
Unit cost per board foot			= $.276 or $.28

Roof boards (flat roof)

1,000 FBM 1″ x 8″ matched boards	@	$160.00 = $160.00
30 lbs nails	@	.20 = 6.00
14 hours carpenter labor	@	6.00 = 84.00
Unit cost per 1,000 FBM		= $250.00
Unit cost per board foot		= $ .25

Unit costs for plywood and composition materials

The following basic formula may be used to develop unit costs for plywood and composition materials used for sheathing, subflooring and roof decking. It is based on 1,000 square feet of material plus the labor to install or apply that quantity. Nails are omitted because their effect on the unit cost is minor.

Basic Formula

1,000 sq ft material	@	$	= $
Hours carpenter labor	@	$	= $_____
Unit cost per 1,000 sq ft			= $
Unit cost per sq ft	$ / 1,000		= $_____

Illustrative Examples
(Prices and Wages Assumed)

5/16″ Plywood sheathing

1,000 sq ft 5/16″ plywood sheathing	@	$140.00 = $140.00
14 hours carpenter labor	@	6.00 = 84.00
Unit cost per 1,000 sq ft		= $224.00
Unit cost per sq ft		= $.224 or $.23

½″ Plywood roof decking on cut-up roof

1,000 sq ft ½″ plywood decking	@	$140.00 = $140.00
18 hours carpenter labor	@	6.00 = 108.00
Unit cost per 1,000 sq ft		= $248.00
Unit cost per sq ft		= $.248 or $.25

5/8″ Plywood subflooring

1,000 sq ft 5/8″ subflooring	@	$170.00 = $170.00
12 hours carpenter labor	@	6.00 = 72.00
Unit cost per 1,000 sq ft		= $242.00
Unit cost per sq ft		= $.242 or $.24

½″ Insulation board sheathing

1,000 sq ft insulation board	@	$ 60.00 = $ 60.00
14 hours carpenter labor	@	6.00 = 84.00
Unit cost per 1,000 sq ft		= $144.00
Unit cost per sq ft		= $.144 or $.15

The unit costs for other materials may be computed by using the same formula and inserting the cost of material, the number of pounds of nails from Table 9-14, and the hours of labor from Table 9-15. The hours of labor should be carefully selected from the Tables depending on the size and type of the material used.

Application of unit costs

In the following illustrative example, the small one story dwelling in Figure 9-13 requires rough flooring, sheathing and roof boards.

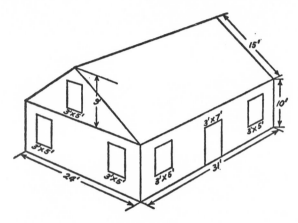

Figure 9-13

Quantities are obtained as shown, and the unit costs per square foot used are the same as those developed for illustrative purposes. For example, subflooring laid diagonally $.26, sheathing applied horizontally $.25, roof boards on plain gable type $.28.

Roofboards
(1″ x 6″ matched boards)
One side 15′ x 31′ =	465 sq ft	
Two sides	x 2	
	930 sq ft	
Add 20% waste	186	
	1,116 FBM @ $.28 = $312.48	

Exterior wall sheathing
(1″ x 6″ matched boards)
Sides	2(10′ x 31′) =	620 sq ft
Ends	2(10′ x 24′) =	480 sq ft
2-Gables	9′ x 24′ =	216 sq ft
		1,316 sq ft

Deduct openings
(Same for sides not shown)
10—3' x 5' wds. = 150 sq ft
 2—3' x 7' drs. = 42 sq ft
                    192

                       192 sq ft
               1,124 sq ft
Add 20% waste   225
               1,349 FBM@$.25 = $337.25

Rough flooring
(1" x 6" matched boards diagonal)
24' x 31'   = 744 sq ft
Add 25 % waste = 186
               930 FBM@$.26

                              = $241.80
                                $891.53

ESTIMATING WOODEN FENCES

The cost to erect fences generally is developed on a lineal foot basis. The cost of the labor and material for a section between two posts is first obtained; then the spacing in feet between posts is divided into this cost to determine the unit cost per lineal foot. This unit cost is then multiplied by the total number of lineal feet of fencing required. When determining the cost of one section, the second post is omitted.

Illustrative Example

To find the cost per linear foot of a solid board fence using 4" x 4" posts, 2" x 4" rails and 1" x 6" matched boards. The posts are 8 feet on center and the fence is 5 feet above ground with posts 3 feet in the ground. Material prices and wage rates are assumed.

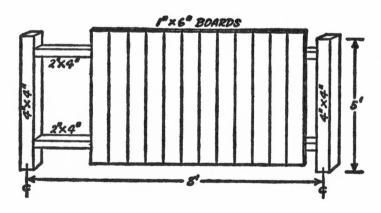

1—4″ x 4″ x 8′ post	= 11 FBM
2—2″ x 4″ x 8′ rails	= 11 FBM
	22 FBM @$160.00 per MBF = $3.52
5′ x 8′ - 1″ x 6″ matched boards	40
add 20% waste	8
	48 FBM@$150.00 per MBF = 7.20
2 lbs nails (estimated)	@ .20 per lb = .40
	Total material cost = $11.12

Labor for one section

Allow ½ hour to dig hole, plumb and set one post	@$5.00 per hr = $ 2.50
Allow ⅓ hour to cut and fit rails	@$6.00 per hr = 2.00
Allow labor for boards at the same rate as sheathing 16 hours per MBF	
16 x $6.00 = $96.00 per MBF	
40 FBM@$96.00 per MBF	= 4.61
	Total labor cost 9.11
Total labor and material for 8 foot section	=$20.23

Cost per lineal foot $= \dfrac{\$20.23}{8} = \$2.53$

Note: Waste allowance should be increased depending on height of fence. Painting is to be added as required.

If a yard is to be enclosed with this type of fence, and the measured distance is 120 lineal feet, the cost would be 120′ x $2.53 = $303.60.

This principle applies to picket fences, wire fences, and straight post and rail fences. The material costs will be governed by the kind of materials and the sizes specified. The labor to dig post holes will vary with the type of soil as to whether it is sandy, loamy, rocky or hard clay. When many post holes are to be dug by machine, the cost can be greatly reduced.

ESTIMATING SCAFFOLDING

Most contractors and builders keep on hand as part of their equipment, scaffold plank. It is moved from one job to another as it is needed. When a large quantity is required for a particular job, it may be purchased and part of the original cost charged against the specific job. Sometimes the materials find their way into other construction jobs in the form of floor joists or other structural members.

Interior scaffolds such as those used by masons for plastering ceilings, or by brick layers to lay up brick walls, are made by placing 2″ plank on wooden horses. The time for setting up and shifting such scaffold is included in the labor rate of the mason or bricklayer's helper.

Sometimes an exterior scaffold one or two stories high is needed for workmen to install cornices, lay up veneer brick or stone, or apply exterior siding. These are erected by carpenters and consist of posts, ledgers, and bracing. Scaffold plank is laid loose, or is nailed to horizontal ledgers. The posts are usually two 2" x 4" spiked together. The ledgers are 1-inch or 2-inch stock and the braces are 1-inch boards.

The cost of this type scaffold may be estimated by the lineal foot, or computed for the entire scaffold if a moderate amount is required.

Illustrative Example

The cost of erecting the scaffold shown in Figure 9-15 would be estimated as follows. An average price per MBF of material is assumed to be $160.00 and carpenter's wages are arbitrarily taken at $6.00 per hour.

Uprights	3—4" x 4" x 18'	72 FBM
Ledgers	6—2" x 6" x 8'	48 FBM (for two levels)
Braces	4—1" x 6" x 12'	24 FBM
		144 FBM
Planking	8—2" x 10" x 12'	160 FBM
		304 FBM @$160.00. = $48.64

Labor—2 carpenters
  2 hours   = 4 hours@$6.00 =   24.00
                                                                                $72.64

The total cost of erecting this 16-foot scaffold at these prices for material and labor is $72.64 or $4.54 per lineal foot.

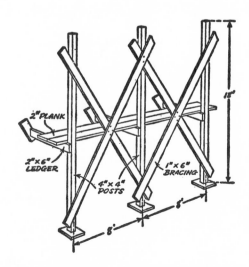

The scaffold will have to be taken down when it has served its purpose and a labor charge should be made for this work. Some of the materials may be used on the job later or, if none of the materials are used, they will go back to the contractor's yard for future use as construction material or scaffolding. Whether or not salvage of scaffold materials should be allowed will depend on the circumstances in each situation.

When a contractor's competitive position permits, he will charge all or as much of the cost of the material as is possible to the job. He charges *all* of the labor for transporting, erecting, and dismantling the scaffold.

Scaffolding made of wood for more than two stories is seldom used. Steel tubular scaffold can be bought or rented and it is more economical. The labor for erecting and dismantling steel tubular scaffolding averages between 2 and 3 hours per 100 square feet of wall surface. To this should be added the cost of rental and the trucking charges to and from the job. The cost of large quantities of tubular scaffold to be erected should be obtained from subcontractors who rent, erect and dismantle on a contract basis.

Figure 9-16 shows workmen on a moveable aluminum "Ladder Scaffold" of Patent Scaffolding Co., Fort Lee, N.J. 07024.

Figure 9-16

Men working from Moveable Aluminum Ladder Scaffold, Courtesy Patent Scaffold Co., Fort Lee, N.J. 07024

Chapter 10.

Finish Carpentry

While there is general agreement among estimators on what constitutes "rough" and what constitutes "finish" carpentry, there are some types of work that could be placed in either one or the other classification. For purposes of convenience in discussion, and for consistency, all carpentry which requires finished materials and perhaps greater skill in installation than rough carpentry, will be classified here as finish carpentry. Such items include interior and exterior trim, doors and windows including frames, screens, wood siding, stair building, cabinet work and finish floors.

A further division is made here by separating interior and exterior finish carpentry. All doors and windows are treated under the classification of interior trim.

FINISH WOOD FLOORS

The top layer of flooring, which is the wearing surface, is referred to as the finish flooring. In dwelling construction the finish flooring may be pine, oak, maple or beech depending on the appearance and quality desired. There are different grades of each kind of wood flooring, so it is important when estimating costs to be certain of the kind, size, and the quality involved.

Finish wood floors can be laid over concrete slabs satisfactorily. The system meets FHA requirements and has been thoroughly tested by the

National Oak Manufacturers' Association. A double layer of 1 x 2 inch wood sleepers are nailed together with a moisture barrier of 4 mil polyethylene film between them. The bottom layer of sleepers is secured to the slab by mastic and concrete nails. The strip hardwood flooring is nailed at right angles to the sleepers, with one nail at each bearing point. The slab when poured should also have a 4 mil or 6 mil polyethylene film over the base or fill.

An alternate method for laying finish wood floors over concrete slabs is shown in Figure 10-1.

1. **Mastic for adhering screeds.** Additional mastic is required for adhering screeds to the slab surface. It is usually applied in "rivers" to a thickness of ¼″ under previously laid out screeds or along lines where screeds are to be placed. If spread over the entire slab, mastic should be leveled with a notched trowel to a depth of 3/32″. Again, use only a hot mastic designed for bonding wood to concrete. Imbed screeds in position immediately after mastic is spread, before it has time to "set." Be sure to observe following instructions for positioning screeds.

2. **Good screeds are important.** For screeds use only flat, dry 2″x4″s. Although FHA requirements specify lengths of 18″ to 30″, lengths up to 48″ have been used satisfactorily. Reject warped pieces. They will distort the finish floor. Use random lengths rather than pieces of all one length since this helps in staggering end joints. Use screeds treated beforehand with an approved wood preservative to prevent rot or termite damage. Avoid use of creosote or other preservative material that might stain the finish floor if it bleeds through the nail holes.

3. **Positioning the screeds.** Lay out screeds in staggered rows, as shown above, on 16″ centers at right angles to the proposed direction of the finish flooring. Lay them flat side down and embed each one firmly in the mastic. Ends must overlap at least 4″. Alternate short and long length screeds to make sure that end joints are staggered as much as possible. These three practices— a staggered pattern of screeds, overlapping end joints, and alternating random lengths—are important in producing the soundest possible nailing surface for the flooring.

4. **Screeds near baseplate.** It's important to allow for expansion of the finish floor when positioning screeds near the base plate. Leave a gap of at least 1″ between the ends of the screeds and the base plate around edges of the room. This provides for normal expansion of the finish flooring that accompanies weather and humidity changes. It helps prevent buckling or cupping of the floor and pressure on the walls. This photograph shows proper positioning of screeds near the base plate and how random lengths are used to stagger end joints.

Figure 10-1

2″ x 4″ screeds set in mastic on a concrete slab in preparation for laying strip hardwood finish flooring. Courtesy National Oak Flooring Manufacturers Association, Memphis, Tenn.

In buildings of joist construction, the finish wood floors are laid over subflooring, either boards 4 or 6 inches wide, or plyscord of 1/2" minimum thickness. When boards are used they should be 1/4 inch apart to allow for expansion; plywood joints should be 1/8 inch.

When electrical conduit or piping is laid over subfloors, furring strips are nailed to the joists through the subfloor.

A layer of 15-lb asphalt felt is layed over the subfloor.

NAILS

Table 10-1 shows a Nail Schedule for finished flooring. The type nail follows the recommendations of the National Oak Flooring Manufacturers' Association.

TABLE 10-1
NAIL SCHEDULE FOR FLOORING

Finished Size in Inches	Type and Size of Nails	Lbs of Nails Needed Per 1,000 FBM Spaced 10 to 12 Inches
25/32 x 3¼	7d or 8d screw type	35
" x 2¼	or	50
" x 1¼	cut nails*	70
½ x 2 or ½ x 1½	5d screw type, cut or casing nail	20
3/8 x 2 or 3/8 x 1½	4d bright casing (8-inch spacing)	

* Machine driven barbed fasteners are equally acceptable; follow manufacturer's recommendation.

Strip flooring is the most popular style of wood flooring used today. It comes in various widths but the 25/32" x 2 1/4" (nominally 1' x 3") is the most popular. It is tongue and grooved and is available in red or white oak and in yellow pine. There is very little difference in red and white oak as to quality and utility. There are several grades of oak flooring and the price varies with these grades. The same is true of yellow pine flooring.

Maple flooring is seldom used in homes except in areas such as New England where there is a fair supply left. Clear maple flooring is the top grade. Select No. 1 is next.

Prefinished strip oak flooring, a newer development in flooring, is prefinished, buffed and polished at the factory. While the cost of prefinished flooring is a third or more higher than unfinished for the same grade, there is less waste and considerable time is saved in completing the job.

Prefinished block oak flooring is the most expensive type floor and is usually laid in adhesive over wood subflooring or concrete slabs. Standard size is 9" x 9" and 5/16" thick. Other sizes are available and also other thicknesses suitable for nailing. This type of flooring is delivered in exact quantities needed and there is no percentage to be added to the area to be covered.

ESTIMATING QUANTITIES OF FLOORING

Table 10-2 shows the percentage that must be added to the *measured* floor area to obtain the quantity that is needed to cover the area. These figures include both milling and cutting and fitting waste.

TABLE 10-2
ESTIMATING FLOORING QUANTITIES

Nominal Size in Inches	Actual Size in Inches	Milling + 5% Cutting Waste	Multiply Area by	Floor Needed Per 1000 FBM
1 x 2	25/32 x 1½	55%	1.55	1,500
1 x 2½	" x 2	42	1.42	1,420
1 x 3	" x 2¼	38	1.38	1,380
1 x 4	" x 3¼	29	1.29	1,290
1 x 2	3/8 x 1½	38	1.38	1,380
1 x 2½	" x 2	30	1.30	1,300
	½ x 1½	38	1.38	1,380
	" x 2	30	1.30	1,300

The labor rates in Table 10-3 are for quality workmanship done by carpenters—no hammer or nail marks. When subcontractors with experienced floor layers do the work the time can be substantially reduced.

TABLE 10-3
AVERAGE HOURS TO LAY FLOORING

Nominal Size and Type Flooring Layed	Per 100 FBM	Per 1000 FBM	FBM Per Hour
1" x 2" Hardwood strip	4.5	45	22
1" x 3" Hardwood strip	3.0	30	34
1" x 3" Softwood strip	2.5	25	40
1" x 4" Softwood strip	2.0	20	50
3/8" x 2" Hardwood strip	4.5	45	22
3/8" x 3" Hardwood strip	4.0	40	25
3/8" x 2" Prefinished strip	4.5	45	22
9" x 9" Prefinished block	3.0	30	34

Use of automatic hammers also reduces the time to lay floors and do a better job. Figure 10-2 shows the use of a nailing machine on strip flooring. (Courtesy of Rockwell International).

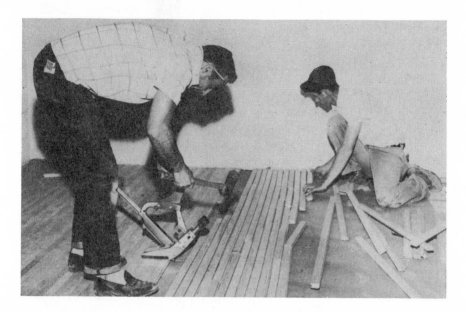

Figure 10-2

Workmen using a Porta-Nailer to nail strip hardwood finish flooring. Courtesy Rockwell International, Power Tool Division, Pittsburgh, Pa. 15208

Unit Cost for Finish Wood Flooring

The unit of material for developing a unit cost for finish flooring is 1,000 FBM of the flooring at the local delivered price. Table 10-1 gives the pounds of nails and Table 10-3 gives the number of hours to lay 1,000 board feet. The following basic formula is used to develop a unit cost.

Basic Formula

1,000 FBM flooring	@ $	= $
Lbs 8d cut nails	@	=
Hours labor	@	=

Unit cost per 1,000 FBM $_____

$$\text{Unit cost per BF} = \frac{\$}{1,000} = \$$$

Assuming prices and wages as indicated, the following are illustrative examples of computing unit costs (7d or 8d screw type nails may be used):

1" x 3" Oak flooring

1,000 FBM oak flooring	@$250.00 =	$250.00
50 lbs 8d nails	@ .20 =	10.00
30 hours labor	@ 6.00 =	180.00
Unit cost per 1,000 FBM =		$440.00
Unit cost per BF =		$.44

1" x 3" Pine flooring

1,000 FBM pine flooring	@$210.00 =	$210.00
50 lbs 8d nails	@ .20 =	10.00
25 hours labor	@ 6.00 =	150.00
Unit cost per 1,000 FBM =		$370.00
Unit cost per BF =		$.37

1" x 4" porch flooring

1,000 FBM fir flooring	@$240.00 =	$240.00
35 lbs 8d nails	@ .20 =	7.00
20 hours labor	@ 6.00 =	$120.00
Unit cost per 1,000 FBM =		$367.00
Unit cost per BF =		$.367 or $.37

Application of Unit Costs

The area over which finish flooring is to be laid is measured and the number of square feet is obtained. Openings such as stairwells are deducted. Areas to be patched should be measured liberally to allow for cutting back so that all joints will not be along one line or joist.

Illustrative Example

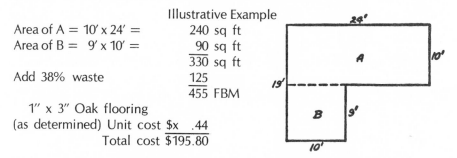

Area of A = 10' x 24' =	240 sq ft	
Area of B = 9' x 10' =	90 sq ft	
	330 sq ft	
Add 38% waste	125	
	455 FBM	
1" x 3" Oak flooring		
(as determined) Unit cost $x	.44	
Total cost	$195.80	

SANDING FINISH FLOORING

After finish floors are laid, they are usually sanded by machine before varnish, shellac or other finishes are applied. This work is done either by the floor layer if he is equipped to do it, or by a floor sanding sub-contractor. In most sections of the country quotations can be obtained for the cost per square foot. The cost varies with the kind of flooring, the size of the rooms and the area involved.

When floors that are ridged or cupped by water require sanding to resurface them, the cost of the work is dependent on the degree of damage and finish to be cut away. Floors that are severely cupped will require more sanding to produce a smooth surface. Old paint or many coats of varnish are difficult to remove.

Floors that are repaired by inserting patches are usually sanded over the entire room and given a uniform finish.

While it is recommended that a unit cost per square foot be obtained from local sources, there are occasions when it is not possible and the unit cost then must be estimated. The basis for determining an approximate cost would be the number of hours of labor required plus the cost of the abrasive paper and rental of a sanding machine. An operator can average 100 square feet per hour on new floors including the finishing along the base. On old floors and ridged floors, the rate may drop down to as little as 50 square feet per hour. Good judgment should be used in any case.

Assuming the following wage rate, machine rental cost, and abrasive paper cost, the unit cost per square foot may be developed.

Cost of floor sanding per hour

Operator's wages	$6.00 per hour
Machine rental	1.00 per hour
Abrasive paper	.50 per hour
	$7.50 per hour

If an operator sands at a rate of 100 square feet per hour, the unit cost would be $7.50 ÷ 100 = $.075 per square foot.

At a rate of 50 square feet per hour the unit cost would be $7.50 ÷ 50 = $.15 per square foot.

ESTIMATING WINDOWS, STORM SASH AND SCREENS

The principal parts of a window in a building are the *frame, sash, interior trim* and the *hardware*. These components may be purchased individually, or a window can be delivered to the job as a completely assembled unit ready to install.

Of the several different styles of windows in use, the following ones are most common:

> *Double Hung*—The sash slide up and down in the frame.
> *Casement*—The sash are hinged to the side of the frame.
> *Jalousie*—The sash pivot and swing up and down like louvres.
> *Stationary*—The sash is fixed permanently in the frames.
> *Gliding*—The sash slide horizontally.

Each type is available in either wood or aluminum. Casement, gliding, and stationary windows are also available in steel. Before estimating the cost of windows, or the particular parts that need to be replaced, it is necessary to determine the style, quality, dimensions and kind of material.

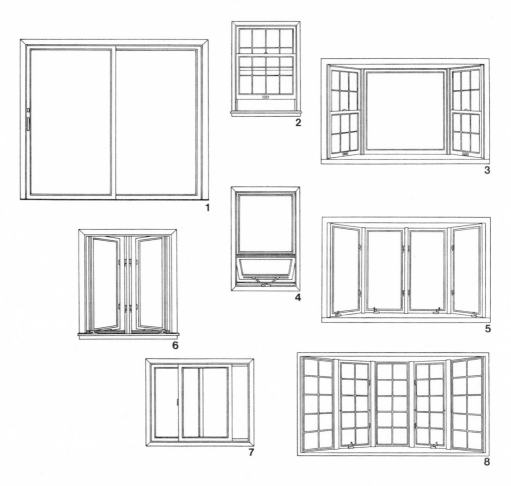

Figure 10-3

Typical window styles readily available in residential construction: 1. Gliding door; 2. double-hung window; 3. double-hung angle bay window; 4. awning window with fixed sash above; 5. casement angle bay window; 6. casement window; 7. gliding window; 8. bow window. Windows and gliding door shown are available in low-maintenance Perma-Shield line from Andersen Corporation, Bayport, Minnesota 55003.

Figure 10-3 show several typical window styles available for residential construction. (Courtesy of Andersen Corporation, Bayport, Minnesota 55003)

Sash should be described according to size, number of lights of glass, thickness and so forth. When wooden sash is broken or burned, new sash can be installed without replacing the window frame if it is undamaged. Expensive non-stock wooden sash can often be repaired by replacing the dividing bars, or a side of the sash-frame as may be required, at less cost than ordering new sash. Such matters need to be investigated carefully before concluding that a damaged window is a total loss.

Table 10-4

Average Hours Required to Install Windows, Storm Sash and Screens

Type of Work Done	Hours Required
Assemble wooden frames on the job	1
Install single wooden frames in frame building	.5
Install single wooden frames in masonry building	1
Additional time for double frames with mullion	.5
Fit and hang wood sash complete (pair)	1.5
Fit and hang casement sash complete (each)	1.5
Interior window trim installed—softwood	1
Interior window trim installed—hardwood	1.5
Complete window, wood frame and sash—factory assembled	2
Complete window, wood frame and sash—knockdown	4
Complete basement window	.5
Wood window screens	1
Wood storm windows	1
Wood shutters—stationary—per pair	2
Wood shutters—hinged—per pair	3
Complete D.H. Aluminum window	2
Complete Aluminum window with sliding sash	2
Picture window, wood frame, trim and sash	4
(Thermopane same labor plus cost of material)	
Aluminum combination screens and storm sash	2
Installing factory steel sash (each)	3
Glazing factory steel sash (see Chapter 11)	5
Weather stripping D.H. windows	2
Placing back band per window	.75

Table 10-4 shows the average number of hours of labor required to install the several parts of different windows, as well as the number of hours to install complete windows, factory sash, screens and storm sash.

UNIT COST FOR WINDOWS

Assuming prices and wages shown, an example of developing a unit cost for a complete window assembled on the job would be as follows:

Illustrative Example

Material

1 2'-8" x 4'-10" pine window frame	$6.00
1 stool	.50
1 pair 6/6—1 3/8" sash	6.00
1 set coil spring balances	2.50
1 sash lock	.40
1 pair sash lifts	.60
1 set interior pine trim	3.00
	$19.00
4 Hours labor (from Table 10-4)@$6.00	$24.00
Unit cost of 1 window installed complete	$43.00

The unit cost for different types of windows, their parts, or accessories, can be computed in a similar manner. Determine the local prices for the material delivered, and add the labor cost, using local wages and the hours required from Table 10-4.

Illustrative Examples

Window trim

1 set pine window trim	$ 3.00
1 hour of labor @ $6.00	6.00
Unit cost for interior pine trim	$ 9.00

Wood window screens

1 2' x 10" x 4'-10" copper screen	$ 3.15
1 hour of labor	6.00
Unit cost for 1 screen	$ 9.15

Wood shutters (hinged)

1 pair 2'-6" x 4'-6" x 1 1/8" pine paneled	$10.00
3 hours labor @ $6.00	18.00
Unit cost per pair of shutters	$28.00

ESTIMATING INTERIOR AND EXTERIOR DOORS, SCREENS ETC.

The component parts of a door are the jambs (called the bucks in exterior doors), the door, door stop, casings, hardware and, for some exterior doors, a saddle. The parts are generally purchased separately, although some suppliers will furnish doors as a complete unit with all parts including hardware.

Doors should be described as to kind of wood, size and style of door, thickness, and also style of casings. Hardware for doors varies in cost depending on type, material and manufacturer. Sometimes only part of a door assembly may be damaged such as a casing, the door itself, the hardware; in other situations the entire assembly may be involved. Table 10-5 shows the average hours of labor required to install the parts of a door, or the entire unit. It also shows the hours required to fit and hang garage doors, screen doors, storm doors, and to weather-strip exterior doors.

Table 10-5
Hours Required to Install Interior and Exterior Doors,
and Their Parts

Type of Work Done	Hours Required
Set interior or exterior jambs (frames)	.75
Install mortise lock set	.75
Cylinder lock on front door	2
Fit and hang interior door—softwood	1.5
Fit and hang interior door—hardwood	2
Put on stops and casings (two sides)—softwood	1
Put on stops and casings (two sides)—hardwood	1.25
Install frame, fit and hang door trim complete—interior	4
Install frame, fit and hang door trim complete—exterior	5
Make and hang 3′ x 7′ batten door	3
Garage swinging doors 8′ x 8′ fit and hang per pair	4
Garage overhead door 8′ x 8′ x 1 3/8″—hang complete	6
Garage overhead door 15′ x 7′ x 1¾—hang complete	8
Storm door—fit and hang complete	2
Screen door—fit and hang complete	2
Weather-stripping exterior door	2
Trimmed opening, set jams and trim 7′ x 8′	2
Placing back-band per one side door opening	.75
Placing back-band per one side trimmed opening	1

UNIT COSTS FOR DOORS

The unit cost for doors and their parts is developed in the same manner that unit costs are developed for windows. Assuming the prices for materials and the wage rate indicated, a typical dwelling door installed complete with frame, door and trim, would be as follows:

MATERIAL

1 2′-8″ x 6′-8″ x 1 3/8″ 5-panel pine door	$10.00
1 set jambs (frame)	4.00
2 sides 1″ x 3″ pine trim	4.50
Stop molding (18 lin ft @ .05)	.90
1 pair 4″ x 4″ butts (hinges)	1.00
1 mortise lock set	2.75
Total material cost	$23.15
Labor 4 hours (Table 10-5) @ $6.00	24.00
Unit cost for 1 door installed complete	$47.15

The unit cost for other types of doors or their component parts may be computed by obtaining local material costs and adding the labor cost based on the average hours shown in Table 10-5 multiplied by the local wage rate.

ESTIMATING MISCELLANEOUS INTERIOR TRIM

There are several items of interior trim, the materials for which are bought by the lineal foot, and the unit cost is figured in the same unit. These items consist mainly of the various base moldings, baseboards, floor moldings, picture moldings, chair and plate rails.

Baseboard may consist of one of the standard single-member molded stock styles, ranging from 2 to 4 inches high, or it may be a so-called three-member base found in older or more expensive homes. This type base is made up of a 4-, 6- and sometimes 8-inch board 7/8-inch thick. A quarter round floor molding, and a wall molding on top of the baseboard completes the three members. Occasionally a two-member base is found with the wall molding omitted.

Plate rails are usually found in older homes that were built in a period when it was a custom to display attractive plates and platters on a plate rail 5 or 6 feet from the floor and extending around the dining room wall. The rail consists usually of two members, and occasionally three members.

Ceiling Molding, usually a cove type; and *picture moldings,* are one-member units.

Chair Rails are either one-member or three-member units applied to dining room walls at chair-back height to prevent the chairs from marring the walls.

Unit Costs for Miscellaneous Interior Trim

The basic formula (with assumed figures below) for developing the unit costs for interior trim items, is expressed in terms of cost per lineal foot.

Basic Formula			
100 lin ft of material	@	$	= $
Number of hours to install 100 lin ft	@		=
Unit cost per 100 lin ft			= $

This cost divided by 100 equals the cost per lineal foot. Nails are relatively a minor item of cost when installing 100 lineal feet of trim and for that reason are omitted in figuring the unit cost.

Table 10-6 shows the average number of hours of labor that are required to install 100 lineal feet of different types of interior trim. By substituting these hours in the formula, and multiplying by the local wage rate, the labor cost per 100 lineal feet can be obtained.

Table 10-6
Hours Required to Install 100 Lineal Feet of Miscellaneous
Items of Interior Trim

Type of Work Done		Kind of Wood	Hours Required Per 100 Lineal Feet	Lineal Feet Per Hour
Baseboard	1 member	Softwood	6	17
Baseboard	2 member	Softwood	7	14
Baseboard	3 member	Softwood	8	13
Baseboard	3 member	Hardwood	10	10
Chair rail	1 member	Softwood	5	20
Chair rail	1 member	Hardwood	6	17
Plate rail	2 member	Hardwood	12	8
Plate rail	3 member	Hardwood	16	6
Picture molding		Softwood	5	20
Ceiling molding		Softwood	6	17
Wood paneling strips		7/8" Hardwood	8	13

Note: On cheap work, hours may be reduced.

Illustrative Examples

1-member baseboard—softwood
   100 lin ft 1" x 3" pine base   @$ .09 = $ 9.00
      6 hours labor   @ 6.00 = 36.00
         Unit cost per 100 lin ft   = $45.00
         Unit cost per lin ft   = $ .45
3 member baseboard—softwood
   100 lin ft 1" x 6" pine base   @$ .17 = $17.00
   100 lin ft ¾ " rnd. floor mldg.   @   .03 =   3.00
   100 lin ft 1" wall molding   @   .04 =   4.00
      8 hours labor   @ 6.00 = 48.00
         Unit cost per 100 lin ft   = $72.00
         Unit cost per lin ft   = $ .72
Chair rail—hardwood
   100 lin ft 1" x 3" pine chair-rail   @   .17 = $17.00
      5 hours labor   @ 6.00 = 30.00
         Unit cost per 100 lin ft   = $47.00
         Unit cost per lin ft   = $ .47
Picture molding—softwood
   100 lin ft 1-1 ½" molding   @$ .05 = $ 5.00
      5 hours labor   @ 6.00 = 30.00
         Unit cost per 100 lin ft   = $35.00
         Unit cost per lin ft   = $ .35

Application of Unit Costs for Interior Trim

When estimating the number of lineal feet of baseboard, plate rail, picture molding, and so forth, to be installed in a room, it is customary to ignore window and door openings less than 8 or 10 feet wide as their inclusion compensates for waste in cutting and fitting. Generally such items extend around the perimeter of a room and the lineal feet of perimeter is the quantity needed.

Illustrative Example

In a room 14' x 22' a three member baseboard, a chair rail, and a picture molding are to be installed. Unit costs are those developed.

The Perimeter is 2(14' + 22') = 72 lin ft

Applying unit costs

72 lin ft 3 member base	@$.72 =	$ 51.84
72 lin ft chair rail	@ .47 =	33.84
72 lin ft picture molding	@ .35 =	25.20
	Total cost =	$110.88

ESTIMATING INTERIOR STAIRS

Stair building is a specialty in carpentry, and today most of the residential staircases are factory built. They are delivered knockdown, or with the stringers, risers and treads assembled.

The main parts of a staircase are the *treads,* the vertical *risers* and the *stringers,* which are the side pieces to which treads and risers are attached. When one or both sides of the stairs are open, a balustrade, or handrail and balusters is installed.

An open, or plain stringer is one where the stringers are notched out and the treads and risers are nailed to them. This type of stairs is found in less expensive construction. The rough, open-string basement stairs is a typical example.

Box stairs are those which have the stringers routed out, or housed out. The treads and risers are slid into place from the back side, wedged and glued with hardwood wedges. This type of stair construction is sturdy and squeak-proof. It is more expensive to build than open-string stairs.

Because of the many different architectural designs in staircases, and the different kinds of wood used in their construction, it is recommended that local stair builders be consulted for accurate estimates. Where an approximate cost is required, the materials should be listed and priced. Table 10-7 shows the average hours of labor required to cut and assemble and install the various parts of staircases.

Table 10-7
Hours Required to Cut and Erect Staircases

Type of Work Done	Hours Required Per Tread
Cutting out open stringer	.3
Housing out box stringer	.8
Erecting box or open stringers	.8
Placing treads or risers on open stringer	.1
Placing treads or risers on housed stringer	.4
Placing rail and balusters	1.0
*Complete erection of open staircase	1.5
*Complete erection of boxed stairs	2.5
Erection of plank cellar stairs without risers	1.0

*Add handrail and balusters as required.

Many estimators compute the cost of the labor and material required in a typical staircase and from that obtain the cost per tread. This unit cost per tread then may be increased or decreased depending upon the quality and design of a particular staircase under consideration.

Illustrative Example

To estimate the cost of labor and material to cut out and erect an oak staircase in a residence with 8' 6" ceiling assuming material prices and labor wages as shown.

Material

11 10½" x 1 1/8" x 3'-0 oak treads	@	$3.00 =	$ 33.00
12 ½" x ¾" x 3'-0 oak risers	@	1.50 =	18.00
2 End Newels	@	6.00 =	12.00
22 Balusters	@	.50 =	11.00
14 Lin ft hand rail	@	.80 =	11.20
48 Lin ft scotia molding	@	.05 =	2.40
1 5/4" x 12' x 14' wall stringer	@	300/M =	5.40
1 1" x 12" x 14' outside stringer		300/M =	4.20
2 2" x 12" x 14' rough stringer		130/M =	7.28
			$104.48

Labor (From Table 10-7)

Complete cutting out and erecting   (Forward) $104.48
11 treads @ 2½ hrs each = 27½ hrs   @ $6.00 = $165.00
Placing rail and balusters
11 treads @ 1 hr each = 11 hrs   @ 6.00 =   66.00
                                                                    $335.48

$$\text{Cost per tread} = \frac{\$335.48}{11} = \$30.50$$

EXTERIOR WOOD TRIM

Water table	Wooden gutters
Drip cap	Cornice trim
Corner boards	Verge or barge trim

Most items of exterior trim are both ornamental, and functional. The woods used are usually fir, pine, cedar, cypress and redwood. For less expensive construction local woods like spruce and a lower grade yellow pine may be used. Quantities of trim are customarily estimated by the lineal foot per member. Unit costs are developed from quantities of 100 lineal feet. Because the cost of nails used to apply exterior trim is trifling when compared to the cost of the labor and lumber, it is inconsequent to include them in unit costs.

Water Table is the baseboard at which exterior siding begins. It is at the base of the building where the foundation ends and the frame wall begins. Actually a water table is a projection which permits rain water that runs down the outside surface of the building to drip away from the foundation. Frequently it consists of two parts; a 1" x 6" or 1" x 8" board and a beveled cap on top called a *drip cap.*

Corner Boards, as their name implies, attach vertically to the outside and inside corners of a house, forming the finish up against which the exterior horizontal siding butts. The stock used is generally 1 1/8 by 3 or 4 inches in width. In some cases, a 4-inch board laps a 3-inch board at the corner, forming a square edge. In other cases two equally wide boards meet at their inside edges, and a quarter round molding is nailed where they meet to form a rounded corner.

Wooden Gutters are used in many sections of the country in place of galvanized or copper gutters. They are less expensive and, for some types of construction, are better suited architecturally. They are purchased in lengths at the lumber yard. Wood gutters are attached with galvanized screws or nails.

Cornices can be generally classified into three types:

1. Closed or Flush cornices.
2. Open cornices.
3. Box cornices.

A cornice which actually has no overhang, the rafters being cut off flush with the exterior sheathing, is called a *closed or flush cornice.* The members consist of a simple frieze board and a crown molding.

A cornice which has the open or exposed rafters projecting is known as an *open cornice.*

There are two common kinds of *box cornice*. Both are projecting cornices that are boxed in with a frieze board on the face of the house. There is a soffit underneath the cornice, and a facia board on the face of the cornice. The soffit and the facia are nailed to the rafter ends. When the cornice has a wide overhang, it is necessary to nail cleats from the end of the rafter to the wall of the house for the soffit boards. These cleats are called lookouts. Examples of typical cornices are shown on pages 269 and 270, Figures 10-4, 10-5 and 10-6.

When estimating the cost of cornice work, the lineal feet of the members are listed individually. A unit cost is developed for each member using 100 lineal feet as a base. The number of hours of labor for installing the members are shown in Table 10-8. In all cases, the kind of wood, and the exact size of each member should be measured with a rule to determine its thickness and width.

Verge or Barge Board is the member used to finish off the gable end of a building. It may be planted on top of the siding.

Estimating Quantities of Exterior Trim

Each member of trim should be measured for thickness and width. The thickness varies with the type of trim and quality of construction. The

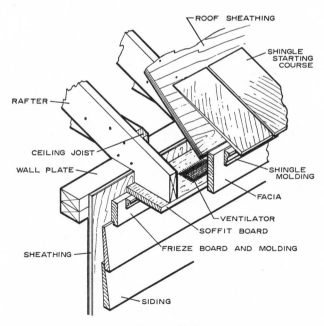

Figure 10-4

Construction details of a "narrow box cornice" (without lookouts).

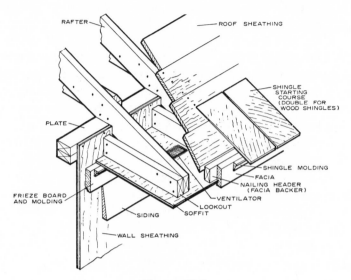

Figure 10-5

Construction details of a "wide box cornice" (with horizontal lookouts).

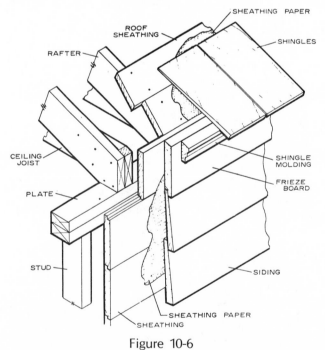

Figure 10-6

Construction details of a "close cornice".

number of lineal feet required should be measured making appropriate allowance for cutting waste.

Unit Costs for Exterior Trim

A unit cost for exterior trim is developed from a quantity of 100 lineal feet of material for each member. The labor to install the material is shown in Table 10-8.

Table 10-8
Approximate Hours of Labor to Install 100 Lineal Feet of Various
Kinds of Exterior Work

Trim	Hours Labor Per 100 Lin Ft	Lineal Feet Per Hour
Water table and drip cap	4	25
Corner boards (2 members)	5	20
Verge board	4	25
2-member closed cornice	8	13
3-member boxed cornice	12	8
Wood gutters	10	10

Basic Formula

100 lin ft trim	= $
Hours carpenter labor @ $	= _____
Unit cost per 100 lin ft	$

Illustrative Example

Assume carpenter wages at $6.00, and the price of materials as used in the examples

Unit Cost for 1" x 8" Water Table

100 lin ft 1" x 6" pine	@$ .25 = $25.00
100 lin ft drip cap	@ .06 = 6.00
4 hours labor (Table 10-8)	@ 6.00 = 24.00
	Unit cost per lin ft = $55.00

$$\text{Unit cost per lin ft } \frac{\$55.00}{100} = \$ \ .55$$

Unit Cost for 2-Member Corner Board

100 lin ft 1 1/8" x 4"	@$ .20 = $20.00
100 lin ft 1 1/8" x 3"	@ .15 = 15.00
5 hours labor (Table 10-8)	@ 6.00 = 30.00
	Unit cost for 100 lin ft = $65.00

$$\text{Unit cost per lin ft } \frac{\$65.00}{100} = \$ \ .65$$

Unit Cost for Closed Cornice
Frieze Molding-(2 member)

100 lin ft 1 1/8" x 8"	@ $ .40 =	$ 40.00
100 lin ft 3" crown molding	@ .20 =	20.00
8 hours labor (Table 10-8)	@ 6.00 =	48.00
	Unit cost for 100 lin ft =	$108.00

$$\text{Unit cost per lin ft } \frac{\$108.00}{100} = \$ \ 1.08$$

Unit Cost for Boxed Cornice
Frieze, Soffit and Facia-(3 member)

100 lin ft 1 1/8" x 10" frieze	@ $ .36 =	$ 36.00
100 lin ft 1" x 10" soffit	@ .30 =	30.00
100 lin ft 1" x 6" facia	@ .18 =	18.00
12 hours labor (Table 10-8)	@ 6.00 =	72.00
	Unit cost for 100 lin ft =	$156.00

$$\text{Unit cost per lin ft } \frac{\$156.00}{100} = \$ \ 1.56$$

DAMAGEABILITY OF ROUGH AND FINISHED CARPENTRY

The types of damage to carpentry items which are most frequently encountered, may be placed in three broad classifications.

1. Damage by fire.
2. Damage by water.
3. Damage by breakage or marring.

Damage by Fire

Wood is combustible, and its rate of burning varies with the density of the species, the moisture content, the bulk or size of the piece, the rate at which heat is supplied, and the amount of air available. Cedar ignites at about 375°F. while long leaf yellow pine ignites at approximately 425 F. Most woods will ignite within this range.

When wood is exposed to heat, it loses moisture which causes it to shrink, warp, or check (split). These conditions may occur even though the heat is insufficient to actually burn or char the wood. The damaging effect of fire or heat on wood will be one of either appearance, structural, or both. A scorched or badly smoked timber, not injured structurally, may require nothing more than scraping and painting where appearance is unimportant. The scorching or slight charring of millwork and trim, however, will usually require replacement, because the spoiled appearance is more important than lack of any structural injury.

Damage by Water

While wood will shrink as it loses moisture, it also swells when it absorbs moisture. Air dried lumber normally contains 12 to 15 percent moisture by weight. Kiln dried lumber, intended for interior millwork, has approximately a 7 percent moisture content. Unprotected wood that is exposed to water or heavily moisture-laden air in a building, expands across the grain. The manner in which the wood is attached or held in place will determine to some extent the amount of deformation. Window and door casings, baseboard, shelving, stair treads and risers may show no signs of warping or twisting when wet. Much will depend on the amount of water and the duration of exposure or submersion. Finish floors, particularly edge-grain, having little or no room to expand sideways, will frequently *ridge* along the edges of each piece producing a washboard effect. In some instances the expanding can result in large areas of the finish floor being thrust upward in great mounds or waves. A floor that is highly finished with paint, varnish, or oil, has less tendency to deform because the water cannot penetrate to the raw wood. If the water comes in contact with the raw wood flooring from the underside, however, there is little to prevent it from soaking into the cells. Because of the susceptibility of finish floors to water damage, prompt measures should always be taken immediately to remove excess surface water by mopping or spreading sawdust. The circulation of air by opening windows or using fans, and also furnishing a moderate amount of heat, generally speeds up the drying out process in order to reduce damage.

The deformation of wood by swelling begins to show up within a matter of hours after exposure, and the ultimate damage may take a few days to disclose itself. For this reason it is well not to attempt to write the specifications for repair until a sufficient period of time has elapsed. If no damage is visible within a week after the water has been removed, it is unlikely that any will occur after that.

Plywood made for interiors does not have waterproof glue between the plies and the veneer and frequently separates if it is permitted to remain directly in water very long, or if it is wet down continuously. Plywoods designed for exterior application are not so affected.

Interior plywood doors are especially susceptible to damage at the bottoms or tops where the water penetrates the unfinished edges.

Exterior woodwork and trim are not usually damaged by water unless submerged, as might be the case during flooding. Damage is then mainly the result of water coming in contact with the unfinished back of the wood.

Wood that is constantly submerged or is continuously kept away from water or moisture will not be attacked by decay or rot. However, intermittent wetting and drying as often occurs to wood near the ground, or to framing members of a roof or to flooring exposed to periodic leaks, will eventually cause rot.

Water in contact with wood may also cause staining which affects the appearance of paneling, trim, and other millwork that has a light or natural finish. The staining may be caused by clean water, or water which carries with it colors or dirt from contents or other parts of the building.

Damage by Breakage or Marring

Breakage of carpentry items includes damage caused by numerous perils other than fire or flood. This includes lightning, explosion, collapse, settlement, slides, earthquake, vehicles, sonic boom, falling trees, windstorm, hail, animals, or occupants. The extent of damage may range from that done by a fireman who chops through a door while gaining entrance to a burning building, to the complete destruction of a frame building by an explosion or a tornado.

METHODS OF REPAIR

Carpentry items that have been damaged will require repairing or replacing, and the specifications will fall into one of the following categories:

1. Replacing completely.
2. Replacing part or parts.
3. Reinforcing.
4. Concealing.
5. Resurfacing.

When taking off the specifications to restore items that are damaged, it is very important to recognize situations where it will be necessary to include the removal and replacement of undamaged adjacent or contiguous structural items. For example, when charred floor joists are to be removed, the subflooring and finished flooring, though slightly or not at all damaged, will also have to be replaced. When rafters are charred, and must be replaced, the roof boards and the roofing will also have to be replaced. Frequently, damaged flooring which runs under the baseboard cannot be properly relaid unless the baseboard is removed and replaced. When charred studding must be replaced, the exterior sheathing and siding, or stucco, though undamaged will usually have to be replaced also. Each situation should be examined carefully to avoid omitting items.

Replacing Completely

The complete replacement of damaged carpentry needs no special explanation, except to point out the need to include all materials and labor necessary to effect proper replacement. For example, when a wall plate or sill has been charred, requiring replacement, and shoring necessary, the removal and putting back of siding or cornice, should be included in the estimate. Unless damaged structural members such as joists, girders, studding and rafters can be satisfactorily concealed or reinforced, it will be necessary to replace them. Millwork and trim items or finished floors that cannot be properly repaired to restore them to their original condition should be replaced.

Replacing Part or Parts

Whenever a door, window, cabinet, or some other type of millwork can be satisfactorily repaired by replacing a stile, panel, or part at less cost than replacing the entire unit, the estimate should be made accordingly. A door, for example, with a broken panel or stile frequently can be repaired for less than the cost of a new door, especially if it is an expensive one.

When the casings, or sash bars in a window have been damaged, the cost of replacing those parts should be compared to the cost of replacing the entire unit.

Finished floors, paneling, or other surface materials that are damaged over a part of the area can usually be repaired by carefully cutting out the damaged section and replacing with new materials.

Reinforcing

The reinforcement of damaged carpentry items applies to structural items. Split or moderately charred timbers frequently can be reinforced by splicing new members alongside those that are damaged. This method of repair is generally considered wherever the cost of replacing damaged timbers greatly exceeds the cost of reinforcing and where such repairs are not objectionable.

Concealing

Many times exposed structural members in a building are heavily smoked, slightly charred or otherwise injured in appearance but not in structural strength. A method of repair which is frequently satisfactory is to conceal the members by scraping and painting, or by covering with a suitable wallboard. Sometimes, when exposed timbers are reinforced, the damaged area is also concealed in this manner.

Damaged rafters and joists lend themselves to this treatment. When the timbers are smoked or charred, the surface is scraped and whitewashed or painted before applying wallboard in order to eliminate any possible odor that might seep through later.

Heavy girders, posts, and truss cords may be scraped and painted when slightly charred, and then boxed-in with 7/8 inch boards to conceal any unsightly appearance.

Resurfacing

Resurfacing by sanding is generally a satisfactory method of repairing floors that have been cupped or ridged by water. Before the sanding operation, it is preceded by any replacement of flooring or patching which is necessary.

Trim and other millwork that has had the surface damaged by heat, smoke, or water may also be sanded down prior to refinishing.

Specific Methods of Repair

The following check-list will serve as a guide when contemplating the repairs to specific carpentry items:

Framing
1. Remove and install new framing members.
2. Reinforce by installing new members alongside.
   (a) Scrape and paint or whitewash charred areas.
   (b) Cover area with wallboard or similar material.
3. Cut out section damaged (as with corner post, sill, plate, etc.) and splice in new member.

Wood Trusses
1. Remove and replace entirely.
2. Remove member (cord, rafter, strut, or brace) and replace with new member.
3. Scrape char (if no structural damage), paint, or box with 7/8 inch boards.
4. Paint or whitewash when only smoked.

Sheathing, Rough Flooring and Roof Boards and Siding
1. Remove and replace entirely.
2. Cut out damaged area and replace with new wood.
3. Paint or whitewash when only smoked.

Windows
1. Remove and replace entirely.
2. Replace sash and hardware only.
3. Replace trim only.
4. Glaze sash only.
5. Refit and hang sash.

Doors
1. Remove and replace frame, door, trim and hardware completely.
2. Replace door and hardware only.
3. Replace door stile or panel.
4. Replace trim only.
5. Glaze door only.
6. Refit door.

Finish Floors
1. Remove and replace entirely.
2. Cut out and patch damaged areas.
3. Sand entire area.
4. Paint, shellac, or varnish.

Baseboard, Picture Molding, Chair Rail
1. Remove and replace.

Built-in Cabinets
1. Remove and replace completely.
2. Replace damaged doors, and hardware as required.

Paneling and Wainscot
1. Remove and replace entirely.
2. Remove and replace damaged stiles, rails, and panels as required.

Stairways
1. Remove and replace entirely.
2. Replace damaged treads or risers.
3. Replace hand rail and balusters as required.
4. Sand treads.

Miscellaneous Exterior Trim
1. Remove and replace units damaged.

Chapter 11.

Interior Wall Finish

INTERIOR WALL BOARDS

This chapter deals with the various types of interior wall finish commonly encountered, excluding wet plaster finishes which are discussed in Chapter 15, under Lathing and Plastering. These wall finishes include gypsum board; the different fibre boards made of wood fibre, cane fibre and wood pulp; plastic coated masonite; the several kinds of plywood finishes; and solid wood paneling.

Most wallboards are available in sheets 4 feet wide by 8 to 10 feet long. Plywoods generally run 8 feet long, and gypsum board runs 6 to 14 feet in length.

Application to studs or furring strips is by nailing, adhesives, or both.

ESTIMATING MATERIALS

An estimate of the materials required to apply composition wallboards should include the furring and blocking that is needed. In many cases where repairs are contemplated, existing furring may be undamaged or simply may require renailing.

The area to be covered is computed after deducting door and window openings to obtain the actual number of square feet of wall and ceiling area. An allowance is then made for fitting and cutting waste; generally 10 percent is adequate for average conditions. Much will depend on the shape, number of openings, size and layout of the room. In actual practice a sketch of each wall is made, from which the number and size of each piece of wallboard may be determined.

ESTIMATING LABOR

Consideration must be given to several factors in estimating the labor necessary to apply wallboard. For example, a man can put on wallboard at a faster rate in a large room than in a small one; cutting and fitting around doors and windows will slow down his rate; and he will put on large sheets at a faster rate per square foot than small sheets, or tile-size pieces. Wallboards with high-grade finishes require more care and consequently more time than the less expensive types where joints are to be covered with wood strips that conceal poor joining or hammer marks. Placing wallboards on ceilings takes more time per square foot than on sidewalls. Light insulating type wallboards are easier to handle than heavier kinds, some of which take two men to carry and put in place.

DRYWALL CONSTRUCTION

Interior walls and ceilings that are finished with gypsum wallboard in place of lath and wet plaster is called "drywall" construction in the building trade. More than seventy-five percent of all new residential buildings have drywall construction.

Sheetrock SW Wallboard

Sheetrock, or gypsum board, is essentially the same type of material described under gypsum lath in Chapter 15, except the paper on the finished side is less porous and is harder and smoother for wearing quality and appearance. It is made in four thicknesses:

5/8", recommended for the finest single layer drywall construction. The greater thickness provides increased resistance to fire exposure and transmission of sound.

½", for single layer application in new residential construction.

3/8", lightweight, applied principally in the double wall system and in repair and remodel work. Not recommended over metal framing.

Width: 4'; length; 8', 9', 12' or 14'; edges: rounded, tapered; finish: ivory manila paper, suitable for paint or other decoration.

Tapered Edge SHEETROCK has long edges tapered on the face side in order to form a shallow channel for the joint reinforcement which provides smooth, continuous wall and ceiling surfaces. Made in the same thicknesses as SHEETROCK SW Wallboard and in one other thickness:

¼", a lightweight, low cost, utility gypsum wallboard, for use over old wall and ceiling surfaces.

Width: 4'; length: 8', 10', edges: tapered; finish: ivory manila paper, suitable for paint, wallpaper or other decoration.

SHEETROCK W/R Gypsum Wallboard is a water-resistant gypsum wallboard that provides an excellent base for the adhesive application of ceramic, metal, and plastic tile. It is water-resistant all the way through: (1) multi-layered and back paper is chemically treated to combat penetration of moisture; (2) the gypsum core is made water-resistant with a special asphalt composition. It was developed for application in bathrooms, powder rooms, kitchens, utility rooms, and other high moisture areas. SHEETROCK W/R Gypsum Wallboard is easily recognized because of its distinctive green face.
In addition to its use as a superior tile base in new construction, SHEETROCK W/R Wallboard is a cost-saver in modernization work. It permits new tilework to be installed over existing surfaces without tearing out old walls.
This back-up for wall tile is available in plain core, ½" and 5/8" thickness.

WALLBOARD TRIM ACCESSORIES

In order to provide true and straight lines for smooth finishing at outside wallboard corners, metal trim accessories are available as corner reinforcements, casing beads, metal trims, control joints and decorative moldings.

PREDECORATED PANELS

Gypsum panels are available, predecorated with a vinyl faced lamination in a variety of patterns making on the job decorating unnecessary.

APPLICATION

Gypsum board may be applied to sidewalls vertically or horizontally; the latter is recommended in rooms where full-length sheets can be used, as it reduces the number of vertical joints. Nails for 3/8 and 1/2-inch sheets are five penny (1 5/8 inches). Spacing of nails is 6 to 8 inches. Where the adhesive nail-on application is used, nail spacing is considerably more. Table 11-1 shows nails needed.

TABLE 11-1
NAIL SCHEDULE FOR GYPSUM BOARD

Thickness	Type Nail	Lbs per 1000 Sq Ft
1/4"	1 1/4" coated	3
3/8" and 1/2"	1 5/8" coated	5
5/8"	1 7/8" coated	5

JOINT TREATMENT

Joints are concealed by perforated tape and joint compound. On large jobs the work is done with a taping machine whereas on small jobs it is done by hand. Approximately 50 lbs. of joint compound and 370 feet of PERF-A-TAPE are required to finish 1,000 sq. ft. of SHEETROCK Gypsum Wallboard (See Table 11-2 for quantities needed).

TABLE 11-2
TAPE AND COMPOUND SCHEDULE

Square Feet Wallboard	Joint Compound	Rolls of Tape
100-200	1 gallon	2-60 ft rolls
300-400	2 gallon	3-60 ft rolls
500-600	3 gallon	1-250 ft roll
700-800	4 gallon	1-250 ft roll and 1-60 ft roll
900-1000	5 gallon	1-500 ft roll

Note: An alternative of 50 lbs of powder joint compound per 1,000 sq. ft. may be used.

ESTIMATING LABOR

An experienced man should place 3/8" or 1/2" gypsum wallboard at the rate of 100 sq. ft. per hour on side walls, and at 100 sq. ft. per 1 1/2 to 2 hours on ceilings in average size rooms. In small rooms or where much cutting and fitting is necessary, the rate of application will range from 50-75 sq. ft. per hour on side walls and ceilings. A rule of thumb for average work is 2 hours per 100 sq. ft. including taping on walls and ceilings.

Most prefinished gypsum wallboards are estimated at the same labor rates as unfinished depending upon the amount of cutting and matching necessary.

TWO-PLY GYPSUM WALLBOARD

Two-ply gypsum wallboard consists of two layers of 3/8" board with the second layer applied with an adhesive. Labor costs can be estimated at about twice that of single-ply.

Unit Costs for Drywall

A unit cost is developed for drywall by adding to the cost of 100 square feet of material, the cost of labor based on local hourly wages. Adjust the rate of installation for unusual job conditions.

Illustrate Examples
(Labor and Material prices assumed)

½-inch Gypsum wallboard on studding

100 sq ft wallboard	@$6.00 =	$ 6.00
1 hour labor	@ 6.00 =	6.00
Unit cost per 100 sq ft	=	$12.00
Unit cost per sq ft	= $	.12

Taping and Finishing Joints

Material per 100 sq ft	@$1.00 =	$ 1.00
1½ hours labor per 100 sq ft	@ 6.00 =	9.00
Unit cost per 100 sq ft	=	$10.00
Unit cost per sq ft	= $	.10

Average For ½-inch Gypsum Wallboard on Walls and Ceilings Including Taping Joints

100 sq ft wallboard	@$6.00 =	$ 6.00
2½ hours labor	@ 6.00 =	15.00
Unit cost per 100 sq ft	=	$21.00
Unit cost per sq ft	$	.21

When the adhesive nail-on method is used, the adhesive is applied to joists or studs before positioning the wallboard. Approximately 8 quart size tubes per 1,000 sq. ft. is needed.

DRYWALL CEILING FINISH

A number of products are available for finishing drywall ceilings economically and artistically. These are either sprayed on, or applied with a roller and stippled, swirled or raked to form pleasing patterns. They come in powdered form and are mixed with water.

Perfect Spray contains shredded polystyrene aggregate and is packaged in 50 lb bags. Mixed with 8 gallons of water this will cover 8 to 10 sq. ft. per lb or 400 to 500 sq. ft.

Quick Spray—Bestex C is sprayed on and comes in medium-fine, medium and coarse powder form.

Texture contains no aggregate and can be sprayed or applied with a roller.

STILTS AND TAPING MACHINES SAVE TIME

Adjustable stilts for workmen which replaces scaffolding are shown in use on page 285. Workmen quickly become accustomed to these and they considerably reduce the time to put on sheetrock. A taping machine is also shown which greatly reduces the hours to apply the tape and compound.

FIBER BOARD

The various types and finishes on fiber base wallboards are too numerous to mention. The sheets are prefinished and are usually 4 ft. wide and 8 ft. long with 6, 10, and 12 ft lengths available on order. Thicknesses range from 1/8" to 3/8" with insulating type sheets 1/2" thick. Some of the well known trade names are Celotex, Fir-Tex, Nu-wood, Presdwood, Masonite, Upson-Board, Tile-board, Asbestos-Board, and Hardboard.

Application varies and may be by nails colored to match the finish, by adhesive, or by both.

Matching inside and outside corner moldings, base molding, and joint strips are available in plastic or wood; metal is also used.

ESTIMATING COSTS

Rough estimates of the quantity of fiber wallboard needed can be made by computing the area of wall space and adding a judgment percentage for waste in cutting and fitting. A more accurate determination can be made by laying out each room on paper and determining the exact number of panels and moldings required.

Labor costs will vary with the type and quality of fiber board, the size of room, and the amount of cutting and fitting. On average work a carpenter should apply 100 sq ft in 1 1/2 to 2 hours on side walls. On smaller rooms and where much cutting and fitting is required the time per 100 sq ft should be increased to 2 1/2 to 3 hours. Application of moldings over joints and corners will require an average of 2 hours per 100 lin ft.

WALL AND CEILING INSULATION TILE

Fiber composition tiles are used for insulating as well as decorative purposes on walls and ceilings. They come in sizes ranging from 12" x 12" to 16" x 32", by 1/2" thick. They are applied by use of adhesives or by concealed nails or staples. Numerous designs and finishes are available.

Tiles may be applied directly over old walls with adhesives or over furring nailed to studs, rafters or over old walls and ceilings.

Figure 11-1

Applying tape to drywall working on adjustable stilts. Courtesy Goldblatt Trowel Trade Tools, Kansas City, Kansas 66110

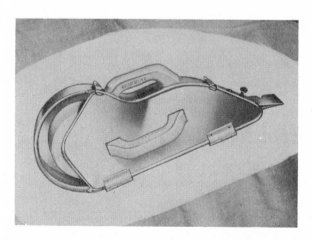

Figure 11-2

Taping Machine used to apply tape to drywall. Courtesy Goldblatt Trowel Trade Tools, Kansas City, Kansas 66110

Furring Strips on Ceilings

Furring strips are usually 1" x 2" nailed across ceiling joists 12 inches on center. A carpenter should average 100 lin. ft. of furring every 3 hours. A unit cost, using the following assumed prices may be developed as follows:

Illustrative Example

1" x 2" Wood furring applied over ceiling joists

100 lin ft furring	@$ .03 =	$ 3.00
3 hours labor	@ 6.00 =	18.00
Unit cost per 100 lin ft	=	21.00
Unit cost per lin ft	=	$ .21

Furring Strips on Studding

Furring on side walls over studs or over old walls can be applied at an average rate of 100 lin ft every 2 hours. If much leveling and shimming is necessary to obtain a smooth surface the time should be increased accordingly.

Unit Costs for Ceiling Tile on Furring

A unit cost may be developed for ceiling tile by adding to the cost of 100 square feet of material, the cost of labor based on the local hourly wages times the number of hours per 100 square feet shown in Table 11-3. The rate of installation may be adjusted for unusual job conditions other than average.

Installing 12" x 12" insulating stapled tile on ceiling furring

100 sq ft insulating tile	@$ .15 =	$15.00
5 hours labor	@ 6.00 =	30.00
Unit cost per 100 sq ft	=	$45.00
Unit cost per sq ft	=	$ .45

PLYWOOD WALLBOARD

Prefinished plywood, a popular interior finish comes in sheets 4 feet wide and 6 to 12 feet long. Thicknesses range from 3/16 to 1-inch but the 1/4 inch is most commonly used.

Plywood may be placed over bare studding, studding with 3/8" or 1/2" gypsum board backing—the joints of which may or may not be taped, or it may be placed over furring on studs or over old walls. Nails, colored to match the finish, are used, or adhesive, or both nails and adhesive.

TABLE 11-3

Hours Required To Install Composition Wallboard, Plywood, and Ceiling Tile.

Type of Work Done	No. Sq. Ft. Per Hour	Hours Per 100 Sq. Ft.
Furring on studding 12″ center		2
Furring on clg. joists 12″ center		3
3/8″ or ½″ Sheetrock on studs	100	1
3/8″ or ½″ Sheetrock on clg. joists	50-70	1½-2
2-Ply, 3/8″ Sheetrock on studs	50	2
Taping and finishing joints	100	1½
Wood strips over joints		1
Fiber board (Celotex etc)		1½-2
Ceiling tile (12″ x 12″) on furring, stapled		5
Ceiling tile (12″ x 12″) on flat surface-adhesive (for large size tile reduce time accordingly)		2
Plywood, prefinished, ¼″ on studs including stripped joints and corners		2 - 3
Plywood, prefinished, ½″ flush joint on studs		6 - 8

Prices per panel vary considerably and it is very important when estimating to be certain of the thickness, and quality of plywood under consideration.

Joints may be flush or V-grooved, but are very often concealed with panel strips at joints and corners.

UNIT COSTS

The following are illustrations of unit costs for prefinished plywood panels using assumed prices for material and wages.

Inexpensive ¼-inch plywood paneling on studding-
  stripped joints
   100 sq ft ¼″ plywood   @$ .20 = $20.00
      2 hours labor   @ 6.00 =   12.00
         Unit cost per 100 sq ft   = $32.00
         Unit cost per sq ft   = $ .32

High grade ½-inch plywood paneling on studding-
  flushed joints
   100 sq ft ½″ plywood   @$ .60 = $ 60.00
      8 hours labor   @ 6.00 =   48.00
         Unit cost per 100 sq ft   = $108.00
         Unit cost per sq ft   = $ 1.08

SOLID WOOD PANELING

Solid wood paneling of knotty pine, cherry, cypress and so forth is usually nailed to furring applied over studding or masonry. There may or may not be an underlayment of composition wallboard for sealing and insulating purposes.

Most paneling consists of 3/4 or 7/8-inch boards 6, 8, or 10 inches wide. They may all be in one width or of random widths. The boards have tongue and grooved or shiplapped edges for joining and concealed nailing. Where it is necessary to facenail, 6 or 8 penny finish nails are used.

The material is estimated by obtaining the net wall area, deducting windows and door openings, and adding the appropriate amount of waste for the width of boards used.

The labor to apply solid wood paneling varies with the size and layout of the area involved, and with the kind of wood used. For average conditions a carpenter should be able to install 1,000 board feet in approximately 40 hours or at the rate of 4 hours per 100 board feet.

Illustrative Example

Installing 1" x 8" Knotty Pine Paneling
1,000 FBM knotty pine	@$300.00 = $300.00
40 hours labor	@ 6.00 = 240.00
Unit cost per 1,000 FBM	= $540.00
Unit cost per BF	= $ .54

Chapter 12.

Exterior Wall Finish

Exterior wall finishes are generally of wood, metal or composition materials. In colder climates the materials are put on over an insulating underlayment of wood or composition sheathing. In temperate climates, on summer bungalows, and secondary buildings, it is common to apply exterior finish directly over the studs, without sheathing.

WOOD SIDING

There are several types of wood siding the most common being *bevel* siding, *drop* siding, *vertical matched* siding, *batten or rough sawed* siding, and plywood of various finishes.

Bevel siding comes in nominal widths from 4 to 10 inches. The wood may be pine, redwood, cedar or fir. As its name implies, bevel siding is bevel-cut, tapering from about 3/16 inch at the top edge, to 9/16 inch or 11/16 inch at the butt edge; the latter is for boards over 6 inches wide.

It is applied horizontally, one board overlapping the one below by 1/2 inch for narrow siding, to 3/4 inch for wider boards. This leaves an exposure which varies from 3 to about 8 inches. The exposure is the distance from the butt edge of one board to the butt edge of the one which overlaps it. Bevel siding is usually applied with 6d or 8d common nails depending on the thickness of the siding. For high grade work, the nails are set with a nail

289

set, and the nailholes are later puttied before painting. Aluminum nails need no setting or puttying.

At the corners of a building, bevel siding may be joined either by mitering, overlapping ends, or by the use of a patented metal clip that conceals the square end-cuts of both boards. The use of corner boards is not as common in applying bevel siding as when drop siding or rustic siding is used. It takes more time to miter corners and, therefore, that method is more expensive than either using metal clips, or installing corner boards against which the bevel siding butts.

The quantity of bevel siding needed to cover an area is determined by computing the number of square feet in the area, deducting the openings, and adding sufficient waste to take care of the loss in milling, lapping, cutting, and fitting on the job. The percent of milling and lapping waste to be added *increases* as the width of the siding *decreases*. Bevel siding 6 inches wide has a waste for milling, lapping and cutting of about 35 percent, while bevel siding that is 4 inches wide has a waste of about 50 percent. Table No. 12-1 shows the quantity of nails required, and the approximate waste to be added for various sizes and types of siding.

Drop siding is made in a number of patterns, and it is known by various names including novelty siding and rustic siding. Log cabin siding is a modification of drop siding which is designed to give a log cabin appearance. Drop siding comes in nominal widths ranging from 4 to 9 1/2 inches, with thicknesses from 9/16 to 3/4 inches. The exposure is fixed by reason of the ship-lapped or tongue and groove edge where the boards are joined.

Corner boards are used with the various types of drop siding rather than mitering or using metal clips as in the case of bevel siding. The nails used are generally 8d common and they should be set and puttied before painting.

As with bevel siding, the quantity of drop siding required is obtained by deducting the window and door openings from the area and adding a percentage for waste. Table No. 12-1 shows the normal percentage to be added.

Vertical matched siding is the application of matched boards vertically. This is usually nailed over horizontal furring strips to provide good nailing. The waste in milling, cutting, and fitting for vertical siding depends on the width of the boards. Table 12-1 shows the approximate percentage to be added. Each board is toe-nailed with 6d or 8d finishing nails. Wide boards which are inclined to cup, are surface nailed and the nails are set before finishing.

Table 12-1
Percentage of Milling and Cutting Waste, and Nails Required
Per 1,000 FBM of Various Types of Wood Siding

Type of Siding	Nominal Size in Inches	Lap in Inches 1" lap	Pounds Nails Per 1,000 FBM	Percentage of Waste
Bevel Siding	1 x 4	1	25-6d common	63
	1 x 6	1	25-6d Common	35
	1 x 8	1¼	20-8d common	35
	1 x 10	1½	20-8d common	30
Rustic and drop siding	1 x 4	Matched	40-8d common	33
	1 x 6	Matched	30-8d common	25
	1 x 8	Matched	25-8d common	20
Vertical siding	1 x 6	Matched	25-8d finish	20
	1 x 8	Matched	20-8d finish	18
	1 x 10	Matched	20-8d finish	15
*Batten siding	1 x 8	Rough	25	5
	1 x 10	Rough	20	5
	1 x 12	Rough	20	5
	1 x 8	Dressed	25	13
	1 x 10	Dressed	20	11
	1 x 12	Dressed	20	10
Plywood siding	¼	Sheets		5-10
	3/8	Sheets	15 per MSF	5-10
	5/8	Sheets		5-10

* For 1" x 10" boards allow 1,334 lineal feet 1" x 2" joint strips for each 1,000 FBM of batten siding. Add 12 pounds 8d common nails.

Batten siding is a type that is very common on farm buildings. Wide 1-inch, square-edge boards are placed vertically. The joints are covered with 1" x 2" strips that are nailed to only one of the vertical boards to permit them to expand and contact without splitting. The amount of waste to be added for milling will depend upon the kind of boards used; i.e., whether they are rough sawed, or square edge dressed. Waste for cutting and fitting will run about 5 to 10 percent. Rough sawed lumber has no milling waste. For milling and cutting waste for dressed square edge boards see Table 12-1.

Plywood siding is becoming more popular as an exterior finish on houses. The sheets of plywood should be a minimum of 3/8 inch thick where the walls are not sheathed. Where studding is spaced 20 or 24 inches on center thickness of 1/2 to 5/8 inch are recommended. Nailing is done with 6d or 8d nails depending on the thickness of the material. The only waste to be computed in figuring the quantity of plywood siding is that which is lost in cutting and fitting. Approximately 5 to 10 percent cutting waste is adequate in most cases. The quantity of siding required is computed by measuring the over-all area and deducting the window and door openings.

UNIT COSTS FOR WOOD SIDING

The quantity of solid siding used to develop a unit cost is 1,000 FBM. The cost of the nails is added, and also the cost of labor, determined from Table 12-2 which shows the number of hours required to install 1,000 FBM.

Table 12-2
Hours of Labor Required to Install 1,000 FBM of Various
Types of Wood Siding

Type of Siding	Nominal Size in Inches	Board Feet Per Hour	Hours Labor Per 1,000 FBM
Bevel siding	1 x 4	25	40
(add 10% for mitered	1 x 6	26	38
corners)	1 x 8	28	36
	1 x 10	30	34
Rustic and drop	1 x 4	32	32
siding	1 x 6	33	30
	1 x 8	36	28
Vertical matched	1 x 4	25	40
siding	1 x 6	28	36
	1 x 8	32	32
	1 x 8	45	22
*Batten siding	1 x 10	50	20
	1 x 12	50	20
	¼" thick	83	12
Plywood siding	3/8" thick	71	14
	5/8" thick	71	14

*Allow 15 hours labor per 1,000 FBM of siding to apply joint strips when 1" x 10" boards are used.

Illustrative Examples
(Prices and Wages Assumed)
½" x 8-Inch bevel siding

1,000 FBM 1" x 8" cedar siding	@ $250.00 = $250.00
20 lbs nails	@ .20 = 4.00
36 hours labor	@ 6.00 = 216.00
	Unit cost per 1,000 FBM = $470.00

$$\text{Unit cost per BF} = \frac{\$470.00}{1,000} = \$ \quad .47$$

6-Inch drop siding

1,000 FBM 1″ x 6″ fir drop siding	@ $220.00 =	1220.00
25 lbs nails	@ .20 =	5.00
30 hours labor	@ 6.00 =	180.00
	Unit cost per 1,000 FBM =	$405.00

$$\text{Unit cost per BF} = \frac{\$405.00}{1{,}000} = \$ \quad .41$$

½″ x 8-Inch redwood siding

1,000 FBM ½″ x 8″ redwood siding	@ $350.00 =	350.00
20 lbs nails	@ .20 =	4.00
36 hours labor	@ 6.00 =	216.00
	Unit cost per 1,000 FBM =	$570.00

$$\text{Unit cost per BF} = \frac{\$570.00}{1{,}000} = \$ \quad .57$$

5/8-Inch Plywood siding

1,000 FBM plywood siding	@ $350.00 =	$350.00
20 lbs nails	@ .20 =	4.00
20 hours labor	@ 6.00 =	120.00
	Unit cost per 1,000 FBM =	$474.00

$$\text{Unit cost per BF} = \frac{\$474.00}{1{,}000} = \$ \quad .47$$

10-Inch Batten siding

1,000 FBM 1″ x 10″ sq edge spruce	@ $150.00 =	$150.00
25 lbs nails	@ .20 =	5.00
1,334 lin ft 1″ x 2″ joint strips		
222 FBM	@ 120.00 =	26.64
12 lbs nails for 1″ x 2″ strips	@ .25 =	3.00
20 hours labor for boards	@ 6.00 =	120.00
15 hours labor for strips	@ 6.00 =	90.00
	Unit cost per 1,000 FBM =	$394.64

$$\text{Unit cost per BF} = \frac{\$394.64}{1{,}00} = \$ \quad .39$$

Application of Unit Costs for Wood Siding

Assuming that the over-all dimensions of the windows in the Figure 12-1 are 3' x 6', and that the door is 4' x 7'; the net area of the two sides shown would be:

Front: 12' x 34'   = 408 sq ft
Less openings:
2(3' x 6') = 36 sq ft
   4' x 7' = 28 sq ft   64 sq ft
                                        344 sq ft

Gable end: 12' x 22'   = 264 sq ft
Plus gable 11' x 8'   =   88 sq ft
                              352 sq ft

Less Openings:
3(3' x 6')   =   54 sq ft

                                          298
    Total net area   = 642 sq ft

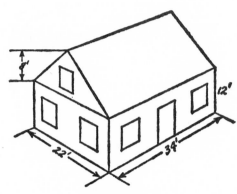

Figure 12-1

Using the unit costs previously developed, the cost of installing wood siding on the two sides of this house would be:

8-Inch bevel siding

Area   = 642 sq ft
Add waste (Table 12-1) 35%   = 225 sq ft
    Amount of siding required   867 FBM
        Unit cost @   .47
        Total cost   = $407.49

6-Inch drop siding

Area   = 642 sq ft
Add waste (Table 12-1) 25%   = 161 sq ft
    Amount of siding required   = 803 FBM
        Unit cost@   .41
        Total cost   = $329.23

5/8-Inch Plywood siding

Area   = 642 sq ft
Add waste 10%   = 64 sq ft
    Amount of siding required   = 706 FBM
        Unit cost@   .47
        Total cost   = $331.82

½ x 8-Inch Redwood siding

Area	= 642 sq ft
Add waste (Table 12-1) 35%	= 225 sq ft
Amount of siding required	= 867 FBM
Unit cost@	.57
Total cost	= $494.19

10-Inch Batten siding

Area	= 642 sq ft
Add waste (Table 12-1) 20%	= 128 sq ft
Amount of siding required	= 770 FBM
Unit cost @	.39
Total cost	= $300.30

WOOD SHINGLE SIDING

Wood shingles are a very popular type of finish for both exterior side walls on frame buildings, and as a roofing surface. In congested areas of many of the larger cities, wood shingle roofing is prohibited by ordinance because of its combustible nature and the possibility of the spreading of a fire by embers, from one residence to another. However, in most suburban and rural areas they are widely used because of their long life (25 to 35 years) and architectural attractiveness.

Wood shingles are manufactured from several kinds of wood, the best of which are western red cedar, redwood and cypress. All of these have a high decay resistance, low shrinkage, and a tendency to lay flat on the surface. The better grades of shingles are of clear vertical (edge) grain wood that consist entirely of heartwood with no defects. The intermediate and the economical grades are clear on the lower 6 inches but may contain defects above such as knots, feather tips, flat grains or some sapwood. These are recommended more for farm buildings, sidewalls of garages, summer homes, and secondary buildings. They are also used for under-courses where double coursing of side walls is contemplated. Grading rules of wood shingles, as with all lumber products, have been formulated by the Bureau of Standards of the U.S. Department of Commerce in cooperation with the industry.

The three standard lengths of wood shingles are 16-inch, 18-inch, and 24-inch. The standard exposure for 16-inch shingles is 5 inches; for 18-inch it is 5 1/2 inches; and for 14-inch, 7 1/2 inches. There are four bundles of shingles to a square. The Red Cedar Shingle Bureau recommends for roofs with a pitch less than 5/24ths, that the shingle exposure be reduced to 3 3/4 inches, 4 1/4 inches and 5 3/4 inches respectively for 16-, 18-, and 24-inch shingles. On all roofs, there should be a minimum of *three* layers of wood at every point to prevent leakage in wind-driven rain storms. On side walls, two layers of wood shingle at every point is considered adequate.

The standard thickness of four butts of 16- and 24-inch shingles is 2 inches. The standard thickness of four butts of 18-inch shingles is 2 1/4 inches.

Table 12-3 shows the covering capacities, for Certigrade Red Cedar Shingles. Since the manufacturer specifies 5, 5 1/2 and 7 1/2 inches respectively for the exposure to apply one square of shingles, any variation of the exposure so specified will result in a square of shingles covering more, or less, than 100 square feet of surface. For example, a 16-inch shingle is designed for a 5-inch exposure and will cover 100 square feet.

Table 12-3
Percentage of a Square of Shingles Required to Cover 100 Square
Feet for Exposures Shown

(1) Shingle Length	(2) Exposure In Inches	(3) Percentage of 1 Square of Shingles For Each 100 Square Feet of Area
	4	125
	*5	100
16 Inch	6	84
	7	72
	**7½	67
	4	123
	*5½	100
	6	92
18 Inch	6½	85
	7	79
	8	70
	**8½	65
	6	125
	*7½	100
	8	94
24 Inch	9	84
	10	75
	11	69
	**11½	66

* Maximum exposure recommended for roofs by the Red Cedar Shingle & Handsplit Shake Bureau.

** Maximum exposure recommended for side walls by the Red Cedar Shingle & Handsplit Shake Bureau.

When the exposure for a 16-inch shingle is increased to 7 1/2 inches one square will cover 150 square feet. Putting it another way, if an exposure of 7 1/2 inches is used, the quantity of shingles needed to cover 100 square feet will be 100/150=67% of a square. It becomes very important, therefore, when estimating the wood shingles required, to make certain of the exposure to be used for the particular length shingle. *Both the quantity of material and the required labor to cover 100 square feet decreases as the exposure is increased and vice versa.*

The length and type nail used to apply wood shingles is important, not more than two nails being used to a shingle. Nails must be rust-resistant, and may be zinc coated or aluminum. For new work the 3d nail is used on 16- and 18-inch shingles, while the 4d nail is recommended for the 24-inch shingle. When new shingles are being applied over old shingles, the 5d or 6d nails are used. For double coursing on side walls the 5d nail is considered best. If shingles are applied on side walls directly over composition sheathing without nailing strips, a special self-clinching nail should be used in accordance with the specifications of the manufacturer.

Shingles are available in either natural wood or stained for roofs, and painted a prime coat for side walls.

APPLYING WOOD SHINGLES TO SIDE WALLS

Wood shingles can be applied on side walls over open sheathing, solid sheathing, or composition types of sheathing.

Open, or spaced sheathing, on centers equal to the exposure used, is common in the warm climates and on seasonal bungalows or farm buildings. The strips may be 1" x 2", 1" x 3", or 1" x 4" nailed to the studs. Saturated felt building paper is first applied between the nailing strips and the studs.

When side wall sheathing is solid boarding, or plywood, a single layer of saturated felt paper is applied before the shingles are put in place.

Composition sheathing requires either stripping, as with open sheathing, or the use of clincher nails if the sheathing is designed for their use. Where composition sheathing has a waterproof surface, it is unnecessary to apply saturated felt paper under the shingles.

There are two general methods of side wall application. The *Single Coursing* and the *Double Coursing*. Because the sides of a building are less exposed to the direct action of rain, snow, and sun, the shingle exposure can be considerably greater (Table 12-3) than on roofs.

In Single Coursing the first or lowest course is doubled, and succeeding courses are applied above, using a tacked lath as a guide for the butts or by striking a chalk-line.

Table 12-4
COVERING CAPACITIES AND APPROXIMATE NAIL REQUIREMENTS OF CERTIGRADE RED CEDAR SHINGLES

Shingle Exposure in Inches	No. 1 GRADE SIXTEEN INCH SHINGLES — NAIL SIZE: 3d, 1¼-inch long — Four-Bundle Square: Coverage in Sq. Ft.	Four-Bundle Square: Pounds Nails	One Bundle: Coverage in Sq. Ft.	One Bundle: Pounds Nails	No. 1 GRADE EIGHTEEN INCH SHINGLES — NAIL SIZE: 3d, 1¼-inch long — Four-Bundle Square: Coverage in Sq. Ft.	Four-Bundle Square: Pounds Nails	One Bundle: Coverage in Sq. Ft.	One Bundle: Pounds Nails	No. 1 GRADE TWENTY-FOUR INCH SHINGLES — NAIL SIZE: 4d, 1½-inch long — Four-Bundle Square: Coverage in Sq. Ft.	Four-Bundle Square: Pounds Nails	One Bundle: Coverage in Sq. Ft.	One Bundle: Pounds Nails
3½	70	2 7/8	17½	¾								
4	80	2½	20	5/8	72½	2½	18	5/8				
4½	90	2¼	22½	5/8	81½	2¼	20	5/8				
5	**100***	2	**25**	½	90½	2	22½	½				
5½	110	1¾	27½	½	**100***	1¾	**25**	½				
6	120	1⅔	30	3/8	109	1⅔	27	3/8	80	2⅓	20	5/8
6½	130	1½	32½	3/8	118	1½	29½	3/8	86½	2 1/8	21½	½
7	140	1 2/5	35	⅓	127	1 2/5	31½	⅓	93	2	23	½
7½	**†150**	1⅓	**37½**	⅓	136	1⅓	34	⅓	**100***	1 7/8	**25**	½
8	*160*		*40*		145½	1¼	36	⅓	106½	1¾	26½	½
8½	*170*		*42½*		**†154½**	1¼	**38½**	¼	113	1⅔	28	½
9	*180*		*45*		*163½*		*40½*		120	1½	30	3/8
9½	*190*		*47½*		*172½*		*43*		126½	1½	31½	3/8
10	*200*		*50*		*181½*		*45*		133	1½	33	3/8
10½	*210*		*52½*		*191*		*47½*		140	1⅓	35	⅓
11	*220*		*55*		*200*		*50*		146½	1¼	36½	⅓
11½	*230*		*57½*		*209*		*52*		**†153**	1¼	**38**	⅓
12	**‡240**		**60**		*218*		*54½*		*160*		*40*	
12½					*227*		*56½*		*166½*		*41½*	
13					*236*		*59*		*173*		*43*	
13½					*245½*		*61*		*180*		*45*	
14					**‡254½**		**63½**		*186½*		*46½*	
14½									*193*		*48*	
15									*200*		*50*	
15½									*206½*		*51½*	
16									**‡213**		**53**	

* Maximum exposure recommended for roofs.

† Maximum exposure recommended for single-coursing on side walls.

‡ Maximum exposure recommended for double-coursing on side walls. Figures in italics are inserted for convenience in estimating quantity of shingles needed for wide exposures in double-coursing, with butt-nailing. In double-coursing, with any exposure chosen, the figures indicate the amount of shingles for the outer courses. Order an equivalent number of shingles for concealed courses. Approximately 1½ lbs. 5d small-headed nails required per square (100 sq. ft. wall area) to apply outer course of 16-inch shingles at 12-inch weather exposure. Plus ½ lb. 3d nails for under course shingles. Figure slightly fewer nails for 18-inch shingles at 14-inch exposure.

From *Certigrade Handbook of Red Cedar Shingles*, published by Red Cedar Shingle Bureau (Red Cedar Shingle and Handsplit Shake Bureau), Seattle, Washington, 1957.

Double Coursing is principally used where a deep shadow line is desired for architectural appeal. Considerably greater exposures are possible with this form of application and, for the under course, a lower grade shingle is used. The starting course should be tripled. The outer course shingles are applied about 1/2 inch lower than the under course. In other words the butt of the outer shingle is about 1/2 inch below the butt of the under shingle.

Where shingles join on an outside corner, they may be mitered or they may be laced by one overlapping the other. Inside corners are finished by jointing the shingles up against a vertical strip which is installed prior to shingling. Mitering of the outside corners takes a little more time than lacing, making the labor cost slightly higher.

Side wall shingles can be applied over old shingles, and also over old stucco or brick, provided horizontal furring or nailing strips are securely attached on centers equal to the shingle exposure.

ESTIMATING MATERIAL

The quantity of shingles needed for the side walls of a building is determined by measuring the actual area to be covered, and deducting all window and door openings to obtain the net area in square feet. By consulting Table 12-3 the portion of a square of shingles needed to cover each 100 square feet can be determined for the corresponding exposure. The quantity should be carried out to the next 1/4 of a square (bundle), as shingles are sold four bundles to the square.

The normal waste allowance for starting courses and for cutting and fitting around windows and doors is 10 percent.

Illustrative Example
To obtain the number of squares of 18 inch shingles required to cover a measured area, using an 8½ inch exposure.

Measured Area: 20' x 30'   600 sq ft
Deduct 2 wds. (3' x 5') 30 sq ft
Deduct 1 dr. 3' x 7'   21 sq ft   51 sq ft
                                    549 sq ft
Add 10% waste   55
                                    604 sq ft

Table 12-3 shows that it takes 65% of a square of 18-inch shingles laid with an 8½ inch exposure to cover 100 square feet.
.65 x 604 = 4.0 squares required

The number of squares of shingles is obtained by multiplying the net area to be covered by the percentage shown in column No. 3, Table 12-3.

LABOR FOR WOOD SHINGLES ON SIDEWALLS

Under average conditions a *carpenter* can apply wood shingles to sidewalls at the rate of about 5 hours per square for 16- and 18-inch shingles, and 4 hours per square for 24-inch shingles. At this rate, and using the exposures recommended by the manufacturer, he will cover 100 square feet of wall area. When conditions are encountered that affect his rate of application, the hours per square may be adjusted accordingly. For example, specifications may call for staggering the butts, or there may be an unusual number of openings, jogs or setbacks to cut and fit around. These factors should be given consideration in the labor rate.

Experienced shinglers can apply shingles as much as 25 percent faster than carpenters who are not accustomed to that type of work. When experienced shinglers do the work, the labor rate may be reduced.

UNIT COSTS FOR WOOD SHINGLES ON SIDEWALLS

Based on the labor rates given, unit costs for shingles may be developed as shown in the following illustrations. Prices and wages are assumed.

*Unit Cost for 16-and 18-Inch Shingles on
Sidewalls with 5 and 5 ½ Inch Exposures*

1 square shingles	@ $24.00 =	$24.00
2 lbs nails	@ .20 =	.40
5 hours carpenter labor	@ 6.00 =	30.00
	Unit cost per square =	$54.40

$$\text{Unit cost per sq ft} \frac{\$54.40}{100} = \$ \ .54$$

*Unit Cost for 24-Inch Shingles on
Sidewalls with 7 ½ Inch Exposure*

1 square shingles	@ $24.00 =	$24.00
2 lbs nails	@ .20 =	.40
4 hours carpenter labor	@ 6.00 =	24.00
	Unit cost per square =	$48.00

$$\text{Unit cost per sq ft} \frac{\$48.00}{100} = \$ \ .48$$

The illustrated unit cost may be applied to the number of square feet of wall area to be covered only when the manufacturer's specified exposure is being used; otherwise the following rules should be noted.

1. When the developed *unit cost per square* is used, the proper number of squares required must be determined from Table 12-3 as illustrated under "Estimating Materials."
2. When the developed *unit cost per square foot* is used, the *unit cost* is multiplied by the factor in Table 12-3 corresponding to the exposure. This adjusted unit cost per square foot may then be applied to the square feet of wall area including the waste to be added.

Illustrative Example

The wall area to be covered, including waste, is 2,000 square feet. The shingles are 18-inch with an 8-inch exposure. Table 12-3, column 3 shows that 70 per cent of a square of 18-inch shingles laid with an 8-inch exposure will cover 100 square feet.

Therefore (.70 x 2,000 sq ft) 1,400 sq ft or 14 squares are needed. The unit cost per square developed is $36.00 which, multiplied by 14 square, is $504.00.

If it is desired to express the unit cost in terms of cents per square foot, the developed unit cost, for standard exposure, of $36.00 per square is $.36 per square foot. This is corrected from Table 12-3 by multiplying the unit cost per square foot by 70 per cent. This adjusted unit cost per square foot may now be applied to the 2,000 square foot area.

$$.70 \times \$.36 = \$.252 \times 2,000 = \$504.00$$

COMPOSITION SIDING

There are various kinds of composition siding. The most commonly used are asphalt shingles, asbestos-cement shingle siding, and rolls of asphalt and felt similar to roll-roofing, manufactured in imitation brick and other patterns. The rigid asbestos-cement siding is used on many new homes. The asphalt shingles and rolls are found mostly on older buildings where it is desired to cover worn wood siding that has served its life.

Asbestos-cement Siding (See also Roofing—Chapter 13)

Asbestos siding, shingles or board, is made by combining asbestos fibers and portland cement under pressure. The material is fire resistive, rigid, and comes in several colors. It may also be painted with a special type of paint.

Asbestos-cement shingles are generally made in a rectangular shape, 12 inches wide and 24 inches long. They are put on like other shingles, lapping 1 1/2 inches. This type of siding is sold by the square. Each square contains three bundles of 57 pieces each. The weight of a square is approximately 180 pounds.

The shingles are applied over solid wood or plywood sheathing and saturated felt paper, or over composition sheathing. Galvanized nails, which are furnished by the manufacturer, come in different lengths. Barbed nails are used to provide better holding in old wood. About 2 pounds of nails will take care of a square of siding. When applied over composition sheathing, patented fasteners are supplied by the manufacturer. A special tool is supplied for cutting, and for punching holes in asbestos siding.

The quantity of asbestos-cement shingles required is computed by measuring the area, deducting window and door openings, and adding 10 percent waste for cutting and fitting around windows, doors, and corners.

The hours of labor required to apply a square of asbestos-cement siding varies with conditions, but for average straight work where there is a moderate amount of cutting and fitting, two men should be able to apply a square in about 2 hours or a total of 4 man hours for a square. When working on the second floor of a two story building from a scaffold, the time required should be increased to about 6 man hours a square.

Illustrative Example

Assuming asbestos-cement shingles cost $20.00 per square, nails $.30 per pound, and wages are $6.00 per hour.

1 square shingles	@ $20.00 =	$20.00
2 lbs nails	@ .15 =	.30
4 hours labor	@ 6.00 =	24.00
	Unit cost per square =	$44.30

$$\text{Unit cost per sq ft } \frac{\$44.30}{100} = \$ \ .44$$

Asphalt Shingle Siding

Asphalt shingles are frequently applied to the exterior of frame buildings, generally over old siding. A more detailed discussion of the characteristics of asphalt shingles, and also asphalt roll siding will be found in Chapter 13 on "Roofing."

Asphalt shingles are applied over saturated felt paper on either sheathing or old wood siding. It requires about 2 pounds of asphalt shingle nails per square. It takes more shingles per 100 square feet than when they are applied as roofing. The reason is that, on side walls, the exposure used is generally less than specified by the manufacturer. It varies between 3 and 4 inches. If the shingles are manufactured for an exposure of 5 inches, then one square will cover exactly 100 square feet at that exposure, because there are 80 shingles to the square and each shingle is 36 inches long.

$$\frac{5'' \times 36''}{144} = \frac{180}{144} \times 80 = 100 \text{ sq ft}$$

Asphalt shingles manufactured for a 4-inch exposure have 100 shingles to the square and cover exactly 100 square feet.

$$\frac{4'' \times 36''}{144} = \frac{144}{144} \times 100 = 100 \text{ sq ft}$$

By reducing the specified exposure when applying a 4-inch or 5-inch shingle, the area that a square will cover is proportionately reduced. The percentage to be added to the area to be covered is shown in Table 12-5.

Table 12-5
Percentage to Be Added to Area to Obtain the Quantity Required
for 36-Inch Strip Shingles

Exposure When Applied	4-Inch Exposure Specified by Manufacturer	5-Inch Exposure Specified by Manufacturer
3	.34	.67
3½	.15	.43
4	— —	.25
4½	— —	.11
5	— —	— —

Illustrative Example

If the net area to be covered is 3,000 square feet and the shingle used has a specified exposure of 5 inches, when applied with an exposure of 3½ inches, the quantity needed to cover an area 3,000 square feet is:

Area   3,000 sq ft
Add 43% 1,290 sq ft
              4,290 sq ft

It is customary to add a percentage to the quantity of shingles computed, to allow for waste in cutting and fitting around windows and doors, and also for the bottom starting course which has two layers of shingles. The percentage varies with each job and depends on the number of openings, corners and so forth. Customarily 10 percent is considered adequate. As with other types of siding, the area of windows and doors are deducted from the gross area.

The hours of labor to apply asphalt shingles to side walls varies, but a good average rate is 3 hours per square.

Illustrative Example

Assume asphalt shingles are priced at $9.00 per square, nails are $.20 a pound and wages are $6.00 per hour.

1 square shingles	@ $9.00 =	$ 9.00
2 lbs nails	@ .20 =	.40
3 hours labor	@ 6.00 =	18.00
	Unit cost per square =	$27.40

$$\text{Unit cost per sq ft} \frac{\$27.40}{100} = \$ \ .27$$

Whatever unit cost is developed, it is applied to the quantity required as adjusted from Table 12-5.

ASPHALT ROLL BRICK SIDING

Rolls of asphalt siding come in different weights, and patterns. The most common is the brick pattern which weighs about 105 pounds to the roll. A roll is 31 inches wide and 43 feet long, and contains 111 square feet. It is intended to cover one square with approximately 11 percent waste for lapping. The nails and asphalt adhesive for joining edges are supplied inside each roll.

The amount of material needed to cover an area is determined by computing the number of square feet in the over-all area, deducting the window and door openings, and adding a percentage for waste in cutting and fitting. While 10 percent is considered adequate for ordinary straight work, each individual job should be examined to judge best the proper amount of waste to be added.

The hours of labor necessary to apply a roll of brick siding varies but under average working conditions two hours per square is considered adequate.

Illustrative Example

Assume that the price of a roll of asphalt roll siding is $5.00, and wages are $6.00 per hour.

1 role siding	@ $5.00 =	$ 5.00
2 hours labor	@ 6.00 =	12.00
	Unit cost per square =	$17.00

$$\text{Unit cost per sq ft} \frac{\$17.00}{100} = \$ \ .17$$

HARDBOARD LAP SIDING

This is a composition of great density for use on the exterior. It is 7/16" x 12" x 16' long and is delivered to the job prime-coated.

Application is the same as for wood bevel siding using metal corner clips. Lapping should be at least 11/2 inches which, on a width of 12 inches would produce an area waste of 12 1/2 percent. The overall area plus cutting and fitting waste will average 20 percent. In other words to cover 100 square feet, 120 square feet of hardboard lap siding is required.

A unit cost would be developed as follows:

Illustrative Example
(Prices and Wages Assumed)

1,000 sq ft hardboard siding	@ $270.00 =	$270.00
20 lbs nails	@ .20 =	4.00
25 hours labor	@ 6.00 =	150.00
	Unit cost per 1,000 sq ft =	$424.00
	Unit cost per sq ft =	$ .42

(Add cost of aluminum metal corner clips)

METAL SIDING

For corrugated metal (aluminum and steel) siding see Roofing, Chapter 13.

Baked enamel *aluminum siding* is available in several styles, different thicknesses, and may be applied vertically or horizontally. Because of variations in cost for materials and the numerous accessories needed around windows and doors, at corners, starter strips etc., it is recommended a subcontractor be engaged to provide an estimate.

Chapter 13.

Roofing

ROOFING

The material that is applied to the surface of a roof to make it water-proof and tight against the weather is called *roofing*. There is a variety of material used for this purpose, but well over 75 percent consists of asphalt roofing products. These are in the form of built-up roofing, roll roofing, and shingles. Other materials that are commonly used for shingles are wood, slate, tile, and aluminum. Flat sheet or corrugated aluminum, and galvanized iron roofing are frequently applied to the commercial, in-dustrial, and rural structures. Corrugated glass-fiber panel roofing has become popular for patios, garages, and structures where the admission of light is desired. Asbestos-cement shingles are also used.

QUALITY OF ROOFING

In estimating, the designation of a particular roofing material by type is not sufficient to determine its cost. There are different grades of each type and the identification of the quality, thickness, or size is equally as important as the kind of material. Asphalt shingles, roll roofing, and felts are made in several weights. Wood and slate shingles are sold in different sizes and grades. It makes considerable difference whether a built-up roof is 3-ply or 5-ply, and whether it has a smooth surface, slag, or crushed marble. Spanish tile may be terra cotta, or it may be made of colored cement to simulate terra cotta. Metal roofs, either corrugated or sheet metal, are made in several thicknesses or gauges, and the cost of the material varies accordingly.

In many instances the quality of a particular kind of roofing material is not easily determined, and it may be necessary to remove a sample to have it identified by a supplier. A properly prepared estimate of the cost of roofing should clearly indicate the type and quality of the particular roofing material.

FACTORS AFFECTING LABOR

While all of the factors which affect labor in general, as discussed in Chapter 2, should be given consideration in estimating the cost of roofing, the following are of special importance:

1. Are the men experienced in applying the particular type of roofing?
2. Is the roof surface of irregular shape, or is it cut up with dormers and intersecting roofs?
3. Is the slope of the roof steep enough to require foot scaffolds, or is it low pitched?
4. Is it necessary to erect scaffolding?
5. Will it be necessary to hoist materials by hand or mechanically? Will materials have to be carried up to the roof from the inside of the building by stairs, ladders, or elevators?
6. Is it a patch job which requires matching in the new roofing at the perimeter of the damaged area?
7. Do the old shingles have to be removed, and does the roof decking have to be repaired, or prepared, for the new roofing?

Experience of Workmen

Some of the experienced roofers, who apply asphalt strip shingles every day, can put them on at the rate of about one square an hour. The average roofer can apply them at the rate of 1 1/2 to 2 hours per square on an ordinary roof without too many irregularities. But a carpenter unaccustomed to such work may take as long as 3 hours per square. Men whose trade is applying asbestos-cement shingles work much more rapidly than a novice at the business who requires time to learn the art of cutting, punching nail-holes, and fitting the material into place. The experience of the workmen who will apply the roofing should be carefully considered.

Shape of Roof

The presence of dormers, intersecting gables, and other architectural features that create a cut-up roof surface will slow down a roofer in his rate of applying roofing; no matter how experienced he is it takes additional time to cut and fit around the valleys, hips, dormers and eaves.

Applying shingles to a Gothic roof, or to the steep sides of a Gambrel or Mansard roof, takes more hours per square than putting them on a building with a low-pitched roof.

Slope of Roof

On roofs that have a steep incline, the roofer generally sets up foot scaffolds to keep from slipping. His rate of work is measurably reduced due to his lack of freedom of movement, and the care he must exercise to keep his materials at hand and maintain his position. The steeper the slope, the slower he works.

Scaffolding

There is usually no need to erect scaffolding on one and two-story buildings, particularly if the roof is not too steep. Sometimes scaffold brackets are used on extension ladders. After the first several rows of shingles are laid, a foot or roof scaffold is used. It consists of a 2" x 4" laid on the roof and supported at each end by stirrups, or by wire loops taken from the shingle bundles. These are nailed to the roof at a point that will be covered by the shingles. As the roofer moves up toward the ridge, the position of the foot scaffold is shifted by slipping the stirrup, or wire loop, off of the nail. The nail which held the scaffold support is then driven the rest of the way in and covered by the shingle.

When conditions require the building of a light scaffold from the ground to the eaves, allowance must be made for its cost, less any anticipated salvage. (See "Scaffolding," Chapter 9.) The cost of the scaffolding should not be included in the *unit cost* of the roofing as it distorts the true cost of application.

Hoisting

Hoisting is an item to be considered, especially in applying built-up roofs when rolls of roofing, asphalt or gravel must be raised from the ground to the area that is being worked on. Circumstances of each job will indicate what additional allowances are to be made for the cost of hoisting materials. In the case of built-up roofs the hoisting charges are included in the unit cost inasmuch as it is a continuous operation during the application of the material.

Patch-roofing

The patching of roofs, either over a small area or spotting-in shingles here and there, cannot be estimated on a unit cost basis. The most reliable method is to estimate the over-all time required, and determine the amount of material it will take to cover the area.

When estimating the labor, ample time should be allowed for lacing-in and matching-in the roofing, particularly shingles, where the new and old join. On shingle roofs of all types such work is time-consuming.

Where the patch-area is in excess of a couple of squares, depending on the conditions, unit costs may be used provided the extra labor is added for joining the new work with the old.

Removing Old Roofing

The cost of removing the old roofing, prior to applying the new, should be estimated as a separate item since it consists only of labor and trucking away from the premises. Some builders and roofers compute the cost to remove the old roofing, and then reduce it to a cost per square or square foot. This cost is then added to make a total unit cost for removing and replacing the roofing.

LABOR RATES TO APPLY ROOFING

Table 13-5 shows the average hours of labor to remove and install various types of roofing. The rates are based on the average roof surface of flat to medium pitch, on which men can work comfortably. The figures in the table contemplate that the roof may have a dormer or two but is not considered cut-up. The rates are for the average workman who is experienced although not necessarily specializing as a roofer. Adjustments in these labor rates are in order where substantial variations from the preceeding conditions are encountered. In the majority of cases, however, the rates shown will be found adequate.

MEASUREMENT OF ROOF AREAS

The unit of measurement for roofing is the *square,* which is an area 10' x 10', or 100 square feet. The quantity of roofing material needed is obtained from the measured area of the roof surface.

The method of obtaining measurements of any roof depends on the shape, and whether or not conditions permit access to it so that the measurements can be taken in the field, as in the case of a flat or low-pitched one-story roof. When it is not possible to get up on the roof, there are several methods used to obtain the necessary dimensions for computing area. Horizontal measurements may be taken from the ground, adding the cornice overhang at the gable or eaves. The distance from the ridge to the eaves, sometimes called the *rafter length,* is frequently obtained by counting the number of shingles from the lower edge of the roof to ridge, and multiplying by the exposure in inches. The product is divided by 12 to obtain the lineal feet.

Illustrative Example

Rows of shingles	36
Exposure in inches	x4¾
	171 inches

$$\text{Rafter length} = \frac{171}{12} = 14'3''$$

In many cases the rafter length is obtained geometrically by measuring at ground level the horizontal distance from the ridge out to the eave line, and obtaining the vertical height from the eave line up to the ridge. By using the Rule of Pythagoras, as illustrated in Chapter 6, the rafter length can be computed. It is not always easy to estimate the vertical height, so the following methods are suggested:

1. If the gable end is finished in shingles, brick, horizontal siding, concrete block, or other materials of uniform courses, the number of courses may be counted and multiplied by the number of inches in one course. The product is divided by 12 to obtain the height in feet.

Illustrative Examples

Common brick courses	36
Inches per course	2¾
	99 inches

$$\text{Vertical height} = \frac{99}{12} = 8'3''$$

2. When the attic is accessible, the vertical height may be measured directly from the interior after establishing any difference in level between the attic floor and the eave line.
3. In some instances the vertical height can be obtained by leaning out of a window in a gable end and measuring down to the eave line, and up to the ridge.
4. Actual measurements can frequently be obtained from inside a garage, barn, shed, or other structure, where rafters and the interior of the roof structure are exposed to view and can be reached.
5. When a structure has been badly damaged by fire or windstorm, and enough of the roof, trusses etc., are available in the debris, measurements of vertical heights may be obtained directly.

6. In buildings of masonry, the wood roof may be completely destroyed, but enough of the outline may remain on the walls to determine its shape and measurements.
7. When a building is completely destroyed, the data for computing the roof area is obtained from existing plans, pictures, and a full description from the occupant or owner as to the number of floors, average ceiling heights, existence of an attic, and shape of the roof.

SPAN, RISE AND PITCH

Roof Pitch—The pitch of a roof is the slope expressed as a ratio of the rise to the span. The rise is the vertical distance between the ridge and the supporting plates. The span is the distance between the sidewalls that support the roof.

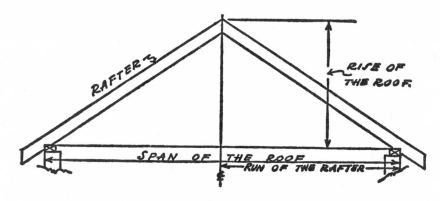

Illustrative Example

Where the rise of a roof is 7 feet and the span is 21 feet.

$$\text{Pitch} = \frac{\text{Rise}}{\text{Span}}$$

$$\text{Pitch} = \frac{7}{21} = \frac{1}{3}$$

When the pitch and the span are known, the rise can be obtained by the following formula:

$$\text{Rise} = \text{Pitch} \times \text{Span}$$

When the pitch and the rise are known, the span is obtained by the following formula:

$$\text{Span} = \frac{\text{Rise}}{\text{Pitch}}$$

Measurement of Pitch From Ground

The pitch of a roof can be fairly well determined from the ground using a carpenter's 6-foot folding rule. The method is shown on page 312, courtesy of the Asphalt Roofing Manufacturers Association, and reproduced from their 1974 Manual.

<u>Pitch</u> - The span, rise and run of a simple gable roof are shown in Fig. 87. The pitch or slope of the roof is most often stated as the relation between the rise and the span. If the span is 24'0" and the rise is 8'0" the pitch will be 8/24 or 1/3. If the rise were 6'0" then the pitch would be 6/24 or 1/4. The 1/3 pitch roof rises 8" per foot of horizontal run, and the 1/4 pitch roof rises 6" per foot of run.

Fig. 87 - Pitch relations

It is possible to determine the pitch of any roof without leaving the ground by using a carpenter's folding rule in the following manner.

Form a triangle with the rule. Stand across the street or road from the building, and holding the rule at arms length, align the roof slope with the sides of the rule, being sure that the base of the triangle is held horizontal. It will appear within the triangle as in Fig. 88. Take a reading on the base section of the rule (note the "reading point", Fig. 88). Then locate on Fig. 89 in the top line headed "Rule Reading"

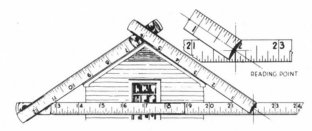

Fig. 88 - Use of Carpenters' Rules to Find Roof Pitch.

the point nearest your reading. Below this point will be found the pitch and the rise per foot of run. In the case illustrated the reading is 22. Under the figure 22 in Fig. 89, the pitch is designated as 1/3, or a rise of 8" per foot of horizontal run.

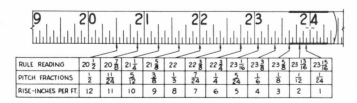

RULE READING	$20\frac{1}{2}$	$20\frac{7}{8}$	$21\frac{1}{4}$	$21\frac{5}{8}$	22	$22\frac{3}{8}$	$22\frac{3}{4}$	$23\frac{1}{16}$	$23\frac{3}{8}$	$23\frac{5}{8}$	$23\frac{13}{16}$	$23\frac{15}{16}$
PITCH FRACTIONS	$\frac{1}{2}$	$\frac{11}{24}$	$\frac{5}{12}$	$\frac{3}{8}$	$\frac{1}{3}$	$\frac{7}{24}$	$\frac{1}{4}$	$\frac{5}{24}$	$\frac{1}{6}$	$\frac{1}{8}$	$\frac{1}{12}$	$\frac{1}{24}$
RISE-INCHES PER FT.	12	11	10	9	8	7	6	5	4	3	2	1

Fig. 89 - Reading Point converted to pitch.

Measurement of roof pitch from the ground using a carpenter's 6-foot rule. Courtesy Asphalt Roofing Manufacturers Association, New York, N.Y.

GABLE ROOF AREAS

The area of a gable roof may be obtained by determining the *slope* length from the ridge down to the lower edge or eave line including the overhang. This measurement is multiplied by the horizontal length of the roof. When the rafter length cannot be physically measured, two methods for computing the length are suggested.

Method 1. The square root of the rafter length is equal to the square root of the sum of the square of the run and the square of the rise.

Illustrative Example

In Figure 13-1, the run is 14 ft and the rise is 8 ft.

Rafter length $= \sqrt{14^2 + 8^2}$

$$= \sqrt{196 + 64}$$
$$= \sqrt{260}$$
$$= 16.12 \text{ ft}$$

RAFTER LENGTH

RUN

RISE

16.12′

8′

14′

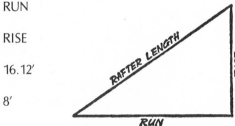 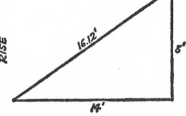

Figure 13-1

Method 2: When the rise and run are known, the rise-per-foot-of-run is obtained by dividing the rise in inches by the run in feet. Table 13-1 shows the factor for rise-per-foot-of-run by which the run is multiplied to obtain the rafter length.

Illustrative Example

In Figure 13-1, the rise is 96 inches (8′ x 12″), and the run is 14 ft.

$$\text{Rise-per-foot-of-run} = \frac{96''}{14'} = 6.85 \text{ inches}$$

The closest factor shown in Table 13-1 for a rise-per-foot-of-run of 6.85 inches is 1.160.

$$\text{Rafter length} = 1.160 \times 14 = 16.24 \text{ ft}$$

Table 13-1
Factor by Which Run Is Multiplied to Obtain Rafter Length When
Rise-per-Foot-of-Run Is Known

No. Inches Rise-per-foot-of-run	*Pitch	Rafter length in inches per foot of run	To get rafter length multiply run in ft by
3.0	1/8	12.37 ÷ 12 =	1.030
3.5			1.045
4.0	1/6	12.65	1.060
4.5			1.075
5.0	5/24	13.00	1.090
5.5			1.105
6.0	1/4	13.42	1.120
6.5			1.140
7.0	7/24	13.89	1.160
7.5			1.185
8.0	1/3	14.42	1.201
8.5			1.230
9.0	3/8	15.00	1.250
9.5			1.280
10.0	5/12	15.62	1.301
10.5			1.335
11.0	11/24	16.28	1.360
11.5			1.390
12.0	1/2	16.97	1.420

* Pitch is determined by dividing the inches rise-per-foot-of-run by 24. (Most roofs of dwellings are constructed with a pitch of 1/8, 1/6, 1/4, 1/3 or 1/2.)

When the rafter length has been obtained by methods 1 or 2, it should be carried to the next half or full foot. The area of the roof is then computed by multiplying the rafter length plus any overhang, by the horizontal length of the roof. Gabled dormers with the same pitch as the main roof, except for their overhang, may be ignored for all practical purposes as the *dormer roof area is the same as that taken up by the dormer in the main roof.*

RAFTER LENGTHS FROM TABLES

Table 13-2 gives the length of rafters for gable roofs with the rise (height) and span (width) shown. Any overhang must be added.

Interpolating is necessary where the width under consideration falls between those given in the table. Heights to the nearest 6 inches should be used in using the table.

Table 13-2
TABLE OF RAFTER LENGTHS

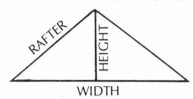

Width of Building in Feet

Height	18		20		22		24		26		28		30		32		34		36	
	Length of Rafters in Feet and Inches																			
4 6	10	2	10	11	11	9	12	10	13	9	14	9	15	8	16	8	17	7	18	7
5 0	10	4	11	3	12	1	13	0	13	11	14	11	15	10	16	9	17	9	18	8
5 6	10	6	11	5	12	4	13	3	14	2	15	1	16	0	16	11	17	11	18	10
6 0	10	9	11	8	12	6	13	5	14	4	15	3	16	2	17	1	18	1	19	0
6 6	11	0	11	11	12	9	13	8	14	6	15	5	16	4	17	3	18	3	19	2
7 0	11	4	12	3	13	0	13	11	14	8	15	8	16	7	17	5	18	5	19	4
7 6	11	8	12	6	13	4	14	2	15	0	15	11	16	9	17	8	18	7	19	6
8 0	12	0	12	10	13	8	14	5	15	3	16	2	17	0	17	10	18	9	19	8
8 6	12	4	13	2	13	11	14	8	15	6	16	5	17	3	18	2	18	11	19	11
9 0	12	9	13	5	14	3	15	0	15	10	16	8	17	6	18	5	19	3	20	2
9 6	13	2	13	10	14	6	15	4	16	1	16	11	17	9	18	8	19	6	20	5
10 0	13	6	14	2	14	10	15	8	16	5	17	3	18	1	18	11	19	9	20	7
10 6	13	10	14	6	15	3	15	11	16	9	17	6	18	4	19	3	20	0	20	10
11 0	14	3	14	10	15	7	16	4	17	0	17	10	18	8	19	6	20	3	21	1
11 6	14	7	15	3	15	11	16	8	17	4	18	2	18	11	19	9	20	6	21	4
12 0	15	0	15	8	16	4	17	0	17	9	18	5	19	3	20	0	20	10	21	8
12 6	15	5	16	0	16	8	17	4	18	0	18	9	19	7	20	4	21	1	21	11
13 0	15	10	16	5	17	0	17	8	18	5	19	2	19	10	20	8	21	5	22	2
13 6	16	3	16	10	17	5	18	0	18	9	19	6	20	3	20	11	21	9	22	6
14 0	16	8	17	3	17	10	18	5	19	2	19	10	20	7	21	3	22	0	22	10
14 6	17	1	17	8	18	3	18	10	19	6	20	2	20	10	21	7	22	4	23	2
15 0	17	6	18	0	18	8	19	3	19	11	20	7	21	3	21	11	22	8	23	6
15 6	17	11	18	5	19	0	19	8	20	3	20	11	21	7	22	4	23	0	23	10
16 0	18	4	18	10	19	5	20	0	20	8	21	4	21	11	22	8	23	4	24	1
16 6	18	10	19	4	19	10	20	5	21	0	21	8	22	4	23	0	23	8	24	5
17 0	19	3	19	9	20	3	20	11	21	5	22	1	22	8	23	4	24	1	24	9
17 6	19	9	20	2	20	8	21	3	21	10	22	5	23	0	23	8	24	5	25	1
18 0	20	1	20	7	21	0	21	8	22	3	22	10	23	5	24	1	24	9	25	6
18 6	20	7	21	0	21	6	22	0	22	8	23	3	23	10	24	5	25	2	25	10
19 0	21	0	21	5	22	0	22	6	23	1	23	8~	24	3	24	10	25	6	26	2
19 6	21	5	21	11	22	5	22	11	23	5	24	0	24	8	25	3	25	10	26	7
20 0	21	11	22	4	22	10	23	4	23	10	24	5	25	0	25	8	26	3	26	11
20 6	22	5	22	10	23	3	23	9	24	4	24	10	25	5	26	0	26	8	27	4

Illustrative Example

To find the rafter length where the height or rise is 8'0" and the width of the building is 31'0".

Rafter length for 8'0" and 30'0" width = 17'0"

Rafter length for 8'0" and 32'0" width = 17'10"

Use the 30'0" width and add 1/2 the difference between the 30' and 32' widths = 17'5". Add any overhang and use next nearest even foot.

DETERMINING ROOF AREA FROM GROUND AREA

The area of a gable, hip or intersecting gable roof can be obtained sufficiently close, for practical purposes, by first measuring the horizontal or ground area. The inches rise-per-foot-of-run is computed, and from Table 13-1 the nearest corresponding factor is selected. The roof area is then obtained by multiplying this factor by the ground area.

Illustrative Example

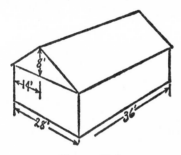

Figure 13-2

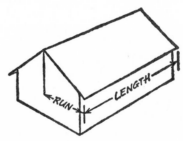

Figure 13-3

In Figure 13-2, the rise-per-foot-of-run is:

$$\frac{96''}{14'} = 6.85 \text{ inches}$$

The nearest factor for 6.85 inches in Table 13-1 is 1.160

The ground area = 28' x 36' = 1,008 sq ft

The roof area = 1.160 x 1,008 = 1,169 sq ft

Where there is an overhang at the eaves, the measurement for the run is taken from the center of the gable to the eave as in Fig. 13-3. *The length measurement of the roof should include the overhang at the gable end.*

INTERSECTING GABLE ROOF AREAS

Where one gable roof joins another, and each has the same pitch or slope, the area of the main roof is computed without deducting the area taken up by the intersecting roof. The area of the intersecting roof is computed by taking its length as the distance from its gable end to the point where it joins the main roof at the eaves. (See Fig. 13-4.)

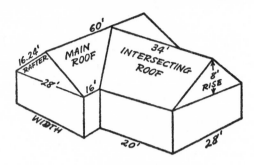

Figure 13-4

Illustrative Example
Roof Area Obtained from Ground Area (Fig. 13-4)

Main roof ground area	$28' \times 60' = 1{,}680$ sq ft
Intersecting roof ground area	$20' \times 28' = 560$ sq ft
Total ground area	$= 2{,}240$ sq ft

Inches rise-per-foot-of-run $\dfrac{96''}{14'} = 6.85$

From Table 13-1, nearest factor $= 1.160$ in
Total area of roof ($2{,}240$ sq ft $\times 1.160$) $= 2{,}598$ sq ft

Actual measurement of the roof in Fig. 13-4, would produce an area of 2,598 sq ft as follows:

One side of main roof:
    $16.24' \times 60'$ $= 974.4$ sq ft
Intersected side of roof:
    $\dfrac{2(16' + 30') \times 16.24'}{2}$ $= 747.0$ sq ft
Intersecting roof:
    $\dfrac{2(20' + 34') \times 16.24'}{2}$ $= 876.9$ sq ft
    Total area $= 2{,}598.3$ sq ft

DORMER ROOF AREAS

The area of a *flat dormer*, excluding overhang, is slightly less than the area which the dormer itself occupies in the main roof. Some estimators determine the gross area of the entire roof without considering the dormer area separately except to make allowance of any overhang. Where exact measurement of area is desired, the area of each segment of the roof is determined.

When the pitch of either a gable or hip dormer is the same as that of the main roof, the area of the dormer roof is equal to that which the dormer itself occupies in the main roof, plus the area of any overhang. In other words, where dormers are encountered, the computation of their roof areas may be ignored except for their overhang. The main roof area is determined as though the dormers did not exist.

HIP ROOF AREAS

The pitch of all sides of a hip roof are usually the same, as a matter of good construction. *The total area of a hip roof is identical to that of a gable roof of the same pitch and covering the same horizontal or ground area.*

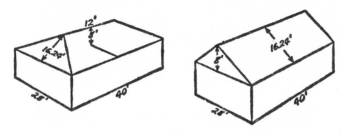

Figure 13-5

Area of hip roof (Fig. 13-5)

2 end areas $= \dfrac{2(28' \times 16.24')}{2}$ $=$ 454.72 sq ft

2 side areas $= \dfrac{2(12' + 40') \times 16.24'}{2}$ $=$ 844.48 sq ft

Total area $=$ 1,299.20 sq ft

Area of gable roof (Fig. 13-5)

2(40' × 16.24') $=$ 1,299.20 sq ft

A quick method of obtaining the area of a hip roof is to treat it as though it were a gable roof. The area is computed by multiplying the longest side by the rafter length, or by determining the ground area (including overhang) and multiplying by the factor in Table 13-1 for the rise-per-foot-of-run.

Illustrative Example

In Fig 13-5, the ground area is:

$$28' \times 40' = 1,120 \text{ sq ft}$$

The rise-per-foot-of-run is:

$$\frac{96''}{14'} = 1.160$$

The roof area = 1.160 x 1,120 = 1,299.20 sq ft

GAMBREL ROOF AREAS

Gambrel roofs are frequently found on barns, and also on older dwellings. They are sometimes called "Dutch Colonial" roofs. In a gambrel roof, the slope or pitch is broken between the plate and the ridge to form a double-pitched roof as in Fig. 13-6.

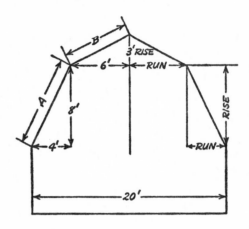

Figure 13-6

While the proportions of this type of roof vary, the outer steeply sloped sides usually have a pitch of 1. This means that the rise is twice the run, which is frequently one fifth of the width of the building. The upper roof has a pitch of 1/4, which means that the rise is one half of the run.

The roof area can best be determined by calculating the length of the rafters A and B, (Fig. 13-6) by measuring the rise and the run of both slopes.

Illustrative Example

Roof 40′ long (Fig. 13-6)

Lower slope length $= \sqrt{4^2 + 8^2}$	$= \sqrt{80}$	$= 8.94$ lin ft
Upper slope length $= \sqrt{3^2 + 6^2}$	$= \sqrt{45}$	$= 6.71$ lin ft
Area of sides $= 2(40′ \times 8.94′)$		$= 715$ sq ft
Area of top $= 2(40′ \times 6.7′)$		$= \underline{536}$ sq ft
		$\overline{1,251}$ sq ft

MANSARD ROOF AREAS

A mansard, sometimes referred to as a French roof, is one that is approximately flat on top, and has four steeply sloped sides. The top deck is usually surfaced with a type of roofing suitable for flat roofs such as roll roofing, built-up roofing or tin. The sides are often finished in some type of shingle.

The sides of a mansard roof appear as trapezoids in profile. The height is the length of the slope from the lower edge to the deck. The lower base of the trapezoid is the distance along the eaves. The upper base is the distance along the deck edge. (Fig. 13-7)

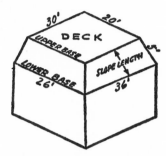

Figure 13-7

The area of the sides is obtained by determining the area of each of the trapezoidal sides, and adding them together.

Illustrative Example

Area side one $= \dfrac{(20' + 26') \times 5'}{2}$ $= 115$ sq ft

Area side two = (opposite side the same) $= 115$ sq ft

Area side three $= \dfrac{(36' + 30') \times 5'}{2}$ $= 165$ sq ft

Area side four = (opposite side the same) $= \underline{165}$ sq ft

Total area $\overline{560}$ sq ft

(or)

Add the perimeter of the lower base to the perimeter of the upper base; divide by 2 and multiply by the slope length.

Lower base—$2(26' + 36') = 124$ lin ft
Upper base—$2(20' + 30') = \underline{100}$ lin ft
$\overline{224} \div 2 = 112 \times 5 = 560$ sq ft

The area of the top deck is obtained by multiplying the length by the width (Fig. 13-7).

$$20' \times 30' = 600 \text{ sq ft}$$

Because of the additional work involved in cutting and fitting around dormers or windows in a mansard, they are usually not deducted when computing the area in order to determine the quantity of shingles.

FLAT AND SHED ROOF AREAS

The areas of flat and shed type roofs are the least difficult to obtain, as in most cases the person making the estimate can get up on the roof to take measurements. The roof area is also relatively easy to determine from the ground measurements of the building.

A shed roof is a single pitch roof, and the length along the slope can be obtained from the rise and the span (Fig. 13-8).

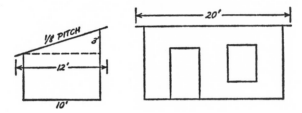

Figure 13-8

Illustrative Example

Span = 12 ft
Rise =   3 ft
Inches-rise-per-foot-of-run = $\dfrac{3 \times 12''}{12'}$ = 3 inches

From Table 13-1 the factor by which the run is multiplied to determine the length of the rafter is 1.03.

Rafter length = 12' x 1.03 = 12.36'
Area of roof = 12.36 x 20' = 247 sq ft
$$\text{(or)}$$
Rafter length = $\sqrt{3^2 + 12^2}$ = $\sqrt{153}$ = 12.36 lin ft
Area of roof = 12.36' x 20' = 247 sq ft
$$\text{(or)}$$

Using the ground area multiplied by the factor 1.03 (Table 13-1):

Ground area = 20' x 12' x 1.03
           = 240' x 1.03
Area of roof = 247 sq ft

GOTHIC SHAPE ROOF AREAS

A gothic roof (Fig. 13-9) is formed by two circles passing through the wall plates and intersecting at the ridge. This type of roof is found principally on barns and is very common in some of the mid-western dairy

states. One of the most frequently used rules to lay out the proportions of a
gothic roof is to scribe two circles, each one with a radius equal to two
thirds of the width of the building, with the centers of the circles at A and
A1 as shown.

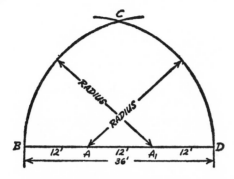

Figure 13-9

To find the roof area, the length of the building including any
overhang at the gables is multiplied by the distance BC + CD.

Distance BC is equal to the circumference of the circle times .2087.
The circumference of the circle is the radius (2/3 of the width of the
building) times 6.2832. Since BC and CD are equal, the formula for obtaining the roof area would be:

$$\text{Area} = 2(\text{Radius} \times 6.2832) \times .2087 \times \text{length}$$

In terms of the width and length of the building, the area for a roof of
these proportions may be expressed for practical application as follows:

$$W = \text{width}$$
$$L = \text{length}$$
$$\text{Area} = 1.75 \ W \times L$$

Illustrative Example

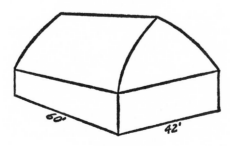

Figure 13-10

Width of building = 42′
Radius of circles = 42′ x ⅔ = 28 lin ft
Using the formula:
Roof area = 2 (Radius x 6.2832) x .2087 x Length
       = 2 (28′ x 6.2832) x .2087 x 60′ = 4,406 sq ft
Using the formula:
Roof area = 1.75 x W x L
        = 1.75 x 42′ x 60′
        = 4,410 sq ft

CIRCULAR ROOF AREAS

The area of a circular roof is obtained by determining 1/2 of the circumference of a circle where the width of the building is the diameter, and multiplying by the length of the roof.

Illustrative Example

Figure 13-11

Where W = Width of building
     L = Length of roof

$$\text{Area of roof} = \frac{3.1416 \times W}{2} \times L$$

(Fig. 13-11) Area of roof = $\frac{3.1416 \times 30'}{2}$ x 80′ = 3,770 sq ft

HEMISPHERICAL ROOF AREAS

The area of a hemispherical roof (Fig. 13-12), commonly found on silos, would be equal to one half the area of a complete sphere. It is also equal to twice the area of its base.

Area = $\frac{(3.1416)}{2}$ x d² (Where d = diameter)

(or)

= 2(3.1416) x r²ˡ (Where r = radius)

Illustrative Example

Area of roof = ½(3.1416) x 20²
              = 1.5708 x 400
              = 628.32 sq ft
                (or)
              = 2(3.1416) x 10²
              = 628.32 sq ft

Figure 13-12

CONICAL ROOF AREAS

Roofs of silos, water tanks, and so forth, are often conical in shape. The area of a conical roof is 1/2 the slope length "s" (Fig. 13-13) x "d" the diameter x 3.1416.

Illustrative Example

Area = ½ s x (3.1416 x d)
            Example
Area = ½(10) x (3.1416 x 20')
     = 314 sq ft

Figure 13-13

SHINGLE ROOFING IN GENERAL

While at one time the word *shingle* referred to a tapered piece of wood, one end being thinner than the other, used to cover roofs, it now includes other materials which are similarly applied. The more commonly used shingles, other than wood, are asphalt, asbestos-cement, slate, tile, or its cement imitation, and aluminum. All shingles, for reasons which are apparent, are suitable only for inclined roof surfaces with sufficient pitch to prevent water from getting under the lapped ends or butts.

In application of the various kinds of shingles, there are several important factors to be given consideration:

Type of Shingle.	Starter Shingles.
Type of Roof Decking.	Ridge and Hip Caps.
Felt Underlayment.	Waste.
Exposure.	

Type of Shingle

Each type of roofing shingle is available in different grades, which is reflected in the cost. The type and grade is governed by the size and shape of the shingle, by the material of which it is made, the color, and the weight. Wood shingles, for example, are made in three sizes: 16-, 18-, and 24-inch, and each size can be obtained in three or more grades. Asphalt shingles are made in several sizes, shapes, and colors; and the weights per square vary from 135 to 350 pounds. Slate shingles are sold in different sizes, thicknesses, colors and grades. To identify a shingle roof solely by the material of which it is made is, therefore, no real indication of its cost.

When a roof has been damaged by hail or windstorm, or partly damaged by a fire, a fairly positive verification of the grade of shingle is possible. When the roof is completely destroyed by fire, the problem is more difficult. Much will depend on the investigative resourcefulness of the individual making the estimate.

Type of Roof Decking

All shingles, with the exception of wood, should be laid over solid roof boards or roof decking. When nominal 1-inch boards are used, they should be matched to provide rigidity and firmness between rafters. Plywood roof decking should be a minimum of 5/16-inch thick, and the joints of the sheets must be reinforced by blocking. The surface should be clean, and all raised nails in old decking should be either replaced or driven in so that they cannot puncture the new shingles.

It is generally recommended that wood shingles be laid over open sheathing. This may be nominal 1-inch square-edge boards spaced an inch or so apart to provide air for the back of the shingles to aid the drying out process when they have been wet.

Felt Underlayment

A layer of roofing paper of asphalt-saturated felt is applied over roof decking before laying all but wood shingles. This seals the roof surface against moisture. Under asphalt and asbestos-cement shingles a 15-pound felt is adequate. Under heavier shingles like tile or slate, a 30-pound felt is recommended.

Exposure

The term *exposure* as applied to roof shingles refers to the number of inches that a shingle is exposed to the weather. It is the distance from the butt edge of one shingle to the butt edge of the shingle above or below it.

Shingles are manufactured and sold by the *square*. When they are laid in accordance with the exposure specified by the manufacturer, one square will cover 100 square feet. If the shingles are laid with an exposure less than specified, a square of them will cover less than 100 square feet. Conversely, as the exposure is increased, the number of square feet one square will cover increases.

The percentage by which the coverage of one square is to be increased or decreased may be obtained by the following formula:

$$\frac{\text{Difference in Specified and Laid Exposure}}{\text{Actual Exposure Laid}} = \%$$

If the shingle is laid with an exposure *greater* than specified, the coverage of one square should be *increased* by the percentage obtained. When the actual exposure of the laid shingle is less than specified, the coverage should be *decreased* by the percentage obtained.

Illustrative Example

Manufacturer's specified exposure   = 5 inches
Actual exposure shingle is laid   = 4 inches

5 inches—4 inches = 1 inch

$$\frac{1}{4} = 25\%$$

The shingle is laid with less exposure and, therefore, 1 square will cover 100 sq ft—25 sq ft = 75 sq ft.

(or)

Manufacturer's specified exposure   = 5 inches
Actual exposure shingle is laid   = 5½ inches

5½ inches—5 inches = ½ inch

$$\frac{.5}{5.5} = 9\%$$

The shingle is laid with greater exposure and, therefore, 1 square will cover 100 sq ft + 9 sq ft = 109 sq ft.

The amount of the exposure is important in estimating the *quantity* of shingles required. The labor factor is automatically taken care of by the quantity. A roofer lays shingles at a particular rate. If his rate of laying shingles is based on the number of hours per square, the over-all labor cost increases or decreases with the number of squares to be laid.

Starter Shingles

Where shingling begins at the eave line of a main roof or dormer, *starting* shingles or a *starting strip* is laid first. The usual practice is to reverse the shingle so that the butt end is toward the ridge. After laying the first course in this manner, the shingling is begun with the first exposed course laid directly over the starters. This method is necessary because wood, asphalt, and asbestos-cement shingles are laid with a spacing between them and the starters seal the roof under the spacing of the first course.

In the case of asphalt shingles a starter roll of similar material and color is available to be used in place of shingles. For asbestos-cement shingle roofs, special starting shingles are made.

Hip and Ridge Caps

To seal the ridge and hips of a roof, and also to give them a finished appearance, they are capped with the same shingles used on the roof, or with special caps provided by the roofing manufacturer.

Asphalt shingle roofs are generally capped with pieces cut out of the roof shingles. Asbestos-cement, and slate and tile shingle roofs are usually capped with pieces of special shape and size made available by the manufacturer.

The quantity of caps to be ordered is determined from the lineal feet of the ridges and hips. When caps are to be made from the roof shingles, the quantity is included in the over-all percentage added for waste.

WASTE

The term *waste* in estimating roofing includes the material lost by cutting and fitting around gables, ridges, hips and valleys, plus starters and caps, when the shingles are used for those purposes. While some roofers estimate waste by adding a square of roofing, more or less depending on circumstances, the common practice is to add a percentage to the net roof area to be covered.

The percentage to be added is a matter of an educated guess based on the shape and size of the roof. More waste is added for a cut-up surface than a plain one. More waste is also required for a hip roof than a gable roof.

Table 13-3 shows the approximate waste to be added to the net area of a roof. The figures include cutting waste, starter waste and an allowance for caps for ridges and hips. Where ridge and hip caps are ordered separately, the waste percentage may be reduced up to one half of the figure shown.

WOOD SHINGLES

Wood shingles are discussed fully in Chapter 12, pages 289 to 305, including Table 12-3 which gives the percentage of a square of shingles

Table 13-3
Approximate Waste to Be Added to the Net Area
of Variously Shaped Roofs

Roof Shape	Plain	Cut-up
Gable	10	15
Hip	15	20
Gambrel	10	20
Gothic	10	15
Mansard (sides)	10	15
Porches	10	—

required to cover 100 square feet. Table 12-4 shows the covering capacities and nail requirements of Centigrade Red Cedar Shingles for various exposures.

Open or solid sheathing under wood shingles on roofs is mainly a matter of choice. Open sheathing is more common as it not only permits the shingles to "breath" in drying out, but also is less expensive because a third to one half of the material is needed; there is a saving in labor as well as material. Open sheathing is applied on centers equal to the exposure of the shingles. Solid sheathing may be of matched or unmatched boards. Plywood may also be used with 5/8 inch thickness on rafters that are 16 inches on center, or 3/4 inch for rafters on 24-inch centers. Wood shingles are not recommended directly on 5/16 inch plywood or on composition roof sheathing.

Rosin-sized building paper is sometimes placed under wood shingles on roofs for insulation, but saturated felt paper is undesirable as it retards the drying of the shingles when they are wet.

The recommended maximum exposure for wood shingles on *roofs* is 5 inches, 5 1/2 inches, and 7 1/2 inches for 16-, 18-, and 24-inch shingles respectively. These exposures are specified by the manufacturer, and a square of 16-, 18-, or 24-inch shingles laid with these exposures will cover 100 square feet of surface. If the exposure is reduced, it will require more shingles to cover 100 square feet, and any increase in the exposure will decrease the quantity of shingles required for 100 square feet (Table 12-3). Shingles on roofs with a steep pitch can have a greater exposure than on roofs with a low pitch.

In applying wood shingles, the first course or row at the eaves is doubled to prevent water or snow from leaking between the shingles; also to induce a slight tension in succeeding courses. Roof shingles should have a space of approximately 1/4 inch between them to allow for swelling when they take up moisture.

To prevent dripping of water from the gables, or the formation of icicles, a roofer frequently nails a strip of 6-inch bevel siding along the edge with butt side out. The shingles are then laid in the usual manner. The result is to tilt the shingles in toward the center of the roof and direct the water away from the edge of the roof.

Ridges and hips are made water-tight by applying "caps." These are available in pre-assembled form, or can be made on the job from shingles of uniform width. This capping gives a finished effect to the roof.

Wood shingles may be laid directly over an old wood shingle roof without removing the old shingles. First, all decayed shingles should be removed and new shingles inserted to form a solid bearing and nailing surface; longer nails are required for penetration, 5d or 6d being recommended. Applying new wood shingles over old is frequently preferred because of the added strength and insulating qualities. Laying wood shingles over composition shingles is not recommended as condensation may accumulate beneath the wood shingles.

ESTIMATING QUANTITIES OF WOOD SHINGLES

The quantity of shingles needed for a roof is determined by computing the square foot area to be covered as discussed under "Measuring Roof Areas." From column 3 in Table 12-3 the percentage of a square of shingles needed to cover each 100 square feet may be obtained for the corresponding exposure and size of shingle. (See example page 299.) The waste to be added to the quantity of shingles for cutting, the starting course, and for ridge and hip caps is normally about 10 percent, unless the roof is considerably cut up with dormers and intersecting gables. Some roofers add 5 percent for cutting waste, and one bundle of shingles (1/4 of a square) for each 25 lineal feet of starters, hip and ridge caps.

ESTIMATING COSTS FOR WOOD SHINGLES ON ROOFS

The method for determining the quantity and cost of applying wood shingles to roofs is, in general, the same as outlined for sidewalls. Carpenters can apply 16- and 18-inch shingles on roofs under average conditions at a rate of about 4 hours per square. The rate for 24-inch shingles is about 3.2 hours per square. Where cut up, irregular roof surfaces are encountered and the time should be adjusted upward accordingly. Where experienced shinglers are employed, the rate of application may be reduced somewhat.

Unit Costs for Wood Shingles on Roofs

Using assumed prices and wages, unit costs may be developed as follows:

Unit Cost for 24-Inch Shingles on Roofs With
7 ½ Inch Exposure

1 square shingles   @ $24.00 = $24.00
2 lbs nails   @   .20 =   .40
3. 2 hours carpenter labor   @   6.00 = 19.20
                                        Unit cost per square = $43.60

$$\text{Unit cost per sq ft } \frac{\$43.60}{100} = \$ \ .44$$

Where the roof shingle exposures are greater or less than those given in the preceding illustration of unit costs, it is important to make an adjustment in either the unit cost or in the area to be covered. This is done by multiplying by the percentage shown in Column 3 of Table 12-3. The method is explained under sidewall shingling, pages 332 to 334.

HANDSPLIT SHAKES

There are few materials used on the roofs or side walls of residences to compare with *shakes* for architectural attractiveness. There are also few that enjoy the length of life of shakes when properly applied.

Handsplit shakes are available in three forms.

* *Handsplit and Resawn Shakes* have split faces and sawn backs, and are produced by running cedar blanks or boards of proper thickness diagonally through a bandsaw to produce two tapered shakes from each blank.

Tapersplit Shakes are produced mainly by hand, using a sharp-bladed steel froe and a wooden mallet. A natural shingle-like taper, from butt to tip, is achieved by reversing the block, end-for-end, with each split.

Straight-Split Shakes are split in the same manner as tapersplit shakes, except that the splitting is done from one end of the block only, producing shakes which are the same thickness throughout.

How Handsplit Shakes Are Made

Selected cedar logs are cut to specified lengths, the resulting blocks or "bolts" being trimmed to remove bark and sapwood. In the case of tapersplit or straight-split types, individual shakes are then rived from the blocks with a heavy steel blade called a "froe." To obtain a tapered thickness for tapersplit shakes, blocks are turned end-for-end after each split.

To produce handsplit-and-resawn shakes, blocks are split into boards of the desired thickness, then passed through a thin bandsaw to form two

* Data for discussion of shakes by courtesy of Red Cedar Shingle and Handsplit Shake Bureau.

shakes, each with a handsplit face and a sawn back. Expert sawyers guide the boards diagonally through the saw to produce thin tips and thick butts.

Roof Application

Roof sheathing may be solid or spaced, 1" x 4" boards on centers equal to the weather exposure the shakes are to be laid, but not greater than 10 inches. At the eave line a 36-inch strip of 30-lb roofing felt is laid. As in regular shingle application, starter courses should be doubled. After applying each course, an 18-inch wide strip of 30-lb felt is laid over the top portion of the shakes, extending onto the sheathing. This roofing felt interlay is not necessary when straight-split or taper-split shakes are applied in snow-free areas at weather exposures less than one-third the total shake length (a 3-ply roof).

Individual shakes should be spaced about 1/4 to 1/2 inch apart to allow for expansion.

Roof Pitch and Exposure

Proper weather exposure is important. As a general rule, a 7 1/2 inch maximum exposure is recommended for 18-inch shakes, and 10-inch maximum exposure for 24-inch shakes. Minimum pitch for handsplit shakes is 1/6 (4-in-12).

Table 13-4 sets forth a Summary of Sizes, Packing Regulations and Coverage for Handsplit Shakes by courtesy of the Red Cedar Shingle and Handsplit Shake Bureau.

Hips and Ridges

Pre-manufactured hip and ridge units are recommended. One bundle covers 16 2/3 lineal feet.

Sidewall Application

Maximum recommended weather exposure on sidewalls with single-course construction is 8 1/2 inches for 18-inch shakes and 11 1/2 inches for 24-inch shakes. Double-course application requires an underlay of shakes or regular cedar shingles. Weather exposures up to 14 inches are permissible with 18-inch handsplit/resawn or taper-split shakes, and 20 inches with 24-inch shakes. If straight-split shakes are used, the double course exposure may be 16 inches for 18-inch shakes and 22 inches for 24-inch shakes.

Nailing

Two 6d rust-resistant nails should be driven at least one inch from each edge, and one or two inches above the butt line of the course to follow.

Table 13-4

Shake Type, Length and Thickness	No. of Courses per Bundle	No. of Bundles per Square	Approximate coverage (in sq. ft.) of one square when shakes are applied with ½" spacing, at following weather exposures (in inches):								
			5½	6½	7	7½	8½	10	11½	14	16
18" x ½" to 3/4" Resawn	9/9 (a)	5 (b)	55(c)	65	70	75(d)	85(e)	100(f)			
18" x 3/4" to 1¼" Resawn	9/9 (a)	5 (b)	55(c)	65	70	75(d)	85(e)	100(f)			
24" x 3/8" Handsplit	9/9 (a)	5		65	70	75(g)	85	100(h)	115(i)		
24" x ½" to 3/4" Resawn	9/9 (a)	5		65	70	75(c)	85	100 (j)	115(i)		
24" x 3/4" to 1¼" Resawn	9/9 (a)	5		65	70	75(c)	85	100 (j)	115(i)		
24" x ½" to 5/8" Tapersplit	9/9 (a)	5		65	70	75(c)	85	100 (j)	115(i)		
18" x 3/8" True-Edge Straight-Split	14 (k) Straight	4								100	112(l)
18" x 3/8" Straight-Split	19 (k) Straight	5	65(c)	75	80	90 (j)	100(i)				
24" x 3/8" Straight-Split	16 (k) Straight	5		65	70	75(c)	85	100 (j)	115(i)		

(a) - Packed in 18"-wide frames.

(b) - 5 bundles will cover 100 sq. ft. roof area when used as starter-finish course at 10" weather exposure; 6 bundles will cover 100 sq. ft. wall area when used at 8½" weather exposure; 7 bundles will cover 100 sq. ft. roof area when used at 7½" weather exposure; see footnote (m).

(c) - Maximum recommended weather exposure for three-ply roof construction.

(d) - Maximum recommended weather exposure for two-ply roof construction; 7 bundles will cover 100 sq. ft. roof area when applied at 7½" weather exposure; see footnote (m).

(e) - Maximum recommended weather exposure for sidewall construction; 6 bundles will cover 100 sq. ft. when applied at 8½" weather exposure; see footnote (m).

(f) - Maximum recommended weather exposure for starter-finish course application; 5 bundles will cover 100 sq. ft. when applied at 10" weather exposure; see footnote (m).

(g) - Maximum recommended weather exposure for application on roof pitches between 4-in-12 and 8-in-12.

(h) - Maximum recommended weather exposure for application on roof pitches of 8-in-12 and steeper.

(i) - Maximum recommended weather exposure for single-coursed wall construction.

(j) - Maximum recommended weather exposure for two-ply roof construction.

(k) - Packed in 20"-wide frames.

(l) - Maximum recommended weather exposure for double-covered wall construction.

(m) - All coverage based on ½" spacing between shakes.

Longer nails should be used if thickness of shake indicates to achieve proper penetration. It takes 2 lbs of nails per square at standard exposures.

Estimating Quantities

Obtain the area of the roof as for other types of roof covering and divide by 100 to compute the number of squares needed at *standard exposures*. Allow for double coursing at eaves. One square will provide about 240 lineal feet of starter course. Allow one extra square for every 100 lineal feet of valleys. Check table for coverage when shakes are applied at exposures less than standard exposures.

Estimating Labor

As with other roof shingle application the labor per square varies with pitch of the roof, how badly cut up the surface is with dormers and intersecting roofs, and the experience of the men.

A fairly experienced shingler can apply shakes at the rate of about 1 square an hour. A highly experienced man can apply 1 1/2 squares an hour on uncomplicated roofs using automatic nailing tools as shown on page 335.

A unit cost per square would be developed from the following formula.

1 square handsplit shakes	$
1 hour labor @ $ per hour	
2 lbs of 6d aluminum nails	———
Total cost per square	$

Illustrative Example

1 Square 24″ handsplit shakes	$48.00
Labor 1 hour @ $6	6.00
Nails 2 lbs @ 50¢	1.00
Cost per square	$55.00
Cost per square foot	.55

ASPHALT SHINGLES

The three major types of asphalt roofing products are the saturated felt used for built-up roofs, shingles, and roll or ready roofing.

Asphalt is a bituminous material found either in native form, or as a by-product of petroleum. The latter source is the one from which the majority of asphalt for roofing is obtained. All three types of asphalt roofing are made by impregnating, with asphalt, a felt paper made from such organic fibres as rag, wood, and jute; asbestos fibres are also used.

Workmen applying handsplit shakes using penumatic nailing tools, courtesy Senco
Products, Inc., Cincinnati, Ohio 45244

Workman applying asphalt shingles with pneumatic stapler. Courtesy Senco
Products, Inc. Cincinnati, Ohio.

To help resist weathering, the asphalt is combined with stabilizers of finely divided minerals. Silica, talc, slate dust, mica, and trap rock are the materials most commonly used. Coarser minerals of the same kind are applied to the surface of shingles and roll roofing to protect the asphalt from the rays of the sun, to improve the fire-resistive qualities of the roofing, and to provide attractive colors for architectural purposes. It has also been found that the light colors reflect the heat of the sun to a greater extent than darker colors, thereby reducing the temperature of the interior of the building during warm seasons.

Figures 13-14 and 13-15 show the weight and quantities per square, the sizes and exposures for Typical Asphalt Shingles and Asphalt Rolls as furnished by the Asphalt Roofing Manufacturers Association.

When asphalt shingles are to be laid directly over an old roof, the galvanized nails should be 1 3/4 inches long. Some property owners consider that putting new asphalt shingles on top of old roofing will provide better insulation. However, more than two thicknesses of roofing is not recommended because of the weight and also the unevenness of the surface which results. If asphalt shingles are laid over old wood shingles, a tapered strip like bevel siding, called a "horse feather" is nailed between the butts of the wood shingle to form a flat surface. The cost of the labor and material to do this work is estimated separately from the cost of shingling.

Estimating Material

The number of squares of asphalt shingles required is obtained by computing the area to be covered, adding the appropriate percentage of waste as shown in Table 13-3, and dividing by 100. The quantity should be carried out to the next half, or third of a square, depending on whether the shingles come two or three bundles to the square.

Large head, rust-resistant nails 1 1/4 inches long are used on new work. The quantity of nails needed per square of shingles varies slightly, but for estimating, 2 pounds is adequate.

When 15-lb saturated felt is used, 100 square feet (approximately 1/4 of a standard roll) will be required for each square of shingles.

Estimating Labor

Table 13-5 shows the approximate number of hours of labor required to lay a square of various types of roofing. Under average working conditions a moderately experienced shingler will lay a square of 3-tab shingles in 2 hours. If the roof is badly cut up it may take slightly longer. A thoroughly experienced roofer will lay 3-tab shingles on a plain roof at a rate close to 1 1/2 hours per square.

1	2	3		4		5	6
		Per Square		Size			
PRODUCT	Configuration	Approximate Shipping Weight	Shingles	Bundles	Width	Length	Exposure	Underwriters' Listing

1	2	3 (Shipping Weight)	3 (Shingles)	3 (Bundles)	4 (Width)	4 (Length)	5 (Exposure)	6
Wood Appearance Strip Shingle More Than One Thickness Per Strip 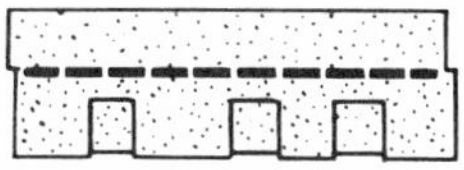 Laminated or Job Applied	Various Edge, Surface Texture & Application Treatments	285# to 390#	67 to 90	4 or 5	11-1/2" to 15"	36" or 40"	4" to 6"	A or C - Many Wind Resistant
Wood Appearance Strip Shingle Single Thickness Per Strip 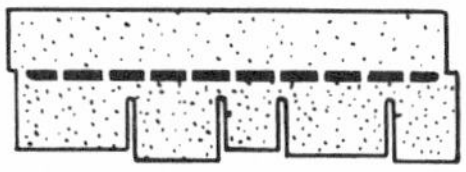	Various Edge, Surface Texture & Application Treatments	Various 250# to 350#	78 to 90	3 or 4	12" or 12.25"	36" or 40"	4" to 5-1/8"	A or C - Many Wind Resistant
Self-Sealing Strip Shingle 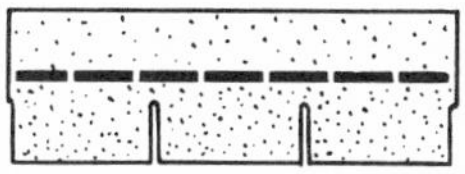	Conventional 3 Tab	205#- 240#	78 or 80	3	12" or 12.25"	36"	5" or 5-1/8"	A or C - All Wind Resistant
	2 or 4 Tab	Various 215# to 325#	78 or 80	3 or 4	12" or 12.25"	36"	5" or 5-1/8"	
Self-Sealing Strip Shingle 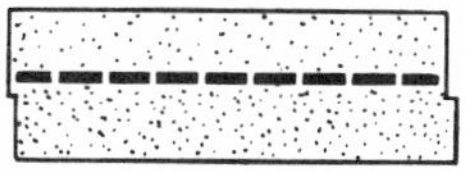 No Cut Out	Various Edge and Texture Treatments	Various 215# to 290#	78 to 81	3 or 4	12" or 12.25"	36" or 36-1/4"	-5"	A or C - All Wind Resistant
Individual Lock Down 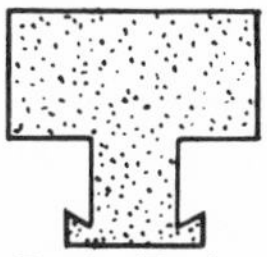 Basic Design	Several Design Variations	180# to 250#	72 to 120	3 or 4	18" to 22-1/4"	20" to 22-1/2"	-	C - Many Wind Resistant

Other types available from some manufacturers in certain areas of the Country.
Consult your Regional Asphalt Roofing Manufacturers Association manufacturer.

Figure 13-14

Typical Asphalt Shingles and Typical Asphalt Rolls. Courtesy Asphalt Roofing Manufacturers Association, New York, N.Y. 10017

1	2		3	4		5		6	7
PRODUCT	Approximate Shipping Weight		Sqs. Per Package	Length	Width	Side or End Lap	Top Lap	Exposure	Underwriters' Listing
	Per Roll	Per Sq.							
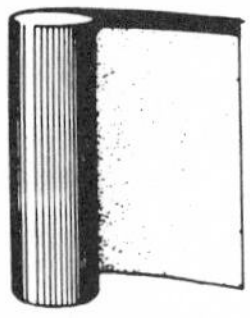 Mineral Surface Roll	75# to 90#	75# to 90#	One	36' 38'	36" 36"	6"	2" 4"	34" 32"	C
	Available in some areas in 9/10 or 3/4 Square rolls.								
 Mineral Surface Roll Double Coverage	55# to 70#	55# to 70#	One Half	36'	36"	-	19"	17"	C
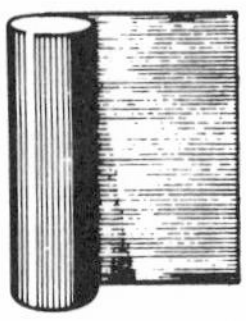 Coated Roll	50# to 65#	50# to 65#	One	36'	36"	6"	2"	34"	None
 Saturated Felt	60# 60# 60#	15# 20# 30#	4 3 2	144' 108' 72'	36" 36" 36"	4" to 6"	2"	34"	None

Figure 13-15

Typical Asphalt Shingles and Typical Asphalt Rolls. Courtesy Asphalt Roofing Manufacturers Association, New York, N.Y. 10017

Individual type shingles require more time to lay as indicated in Table 13-5. Reference is made to the discussion of Factors Affecting Labor, this chapter, page 308.

Table 13-5
Approximate Hours of Labor to Remove and to Lay One
Square of Roofing (Including Underlayment of Saturated
Felt Where Required)

Type of Roofing	Pounds of Nails	Hour of Labor per Square		
		Plain Roofs	Difficult or Cut-up Roofs	To Remove Old Roofing
Asphalt Strip Shingles	2	2	3	1
Asphalt Individual Shingles	4	4	6	1
Asbestos Rectangular Shingles	3	6	8	1½
Asbestos Hexagonal Shingles	3	4	6	1½
3/16″ Slate Shingles	3	6	8	1½
Handsplit Shakes		1	2	1
Wood Shingles	2	4	6	1½
Corrugated Galvanized or Aluminum on Wood	2	2½	3½	1
Corrugated Galvanized Steel or Aluminum	2	6	8	1
3-Ply Built-up Gravel Roof	—	2	—	1½
4-Ply Built-up Gravel Roof	—	2½	—	1½
5-Ply Built-Up Gravel Roof	—	3	—	1½
Roll Roofing	—	1½	2	1

Unit Costs for Asphalt Shingles

The unit cost for asphalt shingles is based on 1 square of material. The hours of labor are taken from Table 13-5.

Unit Cost of 3-Tab 210 Lb. Asphalt Shingles on Plain Roofs

1 sq asphalt shingles	@ $9.50 = $ 9.50
2 lbs asphalt roofing nails	@ .20 = .40
2 hours labor	@ 6.00 = 12.00
	Unit cost per square = $21.90

$$\text{Unit cost per sq ft } \frac{\$21.90}{100} = \$ \ \ .22$$

Unit Cost of 18″ x 20″ Individual Lock Down
Asphalt Shingles on Plain Roofs

1 sq asphalt shingles	@ $12.50 =	$12.50
2 lbs asphalt roofing nails	@ .20 =	.40
4 hours labor	@ 6.00 =	24.00

Unit cost per square = $36.90

Unit cost per sq ft $\frac{\$36.90}{100} = \$ .37$

SLATE SHINGLES

Slate is a laminated rock which is quarried and split into thin slabs by the manufacturer. The size of slate shingles varies between 6 and 8 inches in width, and between 12 and 24 inches in length. While the thickness of standard slate is about 3/16 inch, it is available in thicknesses of 1 inch and over, according to architectural requirements. A popular size slate shingle for dwellings is 10 inches wide and 20 inches long.

Slate is laid on solid roof decking with a 30-lb saturated felt underlayment. They should not be laid over old existing roofing.

The holes in slate for nails are usually punched at the quarry. The *lap* of a slate is the distance by which the upper slate overlaps the head of the second slate below. The standard lap is 3 inches. The exposure of a slate is 1/2 of the length after deducting the lap. A 10-inch shingle with a 3-inch lap will have a 3 1/2-inch exposure $\frac{(10-3)}{2} = 3 1/2$ inches.

Hips and ridges are usually finished by alternately lapping shingles cut to proper size. The first course at the eaves is doubled for a starting course.

Waste from breakage and cutting, and allowance for the starter course, ridge and hip caps, varies between 10 and 20 percent as shown in Table 13-3.

Estimating Materials

The quantity of slate required is determined by adding the appropriate waste to the net roof area, and dividing by 100 to obtain the number of squares. Approximately 1/2 roll of 30-lb saturated felt is required for each square of slate.

Copper nails, 3d or 4d, are recommended. It requires about 3 pounds of nails to lay a square of slate shingles.

Estimating Labor

Experienced slaters are difficult to find in most areas. Two men generally work together and occasionally, a helper is used to fill out a crew.

The labor to lay a square of slate shingles varies with the size of the slate, the experience of the workmen, and the shape of the roof. Table 13-5 shows the approximate number of man-hours per square under average conditions.

Unit Costs for Slate Shingles

A unit cost per square of slate shingles may be obtained by inserting the local material prices and hourly wage rates as shown in the following illustrations. The hours of labor may be taken from Table 13-5.

Unit Cost of 10" x 20" x 3/16" Slate Shingles on Plain Roof

1 square 10" x 20" x 3/16" slate	@ $50.00 =	$50.00
3 lbs 3d copper nails	@ .65 =	1.95
½ roll 30-lb felt	@ 3.00 =	1.50
6 hours labor	@ 6.00 =	36.00
	Unit cost per square =	$89.45

$$\text{Unit cost per sq ft} = \frac{\$89.45}{100} = \$ \ .89$$

Unit Cost of 10" x 20" x 3/16" Slate Shingles on Cut-up Roof

1 square 10" x 20" x 3/16" slate	@ $50.00 =	$ 50.00
3 lbs copper nails	@ .65 =	1.95
½ Roll 30-lb felt	@ 3.00 =	1.50
8 hours labor	@ 6.00 =	48.00
	Unit cost per square =	$101.45

$$\text{Unit cost per sq ft} = \frac{\$101.45}{100} = \$ \ 1.01$$

ASBESTOS-CEMENT SHINGLES

Asbestos-cement shingles are a rigid type, fire-resistive shingle made of a composition of asbestos fiber and portland cement under heavy pressure. They are available in the popular white or gray shingle textured to simulate wood, or in a variety of colors. The shingles are usually rectangular or hexagonal.

Asbestos-cement shingles are laid over solid roof decking with an underlayment of 30-lb saturated felt paper. An advantage of asbestos-cement shingles is that they can be placed directly over old wood shingles if desired.

A shingle cutter is available for sale, or may be rented from a dealer to cut, punch, and notch asbestos-cement shingles.

Galvanized needle-point nails are recommended to lay these shingles. A 3d nail is used on new work, but when shingling over an old roof the 5d or 6d nail should be used. About 3 pounds of nails are required to lay one square.

Estimating Materials

Asbestos-cement shingles are sold by the square which is sufficient to cover 100 square feet when laid in accodance with the specified exposure. Exposures vary between 6 and 9 inches. Starter shingles for the first course, and also hip and ridge shingles are available.

The waste to be added to the net roof area for cutting and fitting, and for starters, hip and ridge shingles is shown in Table 13-3.

The number of squares of asbestos-cement shingles is obtained by adding the appropriate amount of waste to the net roof area and dividing by 100.

Approximately one-half roll of 30-lb asphalt-saturated felt and 3 pounds of nails are needed per square of shingles.

Estimating Labor

The number of man-hours required to lay asbestos-cement shingles varies with the type and size of the shingle. The 8" x 16" rectangular shingle, similar to the slate shingle, takes longer to lay than the larger hexagonal shingle, which is 14" x 30" wide. Table 13-5 shows the approximate number of hours of labor per square under average conditions.

Unit Costs for Asbestos-Cement Shingles

The following material prices and hourly wage rates are assumed; labor is taken from Table 13-5.

Unit Cost of 8" x 16" x ¼" Asbestos-Cement
Shingles on Plain Roof

1 square asbestos-cement shingles	@ $40.00 =	$40.00
3 lb 3d needle-point nails	@ .30 =	.90
½ roll 30-lb sat. felt paper	@ 3.00 =	1.50
6 hours labor	@ 6.00 =	36.00
	Unit cost per square =	$78.40

$$\text{Unit cost per sq ft } \frac{\$78.40}{100} = \$ \quad .78$$

Unit Cost of 14" x 30" Hexagonal Asbestos-Cement
Shingles on Plain Roofs

1 square asbestos-cement shingles	@ $18.00 =	$18.00
3 lb 3d needle-point nails	@ .30 =	.90
½ roll 30-lb sat. felt paper	@ 3.00 =	1.50
4 hours labor	@ 6.00 =	24.00
	Unit cost per square =	$44.40

$$\text{Unit cost per sq ft} = \frac{\$44.40}{100} = \$ \quad .44$$

ROLL ROOFING

Asphalt roll roofing comes in rolls that are 36 inches wide and 36 feet long. A roll contains 108 square feet and, making allowance for lapping, it is sufficient to cover 100 square feet. The 65-lb smooth surfaced, and the 90-lb mineral surfaced rolls are the most commonly used. An underlayment of 15-lb saturated felt is sometimes used, but not required. Lapping cement and 7/8 inch galvanized nails are provided inside of each roll. When laid over old roofing longer nails should be used in order to penetrate the wood decking at least 3/4 of an inch.

Roll roofing is applied to roofs of every pitch. It is used on all types of structures with flat roofs, particularly where initial cost is a consideration. It is the least expensive roofing to apply.

Roll roofing may be laid parallel with the eaves or the ridge of a pitched roof. Caps for ridges and hips are usually made by splitting a roll down the center. For better quality of work, 9-inch starter strips are applied along the eaves and up the *rake* or gable ends.

The ends are lapped 6 inches and the edge laps are 2 inches. Roll roofing may be applied by nailing the upper edge of each strip and cementing the bottom edge. This is called the *concealed nail method*. The more common method of application is to nail the bottom edge of the overlapping strip which is called the exposed nail method.

Estimating Material

The quantity of roll roofing required for a particular roof is determined from the net roof area to be covered. The amount of waste in cutting, lapping at edges of eaves and gables, and for caps, varies with each job. Generally 10 percent is adequate to take care of most situations.

Estimating Labor

Labor factors which affect the rate of laying other types of roofing should be properly considered in estimating labor for roll roofing. The rolls are heavy and more difficult to manage as the pitch of the roof increases. On flat or slightly pitched roofs a man should be able to cover a square in 1 1/2 hours. On steep pitched or difficult roofs, the time should be increased to 2 hours per square.

Illustrative Example

Assuming the local price for 1 roll of 90-lb roofing to be $5.00, and the hourly wage rate to be $6.00, the unit cost applied to a plain flat roof would be:

1 roll 90-lb roofing   @ $5.00 = $ 5.00

1½ hours labor
                                                      @ 6.00 =   9.00
                                        Unit cost per square = $14.00

                                                        $14.00
                                        Unit cost per sq ft ——— = $ .14
                                                         100

BUILT-UP ROOFING

As its name indicates, built-up roofing is one that is applied by *building up* successive layers of saturated felt, lapped and cemented together with hot tar (pitch) or asphalt. The layers of felt are called *plies,* and this type of roof is identified by the number of plies, usually running from 3 to 5 thick. The top may be finished off with a mopping of tar or asphalt, and is referred to as a built-up smooth-top roof. Where slag or gravel is imbedded in the top coating, the roof is commonly spoken of as a tar and gravel, or a slag roof. When the incline of the roof exceeds 2 inches to the foot, the top ply is frequently mineral-surfaced 19-inch Selvage, Double Coverage Roll Roofing. (See Page 343.)

The saturated felt used for built-up roofs is usually the 15-lb type. It is 36 inches wide, 144 feet long, and contains 432 square feet. One roll weighs 60 pounds. On many roofs, particularly bonded roofs, the first ply is a 30-lb felt which is available in rolls of 216 square feet weighing 60 pounds.

On a new roof over wood decking, a layer of rosin-sized building paper or one layer of 15-lb saturated felt is applied to prevent the pitch or asphalt from leaking through cracks, knot-holes, or joints in the boards. This layer of paper is not counted as one of the plies. Successive layers of 15-lb felt are then laid, lapping the sheets 19 inches and mopping between plies with hot pitch or asphalt. Approximately 30 pounds of asphalt or pitch per square is used for each mopping with at least double that quantity (60 pounds) applied to the surface if slag, gravel, or crushed marble is to be imbedded. About 300 pounds of slag or 400 pounds of gravel is required for each 100 square feet of surface covered. The quantity of pitch or asphalt applied to the surface when slag or gravel is not used is about 40 pounds per square.

In the application of lower quality built-up roofing, the first couple of layers of felt are laid dry, that is, without mopping between them. Other means of reducing both cost and quality include the use of less pitch or asphalt. An examination of a cross-section of a sample of the roofing that has been damaged will generally disclose the number of plies and quality of the roofing.

Estimating Materials

The number of squares of built-up roofing required is obtained from the net area to be covered. Waste is taken care of in the rolls of felt by the manufacturer. A roll of 15-lb felt contains 432 square feet and is intended to cover 400 square feet or 4 squares. A roll of 30-lb felt contains 216 square feet and will cover 200 square feet or 2 squares.

Measurements of the roof are taken from the *outside* of the walls. Allowance should be made for flashing against bulk-heads and parapet walls. Normally the flashing is carried up 1 foot, but in many older buildings the entire parapet may be flashed up to the coping.

Skylights, bulk heads, ventilators and so forth, should not be deducted unless the area they occupy is in excess of 100 square feet, and then it is customary to deduct only one half of the area to allow for the additional work of cutting, flashing, and working around them.

The materials used in built-up roofing are saturated felt, asphalt, or coal-tar pitch, and slag or gravel when so surfaced. Different manufacturers and roofers vary the specifications, but for estimating purposes, the quantities shown in Table 13-6 are approximately correct.

Table 13-6
Materials Required to Apply 100 Square Feet Built-up Roofing
Using Asphalt, and Surfacing with Slag or Gravel

Number of Plies	Rolls of 15-Lb Saturated Felt (432 Square Feet)	Pounds of Asphalt	Pounds of Slag	Pounds of Gravel
3	¾	150	300	400
4	1	180	300	400
5	1¼	230	300	400
3 (1 dry ply)	¾	120	300	400
4 (2 dry plies)	1	120	300	400
5 (2 dry plies)	1¼	150	300	400

(For smooth-top finish without slag or gravel, deduct 20 pounds of asphalt per 100 square feet of roofing. When 30-lb saturated felt is used as a first ply, add ¼ of a roll of 15-lb felt.)

Estimating Labor

Built-up roofing is applied by a crew ranging from three to five men, depending on the size of job and working conditions. All materials are raised with a rope and pulley hoist rather than carried up ladders. One man generally tends the asphalt-heating equipment on the ground while the others lay the felt and mop.

Relatively flat roofs are easier to work on than those with slopes, or which are built over crescent or bowstring trusses. The latter frequently require guard-rails, and other safety devices to protect the workmen. As the number of story heights increase, more labor is used in hoisting materials to the roof.

A roof with numerous skylights, ventilators, bulkheads and so forth, or one of irregular shape, requires more labor than a plain roof. Each job should be studied for its individual characteristics before deciding on the number of man-hours to allow per square. The approximate number of man-hours per square for uncomplicated roofs, including a foreman, are:

> 2 hours for 3-ply
> 2½ hours for 4-ply
> 3 hours for 5-ply

Fuel, Mops, Supplies

The fuel required to heat asphalt or pitch, the mops to apply it, brushes for spreading gravel, pails, nails, and so forth, are all expendable supplies. In computing the cost of a roof, a flat allowance may be made to cover these items, or a percentage may be added to the *total cost of the labor and the material* for each job. The amount varies, but ranges between 3 and 5 percent.

Contractors Overhead and Profit

Because built-up roofing is subcontracted to roofers who are properly equipped, and who specialize in that work, it is important to add their overhead and profit to the cost of all material and labor. The amount to be added usually varies between 20 and 30 percent depending on the roofer who is doing the job.

Unit Costs for Built-up Roofing

The following formulas are used to show how to develop a unit cost. Overhead and profit should be added (Prices and wages assumed). Labor rates are from Table 13-5.

Unit Cost of 3-Ply Built-up Slag Surface Roofing Over Wood Decking Using 30-lb Felt for First Layer—All Plies Mopped

¼ roll 15-lb felt laid dry to seal decking	@	$3.00 = $	.75
1 roll 15-lb felt	@	3.00 =	3.00

400 lbs gravel	@ $.25/100 lb. =	1.00
150 lbs asphalt or pitch	@ 2.36 =	3.54
2 hours labor	@ 6.00 =	12.00
		$20.29
	Fuel and supplies 5%	1.01
	Unit cost per square =	$21.30

$$\text{Unit cost per sq ft} = \frac{\$21.30}{100} = \$ \ .21$$

Unit Cost of 4-Ply Built-up Smooth Surface Roofing Over Wood
Decking Using 30-lb Felt for First Layer—All Plies Mopped

¼ roll 15-lb felt laid dry to seal decking	@ $3.00 =	$ .75
1¼ roll 15-lb felt	@ 3.00 =	3.75
180 lbs asphalt or pitch	2.00 =	3.60
2½ hours labor	@ 6.00 =	15.00
		$23.10
	Fuel and supplies 5%	1.15
	Unit cost per square =	$24.25

$$\text{Unit cost per sq ft} = \frac{\$24.25}{100} = \$ \ .24$$

CORRUGATED METAL ROOFING

Both corrugated aluminum and galvanized steel roofing are used extensively on industrial type buildings and on farm and secondary buildings. The sheets are available with either 1 1/4 or 2 1/2-inch corrugations, and come in different widths and lengths. The 26-inch wide sheets are most popular, and lengths ranging from about 6 to 12 feet should be selected according to the area to be covered to avoid as much cutting as possible.

Side laps should be a minimum of 1 1/2 corrugations, and end-laps at least 2 to 6 inches depending on the pitch of the roof.

Ridge and valley rolls, rubber end-seals, and end-wall flashings may be obtained for both galvanized and aluminum corrugated roofing. Special rust-resistant nails and washers are used to apply this type of roofing. A 1 3/4 inch nail is used for new work on wood, and a 2-inch nail should be used when applied over an old roof; allow 2 pounds per square.

Corrugated roofing may be applied to solid wood decking, or over open framing. When applied over steel framing, special fittings are required. Add 15 to 20 percent to the cost of the material for rivets, clips, and so forth.

The price of corrugated roofing varies considerably with its thickness. Galvanized sheets are available in 12 gauge to about 26 gauge. Aluminum sheet thickness is measured in inches and range from .019 inch (approximately 26 gauge) to .032 inch. The 26 gauge galvanized and .019 inch aluminum are frequently used on farm buildings, and the heavier types are used on industrial buildings.

Estimating Material

The quantity of corrugated roofing is obtained from the net roof area to be covered. The amount of *waste* to be added for lapping sides and ends, and for cutting and fitting varies between 15 and 25 percent depending on the amount of lap and cutting required. Generally 20 percent added to the net roof area will be adequate.

Two pounds of nails should be included for each square of roofing. Ridge rolls, valley rolls, and other accessories should be estimated by the number of lineal feet required.

Estimating Labor

The labor rate to apply corrugated roofing varies with the job conditions, to some extent the weight of the sheets, and whether or not it is being applied over wood or metal. Table 13-5 shows the approximate number of man-hours for applying corrugated roofing to wood and to steel.

Unit Costs for Corrugated Roofing (Prices and Wages Assumed)

Unit Cost of .019 Inch Corrugated Aluminum
Roofing Applied Over Wood on Plain Roof

1 square .019 in. roofing	@ $20.00 =	$20.00
2 lbs nails	@ .45 =	.90
2½ hours labor	@ 6.00 =	15.00
	Unit cost per square =	$35.90

Unit cost per sq ft $\frac{\$35.90}{100} = \$ .36$

Unit Cost of .032 Inch Corrugated Aluminum
Roofing Applied Over Wood on Plain Roof

1 square .032 in. roofing	@ $34.00 =	$34.00
2 lbs nails	@ .45 =	.90
2½ hours labor	@ 6.00 =	15.00
	Unit cost per square =	$49.90

Unit cost per sq ft $\frac{\$49.90}{100} = \$ .50$

Unit Cost of 20 Gauge Corrugated Galvanized
Steel Roofing Applied Over Steel Framing
Plain Roof

1 square 20 gauge roofing	@ $22.00 = $22.00
add for fittings 20%	= 4.40
6 hours labor	@ 6.00 = 36.00
	Unit cost per square = $62.40

$$\text{Unit cost per sq ft} = \frac{\$62.40}{100} \$ \quad .62$$

DAMAGEABILITY OF ROOFING

No roofing material may be considered permanent inasmuch as all types are subject to normal weathering, or a shortened life because of improper application. All types of roofing are also subject to damage by fire, and by wind or hail. There are other miscellaneous causes of damage, less frequently encountered such as lightning, collapse, falling trees or their limbs, and explosion.

Normal Weathering

All roofing materials deteriorate from exposure to the weather. The rate of the deterioration depends to a large extent on the kind of material, the degree of maintenance, and the conditions of exposure. Rigid materials such as slate, metal, asbestos-cement, and tile, are less vulnerable to the action of wind, rain and sun than the pliable asphalt roofing products.

Some roof materials in southern climates wear out faster than those in northern latitudes, and generally the south and west exposures of a roof will deteriorate quicker than the north and east exposures. As a rule, most roofing materials on steep slopes will outlast the same materials when applied to those of less pitch. Shaded roofs with wood shingles have a shorter life than those exposed to the sun and wind, which are necessary to keep them dry after wetting by rain, snow or dew.

Summer heat of extreme temperatures dries out the oils in asphalt roofing, and the shrinkage and expansion due to extreme ranges of temperature may cause cracking or alligatoring of the surface.

Maintenance

A roof that receives little or no maintenance weathers and wears out much more rapidly than one that receives proper attention. Wood shingle roofs that are periodically treated with creosote will lay flat and last much longer than untreated shingles. The painting of wood cedar shingles increases their life as the low rate of expansion and contraction does not disturb the film of paint. Heavy oil paints should not be used.

Built-up smooth surface roofs should be coated with bitumen periodically, particularly in areas of high temperatures, to replace the oils lost by exposure to the sun. Built-up roofs covered with crushed stone deteriorate in those places where the gravel, or slag, has been washed or blown away. Such areas should be cleaned off, recoated and the mineral should be replaced.

When the top coating of bitumen is no longer heavy enough to imbed the mineral surfacing properly, the roof should be scraped. A heavy coating of bitumen should be applied and the slag or gravel replaced.

Flashings against parapet walls, bulkheads, and chimneys may crack from shrinkage, particularly if not properly applied. They require recovering or replacing. Low spots in flat-decked roofs accumulate water which freezes in winter and causes breaks in built-up or roll type roofing. Unless repairs are made promptly, water gets under the roofing surface, causing blisters, or rotting of the wood-decking below.

Split wood and cracked slate, or asbestos-cement shingles, should be replaced or patched to prevent water from getting under the surface to cause further deterioration.

IMPROPER INSTALLATION

When roofing is not properly laid, it deteriorates more rapidly than when applied according to manufacturer's specifications.

If the wood decking is laid with green lumber, the curling or cupping causes ridges which press upward to loosen nails and cut through the surface. Inadequate nailing of asphalt shingles or roll roofing exposes the laps to moderate wind action. Nailing asphalt shingles above the thick butt prevents them from lying flat. The use of nails which are too short, or are not rust-resistant, endangers the life of the roofing. The placing of more than two nails in wood shingles causes them to split when they expand and contract. Laying shingles with too great an exposure also exposes them to wind action and frequently causes curling.

Asphalt shingles should be laid over old wood shingles only after filling in the exposed butts with a feather-edge piece to give a solid base. Otherwise they will not lie flat and the surface will be uneven. When laid directly over old wood shingles, asphalt shingles are highly vulnerable to hail damage in the places that are not supported by a solid backing.

When a new asphalt shingle roof is laid over an old asphalt shingle roof, great care is necessary to obtain solid nailing and to select the proper length of nail which will penetrate the roof boards at least 3/4 of an inch.

Aluminum roofing placed in contact with steel or copper in the presence of moisture will corrode or pit from the electrolytic action. Most unprotected metal roofs that are exposed to spray from salt water will also deteriorate rapidly.

FIRE DAMAGE

Wood shingles and asphalt roofing products are combustible, although the degree of heat required to ignite the latter is relatively high. The principal value of asphalt roofing products in the presence of fire is their longer resistance to ignition.

Sparks and embers from exposure fires, although failing to ignite wood or asphalt roofing, can cause scorching and marring to the surface. In more serious exposure fires, the scarring due to embers is sufficient to require complete replacement. Rigid type roofings such as tile, slate, asbestos-cement, or metal, are highly resistant to exposure fires.

Damage by fire to all types of roofing materials is the result, in most cases, of a fire which originates within the structure itself. It either chars or otherwise damages the roof framing and decking to the extent that replacement of the area is necessary, or the fire actually breaks through the roof. Frequently holes are chopped in roofs by firemen to ventilate the fire or to extinguish it.

HAIL DAMAGE

The extent of damage to roofing by hailstones depends largely on the age of the roof, the type of material, and the size of the hailstones. Asphaltic roofing products are more vulnerable to the impact of hail than rigid materials. However, no roofing material is completely safe in storms involving hailstones as large or larger than golf balls, a phenomenon not uncommon in many parts of the country. Spanish tile and slate roofing surfaces have been pulverized in severe hailstorms.

The lighter gauge metal roofs may be badly dented or pocked by pea-sized hail. Wood shingles can be dented and split. Asphalt shingles and roll or built-up roofing may also be dented and frequently hail perforates the materials.

Numerous instances are on record of hail 2 inches or more in diameter passing through the roof decking.

Damage to asphalt shingles applied over old wood shingles is more serious when no feather-edge inserts have been used to fill between the butts of the old shingles.

Built-up roofing surfaced with gravel or slag is much less susceptible to hail damage than smooth top roofing.

WIND DAMAGE

Damage to roofing during a windstorm, hurricane, or tornado may occur as a result of the direct force of the wind getting under the material; by suction; or by the collapse or destruction of the roof.

The effect of the direct action of wind on roofing depends on its velocity, the kind of material, its condition, and method of application. The Red Cedar Shingle Bureau states that: "To lift a shingle 8 inches wide away from a roof covered with No. 1 16-inch shingles laid with a 5-inch exposure requires a pull of 85 pounds—a force so much greater than a hurricane can exert that it can be conservatively stated that properly nailed shingles simply cannot be blown from a roof." Yet an old worn shingle roof or one laid with too much exposure can be seriously damaged by winds of considerably less velocity than those experienced in hurricanes. Asphalt shingles properly nailed and laid with a 4-inch exposure, will fare much better in high winds than shingles nailed too high, or with a 5- or 5 1/2-inch exposure. Shingles applied to steep-pitched roofs suffer less damage by wind than those laid on low-pitched roofs.

Both experience and experiments conducted in wind tunnels indicate that the roof shingles located 5 feet from the rake or edge of a roof, and four or five courses nearest the ridge, are subject to greater damage than other areas of the surface.

Asphalt shingles which are lifted during a strong wind may break along the nailing line. After the wind has subsided, the shingles return to their former position concealing the real damage.

Roll roofing, especially when laid with exposed nailing, is highly vulnerable to winds if the nails are too far apart or too close to the edge.

Slate, asbestos-cement, and tile roofs are extremely wind-resistant when properly laid. Usually they are not damaged except in the high winds of a hurricane or tornado.

Heavy gauge corrugated metal or transite roofs properly applied will withstand winds of high velocities, but the light gauge aluminum or galvanized corrugated roofs are subject to damage when wind gets underneath the edges or laps. Built-up roofing may be damaged by wind of high velocity if a break in the surface exists, or occurs during the storm, and gusts are able to get underneath. The force of hurricane or tornado winds may be sufficient to catch under a small break, and roll back large areas, or lift the roofing, loosening the entire surface.

In some cases, sections of built-up roofing are pulled away from the decking by suction, particularly if the first ply or layer is not adequately nailed. Frequently, after the storm, the surface of the roofing is loose and fluffed like a loose blanket. Damage of this kind usually cannot be repaired and the entire area involved must be removed and re-roofed.

Strong winds often blow the loose gravel or slag into piles on the roof, or blow it off the roof entirely.

METHODS OF REPAIR

The method of repairing or replacing of roofing that has been damaged by fire generally presents no unusual problems. The area damaged is plainly visible, unless destruction is complete, and repairs are determined by observation and measurement.

The extent of injury to roofing by hail or windstorm is not so readily determined without a careful survey. In all cases, when practical, the inspection should be made *on the roof* rather than from the ground. A roof that appears undamaged from ground level may show extensive injury upon close examination. Wind-broken shingles or small pocks or holes caused by hail are not always easy to observe unless seen at close range.

Repairing Asphalt Shingles

Replacing individual asphalt shingles is done by sliding a ripper underneath the shingle and cutting off the nails. A new shingle is then inserted and nailed. Cracked, torn, or otherwise damaged shingles should be removed and replaced. The cost of individual shingle replacement is primarily one of labor. Most roofers have a minimum flat price for a call, though only one shingle is involved. Where a number of them must be replaced, the charge may be based on a flat charge per shingle.

When the damage occurs over several square feet or more, ample allowance of material should be made for lacing in the shingles at the periphery of the damaged area. If the damaged portion of one side of a roof is so extensive, or so scattered, that the labor to patch and repair it equals or exceeds the labor required for complete replacement, it is advisable to estimate recovering the entire side of the roof.

Asphalt shingles that have been peppered by light hail, sufficient to mar the mineral surfacing, are not seriously injured except for appearance and possibly a slightly shortened life. However, when the shingle itself is perforated, even though immediate leaking is not apparent, it is advisable to replace or recover the roof with new shingles.

Matching asphalt shingles. The matching of asphalt shingles is sometimes a difficult problem. The inability to provide a suitable match may be due to the age and fading of the existing shingle, or it may be that the particular color or pattern is no longer available.

Where one or only a few individual shingles are required, it is sometimes possible to cut out a shingle from a piece of roll roofing of similar color. At other times a considerable amount of shopping around is necessary to locate proper matching shingles. Occasionally, where expensive shingles are involved, the manufacturer is able to produce enough to make repairs.

Where larger quantities of shingles are needed to patch a dwelling roof, many property owners are not adverse to having them removed from their garage. The latter is then re-roofed with a suitable shingle. On other occasions, matching shingles have been removed from the rear of a roof and placed on the front. The rear is then re-roofed with a shingle selected to satisfactorily blend or conform.

Patching or replacing small areas of asphalt shingles should be estimated on a time and material basis. The cost of laying single areas of one or more squares may be determined by unit cost, provided ample consideration is given to matching and lacing in.

Repairing Wood Shingles

A wood shingle roof surface that has been peppered by hail, so that the painted or creosoted surface is unsightly, may be repainted or refinished with creosote to restore it. If the surface is only slightly marred even by indentations, the same means of restoration are frequently satisfactory.

When wood shingles are split or broken by hail they must be replaced.

Shingles that have been loosened or dislodged by wind require replacement. When replacing individual shingles or a series of them to patch out a roof, the nails should be cut off with a ripper before the new shingles are slipped underneath the sound ones. The new shingles are then nailed with copper nails.

Patching of very old wood shingle roofs is sometimes done by use of tin pieces which are slipped under the split shingles. Loose nails should be drawn and new ones put in. Older wood shingle roofs are often satisfactorily patched by caulking or by cutting a piece of sutiable size from a piece of 65-lb roll roofing, spreading roofing cement on it, and sliding it under the broken shingle. Copper nails are then driven to secure it.

As with asphalt shingles, all patch jobs should be estimated on a time and material basis. Solid areas of one or more squares may be figured by unit cost if proper consideration is given to lacing in the new shingles where they join the old.

Repairing Rigid Shingles

Tile, slate, or asbestos-cement shingle roofs may be treated very similarly to damaged wood shingle roofs. Individual replacements are made by use of a ripper to cut off the nails under the sound shingles. A new shingle is inserted and nailed with rust-resistant nails. Holes have to be drilled in asbestos-cement shingles for nailing. The nail head may then be covered with a copper strip made for that purpose. Lead plugs are sometimes used in place of the copper strip.

Broken clay tile are removed. The joints are cemented and the new tile is wired in place with copper wire.

Repairing Built-up Roofing

The damage to built-up roofing, outside of replacement, when necessary, may be repaired in one of four ways: 1. Spot-patch. 2. Patch and mop. 3. Cap-sheet and mop. 4. Resurface with gravel or slag.

Before patching built-up roofing, all gravel or slag must be removed from the area, to be replaced after the repairs are complete.

Spot-patch. This method is used when small or medium sized breaks occur in the surface involving one or more plies. The area around the break is cut out and removed. All moisture under the break should be thoroughly dried out. New felts are then laid and mopped between if more than one has been cut through. A coating of asphalt or pitch is then applied to the patch surface.

Patch and Mop. This method of repair is generally used where large patches or many small patches are necessary. The patching is done in much the same manner as in spot-patching. When the patches are complete, the entire surface of the roof is given a mop coat of asphalt or pitch.

Cap-Sheet and Mop. When hail has badly dented or has perforated a built-up roof, a method of repair is to clean off the surface of the roof area to be repaired, lay a 30-lb cap-sheet, and apply 25 or 30 pounds of asphalt or pitch per square.

Resurface with Gravel or Slag. In hurricanes and tornadoes, it frequently happens that no greater damage occurs to a built-up roof than having the loose gravel or slag blown around or off the roof. In some cases replacing the mineral surfacing is all that is required. At other times it is necessary to clean off the balded areas and mop coat with sufficient asphalt or pitch to hold new gravel or slag. Close inspection is the only way of determining which treatment is needed.

Repairing Metal Roofs

Whether or not corrugated aluminum or galvanized roofing has been injured by wind, hail, or fire is readily determined by examination. Badly twisted or dented sheets should be removed and replaced.

Chapter 14.

Painting, Paperhanging and Glazing

PAINTING

Most estimates of building losses include an item for washing down painted surfaces, or redecorating some part of the premises as a result of the loss. While the majority of people have some knowledge of painting or papering as a result of personal experiences in their own homes, and even though the materials and their application are basically simple, there is a greater range in the estimates of cost than is generally recognized. The unit cost per square foot for painting in some areas, for example, may range from 20 to 50 percent higher than in others, in spite of the fact that painter's wages in the lower unit cost areas are considerably higher. There is little if any difference in the price of materials. A unit cost per square foot for painting tends to become uniform and accepted among adjusters and estimators within a specific region more so than is the case with many other building trades. Much of this can be attributed to local customs in methods of estimating which become established over a period without understanding or analysis of the factors that make up the unit cost.

In metropolitan New York, in New Jersey, and in several New England and Southern States, a unit cost for painting is applied to the gross wall and ceiling area of a room. In most of the midwestern and Pacific Coast states a unit cost is also applied to the gross area without deducting openings; but in the latter states a flat sum is added for each window and door opening, and the baseboard. These units are listed as *sides*. This practice tends to inflate the cost of painting, particularly in a room having several windows and doors. If openings are not deducted when applying the unit cost to the gross wall area, it amounts to doubling up on the cost of both labor and material where *sides* are added.

JOB CONSIDERATIONS

There are several different methods used to approximate the number of gallons of paint and the hours required to apply it. Some of these are described in the following pages. Whatever method is used, the main objective should be to compute as accurately as possible the amount of material and, considering all conditions that pertain to the particular job, determine a reasonable number of hours for a man to apply the materials. The following conditions are those which should be given particular attention as they most generally affect the cost of material and labor.

1. Shifting and protecting contents
2. Preparation of the surface
3. Kind of surface
4. Shape or form of surface
5. Type of paint
6. Number of colors
7. Number of coats
8. Method of application
9. Size and layout of area being worked in
10. Degree of workman's skill
11. Working from ladders or scaffold.

1. Shifting and protecting contents

A painter works a little faster in a room that is vacant than he does in one that requires him to shift and cover the contents to conveniently reach the areas to be painted. It takes extra time to move furniture and take down pictures. In mercantile risks it should be determined who is to move stocks on shelves, and away from walls that are to be painted.

2. Preparation of the surfaces

Walls that are badly smoked are customarily washed down to remove grease and dirt before applying paint. In many instances the washing down of smoked walls results in only one coat of paint being needed to cover the surface. Blistered paint from heat or water should be removed and sanded prior to applying paint. Frequently minor cracks in plaster, or in the joints of wall board, have to be filled with patch plaster or spackle, and nail holes in woodwork must be puttied.

3. Kind of surface

Both the labor and the quantity of paint varies with the kind of surface. Rough plastered brick, or cement walls require more paint than smooth plastered walls and woodwork. A painter using a brush can cover more square feet per hour on a smooth surface than he can on one that is porous. Exterior unpainted wood shingles and weather worn siding require more paint and labor than surfaces where the old paint has been well maintained.

4. Shape or form of surface

Stairways, trims, cabinets, windows and doors take longer to paint per square foot than smooth uninterrupted areas such as walls and ceilings, bevel siding, and floors.

5. Type of paint

The covering capacities of some types of paint are greater than others regardless of the kind of surface being painted. Cold water mix paints and calcimine cover less square feet per gallon than oil, rubber or alkyd resin base paints, while shellac and oil stains have greater covering capacity than any of those named. In figuring the quantity of material it is important to know the probable covering capacity of the paint to be used. The type of paint also affects the cost of the material. See covering capacities in Table 14-1.

Painters can apply different types of paint at varying rates. For example, while calcimine covers less area per gallon, a painter can apply it faster than oil paint.

6. Number of colors

When doors, windows and trim in a room or on the exterior must be painted a different color than other areas, a painter will require a little more time than if only one color is used.

7. Number of coats

The number of coats of paint required naturally affects the cost of the work. See Table 14-1 for covering capacity for each coat.

8. Method of application

Brush work is a little slower than using a roller on walls and ceilings. In some jurisdictions a painter who uses a roller is paid a slightly higher rate of wages which may offset the advantage in speed. Where spray painting is permissible, the labor cost is considerably reduced although slightly more paint is usually required.

9. Layout of area

Where a painter is working in a large area, as a loft building, his rate of applying paint on walls is higher than one working in a confined area or a cut-up room. Areas such as stores, offices, or factories that are being painted during working hours have to be estimated at a slightly higher rate of application of paint because employees or customers moving about slow the progress of the painter.

10. Degree of workman's skill

Some painters have more experience and skill than others. A painting contractor will usually put his men at work where their talents and capabilities are most productive.

11. Working from ladders or scaffolds

A man painting from ladders or scaffolding generally works at a slightly slower rate of speed than when he is working at ground or floor level. It also requires extra time to set up and move ladders or scaffolds from one position to another.

COVERING CAPACITY OF PAINT

The number of square feet that one gallon of paint will cover varies with the condition and porosity of the surface, the type of paint, and the method used to apply it. Paint manufacturers usually specify on each container the approximate area that can be covered. A first coat generally covers less area per gallon than succeeding coats primarily because of the absorption. The first coat acts as a sealer. It is seldom that two painters will agree on the covering capacity of paint, but their differences of opinion should not be much more than 10 or 15 percent apart.

For estimating quantities, an average number of square feet per gallon of material is used. For practical purposes there is no advantage to be gained by attempting to use different spreading capacities for the first, second and third coats. One average figure should be used regardless of how many coats are to be applied.

Table 14-1 shows the approximate number of square feet that a gallon of painting materials will cover on the most frequently encountered surfaces. When the surfaces are unusually rough or porous, the number of square feet should be reduced accordingly.

TABLE 14-1
APPROXIMATE COVERING CAPACITY OF PAINTING MATERIALS
USING BRUSH APPLICATION

USE OF PAINT MATERIALS	Sq Ft Per Gallon		
	First Coat	Second Coat	Third Coat
Interior			
Flat paint on smooth plaster	450	450	450
Flat paint on sand finish plaster	300	350	300
Flat paint on texture finish plaster	300	300	300
Gloss or semi-gloss on smooth plaster	400	500	400
Gloss or semi-gloss on sand plaster	300	400	400
Gloss or semi-gloss on texture plaster	250	350	400
Calcimine-Watersize - size	720		
Calcimine	240		
Calcimine-Oilsize - size	225		
Calcimine	240		
Concrete block - Resin Emulsion	200		
Wood veneer - Lacquer	500	540	
Wood veneer - Synthetic resin	600	675	
Penetrating wax	600	675	
Enamel on wood floors	400	400	
Stainwax on floors or trim	500	600	
Shellac (4 lb cut)	500	500	500
Varnish over shellac	500		
Varnish remover	150		
Exterior			
Wood siding and trim-flat or semi-gloss	450	550	600
Staining shingle roofs and siding	200	300	
Oil Paint - asbestos shingles	200	300	
Cedar shakes - Latex	200	300	400
Brick - Latex	200	300	400
Concrete floors and steps-enamel	300	400	400
Link fences	750		
Gutters and leaders	200-250 lin ft per coat		

Note: See pages 362 to 269, Measurement of Areas.

Wall and Ceiling Areas from Table 14-2

Table 14-2 gives the *total* square foot area of the walls and ceiling of rooms with lengths and widths shown and ceiling heights of 8,9, and 10 feet. There has been no deduction for window or door openings and rooms are presumed to be rectangular in shape.

Interpolation can be made for room sizes to the nearest 6 inches.

Illustrative Example

To find the total square foot area of a room 10'6" x 20'0" x 10'0":

Area of room 10' x 20' x 10'	= 800 sq. ft.
Area of room 11' x 20' x 10'	= 840 sq. ft.
Difference	40 sq. ft.
Area of room 10' x 20' x 10'	= 800 sq. ft.
Add ½ the difference	20 sq. ft.
Area of room 10'6" x 20' x 10'	= 820 sq. ft.

If the room size was 10'3" x 20' x 10', then 1/4 of the difference would be added or 10 sq. ft. giving that room an area of 810 sq. ft.

Interpolating ceiling heights. If the ceiling height is 8' 6", add the area shown for 8' and 9' in the Table and divide by 2 to get the average. If the ceiling height is 9' 6", add the area shown for 9' and 10' and divide by 2 to get the average. The answer is accurate.

Illustrative Example

Find the area of a room 10' x 20' with an 8'6" ceiling.

Area of a room 10' x 20' x 8'	= 680 sq ft
" " " " " " x 9'	= 740 " "
	1,420 " "
Area of a room 10' x 20' x 8'6"	= $\frac{1}{2}$ x 1420 sq ft
	= 710 sq ft

MEASUREMENT OF AREAS

The unit of measure for painting customarily is the square yard or square foot. In some parts of the country a *square* (100 square feet) is used as a unit.

All measuring is done in lineal feet, and the areas are first calculated in terms of *square* feet. Since this is true, and because the covering capacity of the materials and the rate of application is expressed in square feet, there would appear to be no reason to convert the figures to square yards or "squares." If it becomes necessary when analyzing or comparing estimates that express areas in terms of square yards or squares, the conversion can readily be done at that time. The unit that will be used in this chapter is the *square foot.*

Table 14-2

TOTAL AREA....4 WALLS AND CEILING (In Square Feet)
ROOMS WITH CEILINGS 8 FEET

	3'	4'	5'	6'	7'	8'	9'	10'	11'	12'	13'	14'	15'	16'	17'	18'	19'	20'	21'	22'
3'	105	124	143	162	181	200	219	238	257	276	295	314	333	352	371	390	409	428	447	466
4'	124	144	164	184	204	224	244	264	284	304	324	344	364	384	404	424	444	464	484	504
5'	143	164	185	206	227	248	269	290	311	332	353	374	395	416	437	458	479	500	521	542
6'	162	184	206	228	250	272	294	316	338	360	382	404	426	448	470	492	514	536	558	580
7'	181	204	227	250	273	296	319	342	365	388	411	434	457	480	503	526	549	572	595	618
8'	200	224	248	272	296	320	344	368	392	416	440	464	488	512	536	560	584	608	632	656
9'	219	244	269	294	319	344	369	394	419	444	469	494	519	544	569	594	619	644	669	694
10'	238	264	290	316	342	368	394	420	446	472	498	524	550	576	602	628	654	680	706	732
11'	257	284	311	338	365	392	419	446	473	500	527	554	581	608	635	662	689	716	743	770
12'	276	304	332	360	388	416	444	472	500	528	556	584	612	640	668	696	724	752	780	808
13'	295	324	353	382	411	440	469	498	527	556	585	614	643	672	701	730	759	788	817	846
14'	314	344	374	404	434	464	494	524	554	584	614	644	674	704	734	764	794	824	854	884
15'	333	364	395	426	457	488	519	550	581	612	643	674	705	736	767	798	829	860	891	922
16'	352	384	416	448	480	512	544	576	608	640	672	704	736	768	800	832	864	896	928	960
17'	371	404	437	470	503	536	569	602	635	668	701	734	767	800	833	866	899	932	965	998
18'	390	424	458	492	526	560	594	628	662	696	730	764	798	832	866	900	934	968	1002	1036
19'	409	444	479	514	549	584	619	654	689	724	759	794	829	864	899	934	969	1004	1039	1074
20'	428	464	500	536	572	608	644	680	716	752	788	824	860	896	932	968	1004	1040	1076	1112
21'	447	484	521	558	595	632	669	706	743	780	817	854	891	928	965	1002	1039	1076	1113	1150
22'	466	504	542	580	618	656	694	732	770	808	846	884	922	960	998	1036	1074	1112	1150	1188
23'	485	524	563	602	641	680	719	758	797	836	875	914	953	992	1031	1070	1109	1148	1187	1226
24'	504	544	584	624	664	704	744	784	824	864	904	944	984	1024	1064	1104	1144	1184	1224	1264

ROOMS WITH CEILINGS 9 FEET

	3'	4'	5'	6'	7'	8'	9'	10'	11'	12'	13'	14'	15'	16'	17'	18'	19'	20'	21'	22'
3'	117	138	159	180	201	222	243	264	285	306	327	348	369	390	411	432	453	474	495	516
4'	138	160	182	204	226	248	270	292	314	336	358	380	402	424	446	468	490	512	534	556
5'	159	182	205	228	251	274	297	320	343	366	389	412	435	458	481	504	527	550	573	596
6'	180	204	228	252	276	300	324	348	372	396	420	444	468	492	516	540	564	588	612	636
7'	201	226	251	276	301	326	351	376	401	426	451	476	501	526	551	576	601	626	651	676
8'	222	248	274	300	326	352	378	404	430	456	482	508	534	560	586	612	638	664	690	716
9'	243	270	297	324	351	378	405	432	459	486	513	540	567	594	621	648	675	702	729	756
10'	264	292	320	348	376	404	432	460	488	516	544	572	600	628	656	684	712	740	768	796
11'	285	314	343	372	401	430	459	488	517	546	575	604	633	662	691	720	749	778	807	836
12'	306	336	366	396	426	456	486	516	546	576	606	636	666	696	726	756	786	816	846	876
13'	327	358	389	420	451	482	513	544	575	606	637	668	699	730	761	792	823	854	885	916
14'	348	380	412	444	476	508	540	572	604	636	668	700	732	764	796	828	860	892	924	956
15'	369	402	435	468	501	534	567	600	633	666	699	732	765	798	831	864	897	930	963	996
16'	390	424	458	492	526	560	594	628	662	696	730	764	798	832	866	900	934	968	1002	1036
17'	411	446	481	516	551	586	621	656	691	726	761	796	831	866	901	936	971	1006	1041	1076
18'	432	468	504	540	576	612	648	684	720	756	792	828	864	900	936	972	1008	1044	1080	1116
19'	453	490	527	564	601	638	675	712	749	786	823	860	897	934	971	1008	1045	1082	1119	1156
20'	474	512	550	588	626	664	702	740	778	816	854	892	930	968	1006	1044	1082	1120	1158	1196
21'	495	534	573	612	651	690	729	768	807	846	885	924	963	1002	1041	1080	1119	1158	1197	1236
22'	516	556	596	636	676	716	756	796	836	876	916	956	996	1036	1076	1116	1156	1196	1236	1276
23'	537	578	619	660	701	742	783	824	865	906	947	988	1029	1070	1111	1152	1193	1234	1275	1316
24'	558	600	642	684	726	768	810	852	894	936	978	1020	1062	1104	1146	1188	1230	1272	1314	1356

ROOMS WITH CEILINGS 10 FEET

	3'	4'	5'	6'	7'	8'	9'	10'	11'	12'	13'	14'	15'	16'	17'	18'	19'	20'	21'	22'
3'	129	152	175	198	221	244	267	290	313	336	359	382	405	428	451	474	497	520	543	566
4'	152	176	200	224	248	272	296	320	344	368	392	416	440	464	488	512	536	560	584	608
5'	175	200	225	250	275	300	325	350	375	400	425	450	475	500	525	550	575	600	625	650
6'	198	224	250	276	302	328	354	380	406	432	458	484	510	536	562	588	614	640	666	692
7'	221	248	275	302	329	356	383	410	437	464	491	518	545	572	599	626	653	680	707	734
8'	244	272	300	328	356	384	412	440	468	496	524	552	580	608	636	664	692	720	748	776
9'	267	296	325	354	383	412	441	470	499	528	557	586	615	644	673	702	731	760	789	818
10'	290	320	350	380	410	440	470	500	530	560	590	620	650	680	710	740	770	800	830	860
11'	313	344	375	406	437	468	499	530	561	592	623	654	685	716	747	778	809	840	871	902
12'	336	368	400	432	464	496	528	560	592	624	656	688	720	752	784	816	848	880	912	944
13'	359	392	425	458	491	524	557	590	623	656	689	722	755	788	821	854	887	920	953	986
14'	382	416	450	484	518	552	586	620	654	688	722	756	790	824	858	892	926	960	994	1028
15'	405	440	475	510	545	580	615	650	685	720	755	790	825	860	895	930	965	1000	1035	1070
16'	428	464	500	536	572	608	644	680	716	752	788	824	860	896	932	968	1004	1040	1076	1112
17'	451	488	525	562	599	636	673	710	747	784	821	858	895	932	969	1006	1043	1080	1117	1154
18'	474	512	550	588	626	664	702	740	778	816	854	892	930	968	1006	1044	1082	1120	1158	1196
19'	497	536	575	614	653	692	731	770	809	848	887	926	965	1004	1043	1082	1121	1160	1199	1238
20'	520	560	600	640	680	720	760	800	840	880	920	960	1000	1040	1080	1120	1160	1200	1240	1280
21'	543	584	625	666	707	748	789	830	871	912	953	994	1035	1076	1117	1158	1199	1240	1281	1322
22'	566	608	650	692	734	776	818	860	902	944	986	1028	1070	1112	1154	1196	1238	1280	1322	1364
23'	589	632	675	718	761	804	847	890	933	976	1019	1062	1105	1148	1191	1234	1277	1320	1363	1406
24'	612	656	700	744	788	832	876	920	964	1008	1052	1096	1140	1184	1228	1272	1316	1360	1404	1448

When estimating the number of square feet of painting, the actual surface to be painted should be measured as accurately as possible because the cost of the material and the number of hours of labor will be determined from the surface area to be covered. The painting or decorating of flat wall surfaces such as ceiling and floor areas, presents no special problem in measurement, but questions frequently arise as to how to obtain the areas for windows, trimmed openings, stairs and balustrades, doors, exterior cornices, lattices and so forth. Rules-of-thumb are employed by most painters and, while there are differences in the methods used, each one strives to reasonably approximate the actual surface to be covered. The following methods for the measurement of flat and irregular surfaces are recommended.

Interior Walls and Ceilings

To obtain the area of the ceiling, multiply the length times the width. The area of the walls is obtained by measuring the distance around the room (the perimeter) and multiplying by the height from the floor to the ceiling. When walls are to be painted only above or below a wainscoting, the height measurement should be only for that portion to be painted.

Whether or not window and door openings are to be deducted will depend on the particular method used to estimate. When openings are deducted, the cost of material and labor to paint openings is added separately. If window and door openings are not deducted, then the additional time for cutting in windows and doors is included in the labor allowance for the room or per square foot area. Actually it requires no additional paint, even though different colors are used for the finish coat. It does require more time than painting a flat surface. (See "Methods of Estimating," this chapter.)

Windows and Doors

Measure the width and height from the outside of the trim of the opening. Add 2 feet to the width and height of a window. Add 2 feet to the width and 1 foot to the height of a door. This will allow for moldings, edges and frames. A window that measures 3' x 6' would be figured as 5' x 8' or 40 square feet. (3' + 2' = 5' and 6' + 2' = 8'.) A door measuring 3' x 7' would be figured 5' x 8' = 40 square feet (3' + 2' = 5' and 7' + 1' = 8'). Many painters use the unit of 40 square feet for all average sized openings when figuring windows and doors.

Windows or doors with sash having more than one light of glass, are figured at 2 square feet for each additional light. A 6-light sash would add 10 square feet to the window area.

The proper measurement of window and door areas is very important when estimating brick, masonry, or other surfaces where the painting of such opening is strictly an individual operation.

Floors

The number of square feet in a floor is obtained by multiplying the length times the width. Openings less than 10' x 10' in the floor are not deducted.

Wainscoting

If a wainscoting is painted on a flat surface such as plaster, the area is obtained by multiplying the length times the width. When the wainscoting is paneled, the area should be multiplied by 1 1/2 or 2, depending on the form of the surface.

Stairs and Balustrades

To obtain the surface area of stairs that are to be painted, allow 1 foot for the width of the tread, and 1 foot for the width of the riser to take care of the nosing and molding. Multiply this width by the length of the treads or risers, plus 2 feet for stair strings. A stairway with treads and risers 3'6" wide would have 3'6" + 2' = 5 1/2 square feet for each tread and riser. Multiply the area of the tread and the riser by the number of each.

A stairway with 18 risers 4 feet wide would have an area as follows:

Risers	= 18 x 1' x (4' + 2')	
	= 18 x 1' x 6'	= 108 sq ft
Treads	(1 less than number of risers)	
	= 17 x 1' x (4' + 2')	
	= 17 x 1' x 6'	= 102 sq ft
		Total = 210 sq ft

Some authorities count risers and multiply by 8 times the width.

The area calculation for a balustrade is obtained by measuring the height from the tread to the top of the handrail. Multiply this height by the length of the balustrade. The area of square edge and relatively simple designed balustrades should be multiplied by four. Turned and fancy balustrade areas should be multiplied by five. This takes care of both sides.

A balustrade that measures 2'6" x 18' has a plane area of 45 square feet. The paint area for a simple balustrade would be 45 square feet x 4 or 180 square feet. For a turned and fancy balustrade the paint area would be 45 square feet x 5 or 225 square feet.

Baseboard, Chair-rail, Picture Moldings, etc.

The paint area of baseboards, chair-rails, plate rails, picture moldings and other individual trim members should be estimated on the basis of one square foot per lineal foot.

Trimmed Openings

The paint area of two sides of a *trimmed* opening (one without a door) is obtained by measuring the lineal feet around the two sides and top of the opening and multiplying by three.

A trimmed opening 8 feet wide and 7 feet high would have a paint area of 8' + 7' + 7' = 22' x 3 = 66 square feet. This area is for both sides of the opening. If it is computed in this manner for one room, care should be taken not to include it in the adjoining room also.

Built-in Shelving, Cupboards, and Cases

To obtain the paint area for built-in shelves, cupboards, book cases and cabinets, measure the front area (width x height). Multiply the front area by 3 for open-front units. Multiply the front area by 5 for units having doors. The total area obtained includes the complete finishing of the interior and exterior.

Exterior Shingles and Siding (Wood, Metal and Composition)

To obtain the paint area of exterior shingles and siding, measure the actual wall area including gables, but do not deduct window or door openings less than 10' x 10'. For narrow clapboard exposures, and for all shingles, add 20 percent to the actual area. For wide clapboard exposures (over 5 inches) add 10 percent to the actual area.

Exterior Cornices and Eaves

Measure the length and width of the cornice to obtain the area. Multiply this area by 2 for relatively simple cornices, and multiply it by 3 for fancy or ornate cornices.

Where cornices are open, with exposed rafters, multiply the cornice area (length times overhang) by 3 to obtain the paint area.

For open cornices or eaves that do not have the rafters exposed, add 50 percent to the area.

Exterior Masonry Walls

To obtain the paint area of brick, stucco, cement block, and other exterior masonry walls, measure the actual wall area including gables. Do not deduct window or door openings less than 10' x 10'.

Fences

To obtain the paint area of a solid board fence, measure the length and height. Multiply the length times the height to obtain the area of *one side*. Multiply by 2 if it is to be painted on two sides.

The paint area of *one side* of a picket fence is determined by multiplying the length times the height by 2. Two sides, multiply by 4. Chain link fences, measure one side, multiply by 3.

Lattice

To obtain the paint area of lattice work, multiply the length times the width. To paint *one side* multiply the actual area by 3.

ESTIMATING MATERIALS

Painting and finishing is one of the least difficult of building trades to estimate, because the quantities of material, and the numbers of hours of labor are both determined directly from the area to be covered.

When the paint area has been measured and calculated, the quantity of material required per coat is obtained by dividing that area by the number of square feet a gallon of paint will cover.

Illustrative Example No. 1

A room 14' x 18' x 9' is to be given *one* coat of flat paint on sheet rock walls and ceiling.

Ceiling area	$14' \times 18' = 252$ sq ft
Wall area	$2(14' + 18') \times 9' = \underline{576}$ sq ft
	Total $= \overline{828}$ sq ft

(From Table 14-1)

$$\text{Paint required for one coat} = \frac{828}{450} = 1.84 \text{ gallons}$$

If only one room is to be painted the quantity would be rounded out to 2 gallons. Where other rooms are to be painted, any adjustment to whole or half gallons is made at the end of the entire painting estimate.

Illustrative Example No. 2

A room 10' x 16' x 8'6" is to be painted *two* coats of flat on plastered walls and ceiling.

Ceiling area	$10' \times 16' = 160$ sq ft
Wall area	$2(10' + 16') \times 8\frac{1}{2} = \underline{442}$ sq ft
	602
	(Two coats) x 2
(Table 14-2 may also be used)	$\overline{1,204}$ sq ft

(From Table 14-1)

$$\text{Paint required} = \frac{1,204}{450} = 2.45 \text{ gallons}$$

ESTIMATING LABOR

The conditions which have an influence on painting labor specifically have been discussed in the first part of this chapter. Those factors which affect labor in general will be found under "Estimating Labor" in Chapter 2.

The labor-hours required to apply paint materials to 100 square feet of surface under a given set of conditions varies, but it may be estimated reasonably closely on the basis of accepted average rates of applying paint. Table 14-3 shows the approximate number of square feet per hour required to apply paint materials for various types of interior work. They represent averages for two or three coats rather than individual first, second, or third coats. Included in the rates of application shown is the time needed for a normal amount of preparation such as sanding trim, patching minor plaster cracks, puttying nail holes, dusting, and so forth. The rates also contemplate working from step ladders to paint ceilings and working from extension ladders on exteriors. It includes time for getting ready and for cleaning up when the job is done.

Any special preliminary preparation such as having to move furnishings or merchandise, wash down walls, or burn off blistered paint, should be estimated separately and added to the over-all cost of the job.

The labor rates shown in Table 14-3 for "Interior" and in Table 14-4 for "Exterior" walls include the windows and doors found in the rooms of an average dwelling. They also include the painting of baseboards, chairrails, and so forth. For example, the rate of 150 square feet an hour shown for smooth plastered walls including openings is an average between 200 square feet an hour for a flat surface with no openings and 125 square feet an hour for painting openings and trim. Whenever window and door opeings and trim items are figured separately from the wall surfaces, the area of the openings should be deducted; otherwise there will be an overlapping of cost which will produce an excessive estimate. When it is desired to estimate openings and trim individually, the cost may be computed as outlined under "Measurement of Areas" or by using 1/16 of a gallon of material and 1/3 hour labor as shown in Table 14-3.

When unusual conditions are encountered, the labor rates shown in Tables 14-3 and 14-4 may be modified according to judgment.

METHODS OF ESTIMATING

There are five basic methods or systems of estimating painting and finishing which are in common use. In some sections of the country one

TABLE 14-3
APPROXIMATE NUMBER OF HOURS TO APPLY PAINTING
MATERIALS FOR TYPE OF INTERIOR WORK SHOWN —
PREMISES UNOCCUPIED

TYPE OF WORK	Sq Ft Per Hour Per Coat	
	Brush	Roller
Washing down smoke-stained walls and ceilings	100-150	
Sanding woodwork, preparatory	150-250	
Sizing plastered walls	300-400	
Burning paint from trim	25-40	
Painting smooth plaster incl openings	150	
Painting smooth plaster excl openings	200-250	300-350
Painting sandfinish plaster excl openings	175-200	350
Painting sandfinish plaster incl openings	125	
Painting texture plaster-semi-gloss	100-125	350
Painting texture plaster-flat paint	180	350
Painting Cement block or Brick	150	250
Spray painting		350-400
Painting Concrete floors or walls	200-250	300-350
Floors-remove finish with liquid	50	
Floors shellac, paint, stain or varnish	200-300	400-500
Floors wax and polish	200	
Paneling-Staining, or finishing	250	
Trim-back priming-paint	350	
*Windows-one side	125-150	
**Doors - one side	125-250	
Moulding etc - 1 to 6 inches wide	175 lin ft	

Note: No scaffolding included in above hour rates.

See pages 362 to 369 Measurement of Areas.

*Average 12 - 14 openings per day, one coat, one side per 8 hours, 6/6 lites.
**Flush-average 16 - 18 openings per 8 hours, one coat, one side.
Panel - average 12 - 14 openings per 8 hours, one coat, one side.
There are some painters who estimate 1/16 gal. of paint per side per opening, and ⅓ hours labor per side per opening.

particular method may become established by custom and usage to the exclusion of any other method. For that reason, and also for purposes of analyzing painting estimates, it is important to understand each system. When properly used, there should be little difference in the results obtained.

To demonstrate the application of each of the five methods, a single problem, which contemplates painting a typical room, is used.

TABLE 14-4
APPROXIMATE NUMBER OF HOURS TO APPLY
PAINTING MATERIALS FOR TYPE OF EXTERIOR WORK SHOWN—
AVERAGE FROM GROUND AND LADDERS

| TYPE OF WORK | Sq Ft Per Hour Per Coat | |
	Brush	Roller
Sanding woodwork lightly	250	
Sanding blistered woodwork	100-200	
Burning off paint from woodwork	25-40	
Wood siding, incl window and door openings	100-125	
″ ″ excl window and door openings	200-225	300-350
*Exterior trim only incl windows and doors	80-100	
Shingle staining - walls	150	200
Brick walls - Oil paint	125-150	200-250
Stucco walls - Oil Paint	100-150	200-250
Concrete block	100-125	250-300
Floors - wood	200	250
Floors - concrete	200-250	300-350
Roofs - Wood shingle stain	200-225	250-300
Spraying		500
Shutters	4-5 units	
Screens	5-6 ″	
Storm sash	4-5 ″	
Fences - solid board	250	400
″ - picket	150-200	250-300
″ - picket (spray)		450-500
″ - link	125	300
Leaders and gutters	100 lin ft	

* Windows - painter will paint (one-coat) 12 - 14 openings per day, one side only.
   Flush doors - Painter will paint (one-coat) 16-18 openings per day, one side only.
   Panel doors - painter will paint (one-coat) 12-14 openings per day, one side only.
Note: See pages 362 to 369 Measurement of Areas.

Illustrative Problem
     To paint the walls, ceiling and woodwork of a dwelling living room 16′ x 20′ x 9′. There are two doors with openings 3′ x 7′, and three windows with openings 3′ x 6′. A 6″ baseboard runs around the room. The walls and ceilings are finished in smooth plaster and no preparation of surfaces is required. The walls and ceiling are to be painted 2 coats of flat white. The woodwork and trim are to receive one coat of flat and one coat of semi-gloss enamel of a color different from the wall paint.

Areas

Ceiling	16' x 20' = 320 sq ft
Walls	2(16' + 20') x 9' = 648 sq ft
	Gross area = 968 sq ft

Openings (actual opening size)

2 Doors	3' x 7' =	42 sq ft
3 Windows	3' x 6' =	54 sq ft
Total	=	96 sq ft
Net wall and ceiling area	=	872 sq ft

Baseboard (openings not deducted)

$$2(16' + 20') = 72 \text{ lin ft}$$

For purposes of illustration painter's wages are assumed to be $6.00 an hour, and both flat and semi-gloss enamel to be $7.00 a gallon.

Method No. 1—Material and Labor Separately Estimated

In this method the total number of gallons of paint, and the total hours of labor are determined from the *gross* wall and ceiling area. The number of gallons of paint is based on the average covering capacity of one gallon. The number of hours of labor is based on the number of square feet a painter can average in one hour.

Table 14-1 shows, for this type of paint and surface, an average covering capacity of 450 square feet for each gallon. By dividing the gross area of the room by the covering capacity per gallon, the number of gallons of paint required for one coat is obtained. For two coats it would require twice as much paint.

$$\frac{\text{Gross Area}}{\text{Coverage}} = \frac{968}{450} = 2.15 \text{ gallons per coat}$$

$$\text{Two coats} = \frac{\times 2}{4.3} \text{ gallons}$$

Table 14-3 shows that a painter can apply an average of 150 square feet an hour on smooth plastered walls, ceilings and trim. The number of hours of labor required for each coat is obtained by dividing the gross area of the room by 150 square feet. The number of hours to apply two coats is double that required for one coat.

$$\frac{\text{Gross Area}}{\text{Rate per hr}} = \frac{968}{150} = 6.4 \text{ hours per coat}$$

$$\text{Two coats} = \frac{\times 2}{12.8} \text{ hours}$$

Estimated Cost in Problem

4.3 gals paint	@ $7.00 =	$ 30.10
12.8 hours labor	@ 6.00 =	76.80
	Total cost =	$106.90

When the gross area of a room is used to compute material and labor, it is contemplated that less paint is used on window openings. More labor, however, is needed to paint the window bars, and also the trim, particularly when a different color is used. The rate of 150 square feet an hour for applying paint takes into consideration the additional time for painting a normal number of openings and the trim in an average room. On a flat, smooth plastered wall surface without openings an experienced painter would be able to cover between 200 and 300 square feet an hour with very little difficulty.

Method No. 2—A Unit Cost Applied to Gross Area

A popular method in some sections of the country is one in which a unit cost per square foot or square yard is developed and applied to the gross area without any deductions for openings. It employs the same principle as method No. 1 in that the rate of application makes allowances for the time required to cut in window and wood trim where different colors of paint are used.

Using a covering capacity of 450 square feet a gallon and a rate of application for a painter of 150 square feet an hour, the unit cost is obtained as follows:

1 gallon paint	@ $ 7.00 =	$ 7.00
$\frac{450}{150}$ = 3 hours labor	@ 6.00 =	18.00
Total cost for labor and material	=	$25.00
Unit cost per sq ft per coat $\frac{$25.00}{450}$	=	$ .0555
Unit cost per sq ft for two coats	=	.111

Estimated Cost in Problem

Gross area of room	=	968 sq ft
Unit cost per sq ft	=	x $.111
Total cost	=	$107.45

It should be noted that the covering capacity of a gallon of paint on smooth plastered walls is 450 square feet as taken from Table 14-1. The rate of application for a painter shown in Table 14-3 is 150 square feet an hour. The number of hours in this instance to apply one gallon of paint is obtained by dividing 450 by 150, which equals 3 hours.

The unit cost developed in this method is $.111, which in actual practice would be rounded off at $.11 per square foot.

Method No. 3—Separate Unit Costs Used for Flat Wall and Ceiling Surfaces, and for Openings and Trim

A method that is quite common in many of the central, midwestern, and Pacific Coast states is one in which the walls and ceilings, the window and door openings, and the trim items are estimated on an individual unit cost. The unit cost that is used takes into consideration the conditions applicable to each type of surface, on the principle that paint coverage, and the rate of application varies with the kind of paint and the form of the surface being painted. This method of estimating is taught in a number of trade schools for painters. Because of the refinement of the estimate, it is considered by many to be more accurate than any other system. It has particular merit where the treatment of the wall or ceiling surfaces and that given to openings and trim are substantially different in both character and in the cost per square foot. If, for example, the walls and ceilings are calcimined or painted with cold water paints and all woodwork is painted a coat of flat and finished with semi-gloss, an average unit cost per square foot to be applied to the gross area would be difficult if not impossible to compute.

The deduction of openings from wall area is important where the cost of painting such openings is added as a separate item to the estimate because if a unit cost is applied to the gross wall area, the cost of the opening is figured twice. An average door opening, for example, is roughly 3' x 7', and if a unit cost of $.10 per square foot is used, the cost of the opening is $2.10. An average bedroom contains 2 doors and 2 windows, or 4 openings. Inclusion of the openings adds $8.80 to the estimate. When 20 percent overhead and profit is added, the cost is further increased to $10.56 for one room. There is no way to modify a unit cost which is to be applied to the gross area of a room so that it will take out openings. They must be physically deducted and a proper unit cost applied to the net wall area. These matters should be thoroughly understood as they become important when analyzing estimates on painting. There are many sections of the country where gross areas are figured for loss purposes at a unit cost, and all openings and baseboards are added on top of that at a flat dollar amount per unit. Where one or two rooms are involved, the difference is not great, but the cumulative effect on a loss with several or a great many rooms is substantial.

Under Method No. 3, it is customary to consider a baseboard as approximately equal to one opening. Because the cost of painting the openings is estimated separately from wall and ceiling surfaces, *all window and door openings are deducted* using the inside measurements of the openings. Different rates for applying paint are used for the walls and for the openings and trim. Table 14-3 shows a rate of 200 square feet an hour for plain walls and ceilings. It also shows a rate of 125 square feet an hour for painting windows, doors, and trim. Using these rates for applying paint, the following unit costs are developed:

Unit Cost for Walls and Ceiling

1 gallon paint	@ $ 7.00 =	$ 7.00
$\frac{450}{200}$ = 2¼ hours labor	@ 6.00 =	13.50
		$20.50
Unit cost per sq ft for 1 coat $\frac{\$20.50}{450}$ = .0455		
Unit cost per sq ft for 2 coats	= $	.091

Unit Cost for Openings and Trim

1 gallon paint	@ $ 7.00 =	$ 7.00
$\frac{450}{125}$ = 3.6 hours labor	@ 6.00 =	21.60
		$28.60
Unit cost per square foot for 1 coat = $\frac{\$28.60}{450}$ = .0635		
Unit cost per square foot for 2 coats	= $	.127

Estimated Cost in Problem

Net wall and ceiling area 872 sq ft x $.091	= $ 79.35
Openings and Baseboard:	
2 windows	
2 doors	
1 baseboard	
6 openings @ 40 sq ft* 240 sq ft x .127	= 30.48
	Total cost = $109.83

*(See Measurement of Areas)

Alternate of Method No. 3

Using Method No. 3 and estimating the cost of openings (as shown in Tables 14-1 & 14-3) by figuring ½ pint (1/16 gal.) of paint for each opening, and ⅓ hour labor, the cost would be obtained as follows:

Estimated Cost in Problem

Net wall and ceiling area 872 sq ft x $.091	= $ 79.35
6 openings:	
Material - 6/16 gal	@ $7.00 = $ 2.62
Labor 6($\frac{1}{3}$) = 2 hours	@ 6.00 = 12.00
	Cost per coat = $ 14.62
	x 2
	Cost for 2 coats = $ 29.24
	Total cost = $108.59

Method No. 4—Material and Labor Per Gallon of Material

The cost of painting can be expressed in terms of the cost of labor and material per gallon. This unit cost per gallon can be multiplied by the number of gallons of paint required to obtain the total cost of labor and material.

Unit Cost Per Gallon

1 gallon paint	@ $7.00 = $ 7.00
3 hours labor	@ $6.00 = 18.00
Cost of material	
and labor per gallon	= $25.00

In the problem being used for illustration the total number of gallons of paint for two coats would be $\frac{968 \times 2}{450} = 4.3$ gals.

Estimated Cost

4.3 gals. @ $25.00 = $107.50

Method No. 5—Material and Labor Per Hour

A method occasionally used by painters is to develop a unit cost per hour for both materials and labor. Using a covering capacity of 450 square feet per gallon and 150 square feet per hour, the unit cost would be developed as follows:

Unit Cost Per hour

$\frac{150}{450} = \frac{1}{3}$ gal of paint	@ $7.00 = $2.33
1 hour labor	@ 6.00 = 6.00
Unit cost of material and labor per hour	= $8.33

Many estimators prefer this method, particularly for rough estimating, because they can look at a room, a house, or a complete job and approximate the number of hours of labor required to do the work.

In the illustrative problem the number of hours required at 150 square feet per hour of application would be $\frac{968 \times 2 \text{ (coats)}}{150} = 12.9$ hours

Estimated Cost: 12.9 hrs x $8.33 = $107.46

Summary of Methods

The choice of the method by which painting is estimated is one of individual preference, but whatever system is used, it should consider all of the conditions peculiar to the particular job. The five methods outlined show very little variation in cost when applied to the illustrative problem.

Methods	Total Cost
No. 1	$106.90
No. 2	107.45
No. 3	109.83
No. 3 (alt.)	108.59
No. 4	107.50
No. 5	107.46

Unit Costs for Painting

A practical way to develop a unit cost is to use a gallon of material as a basis, because the covering capacity is expressed in terms of square feet. The rate at which painting material can be applied is expressed in square feet per hour. The following basic formula may be used in each instance:

Basic Formula

1 gal. paint	@ $	= $
Hours labor = $\frac{\text{Coverage per gal.}}{\text{Rate of application per gal.}}$	@ $	= $_____
Total cost of labor and material per gallon		= $
Unit cost per sq ft = $\frac{\text{Total cost}}{\text{Coverage per gal}}$		= $

Illustrative Example No. 1

In Table 14-1 the covering capacity on exterior siding is given as 500 square feet per gallon. In Table 14-4 the rate of applying exterior paint on smooth surfaces including all openings is 125 square feet per hour. You may develop a unit cost for 1 coat painting on bevel siding as follows:

Exterior paint cost assumed @ $6.50 per gal.
Painter's wages assumed   @ $6.00 per hour

Unit Cost

1 gallon paint @ \$6.50 = \$ 6.50

$\dfrac{500}{125}$ = 4 hours labor @ 6.00 = $\dfrac{24.00}{\$30.50}$

Unit cost per square foot $= \dfrac{\$30.50}{500} = \$.061$ per coat

Unit cost per square foot for 2 coats = \$.122 or \$.12

Illustrative Example No. 2

To develop a unit cost for 1 coat of varnish on prepared floors. Table 14-1 shows the covering capacity of varnish as 400 square feet per gallon. Table 14-3 shows the rate of applying varnish to floors as 200 square feet per hour.

Floor varnish assumed @ \$9.50 per gal.
Painter's wages assumed @ \$6.00 per hour

Unit Cost

1 gallon varnish @ \$9.50 = \$ 9.50

$\dfrac{400}{200}$ = 2 hours labor @ 6.00 = $\dfrac{12.00}{\$21.50}$

Unit cost per square foot $= \dfrac{\$21.50}{400} = \$.05375$

$= \$.054$ coat

Illustrative Example No. 3

To develop a unit cost per opening for windows and doors using ½ pint (1/16 gals.) material and allowing ⅓ hour labor per opening per coat.

Paint cost assumed @ \$6.00 per gallon
Painter's wages assumed @ 6.00 per hour

Unit Cost

1/16 gallons of paint @ 6.00 = \$ .38
⅓ hours labor @ 6.00 = 2.00
Unit cost per opening 1 coat = \$2.38
Unit cost per opening 2 coats = \$4.76

APPLICATION OF UNIT COSTS FOR PAINTING

Illustrative Example No. 1

The walls and ceilings of the following four rooms are to be painted with two coats of flat, and the woodwork is to be painted with one coat of flat with a finish coat of semi-gloss.

Assuming that paint costs \$7.50 per gallon and painter's wages are \$6.00 per hour, obtain the total cost using Method No. 2.

Unit Cost

1 gallon paint	@ $7.50 =	$ 7.50
$\dfrac{450}{150}$ = 3 hours labor	@ 6.00 =	18.00
		$25.50

Unit cost per sq ft for 2 coats $\dfrac{\$25.50 \times 2}{450}$ = $ .11

Estimated Cost

Room size	Gross area of walls and ceiling		Cost
10' x 14' x 9'	572 sq ft	@ 11¢	$ 62.92
12' x 18' x 9'	756 sq ft	@ 11¢	83.16
8' x 10' x 9'	404 sq ft	@ 11¢	44.44
11' x 12' x 9'	546 sq ft	@ 11¢	60.06
		Total cost	$250.58

Illustrative Example No. 2

The exterior bevel siding of a dwelling is to be painted with two coats. The size of the building is 45' x 30' with a 6' roof at the gable end. The height of the walls from foundation to the eaves is 12', and the cornices are finished flush.

Wall area	= 2(45' + 30') x 12'	= 1,800 sq ft
Gable area	= 6' x 30'	180 sq ft
		1,980 sq ft

Unit Cost

Assume that paint cost $6.00 per gallon
Assume that painter's wages are $6.00 per hour

1 gallon paint	@ $6.00 =	$ 6.00
$\dfrac{500}{125}$ = 4 hours labor	@ 6.00 =	24.00
		$30.00

Unit cost per square foot 1 coat $\dfrac{\$30.00}{500}$ = $ .06

Unit cost per square foot 2 coats = $ .12

Estimated Cost
1,980 square feet @ .12 = $237.60

DAMAGEABILITY

Paints, varnishes and other decorative materials can be damaged during a fire by heat, smoke, water, and chemicals used to fight the fire, staining from contents, and by scratching or marring. Hailstones damage exterior paint, water from leaky plumbing or the backing up of sewers causes injury to interior paint, and wind-driven sand can scarify or completely remove exterior paint. Whatever the cause of the damage, it is visible and usually presents no problem of discernment with the exception of isolated cases of questionable damage by acid fumes, smoke, or other stains.

The main concern of an estimator is to determine the requirements and the specifications for restoring the surface to its original condition. His problem can be generally divided into two categories:

1. The preparation of the surface before finishing
2. The number of coats to be applied

PREPARATION OF THE SURFACE

Before paint, varnish, or other materials can be applied, the surface must be cleaned to remove the dirt and grease. Blistered areas on wood must be burned off with a torch and sanded smooth. Scraping blackened and blistered woodwork with a knife will disclose whether or not the wood itself has been burned.

Walls and ceilings that are seriously coated by smoke should be washed before painting for two reasons. Washing removes dirt and grease and, on more occasions than are recognized, washing a surface will make restoration possible by the application of a single coat. Where the soilage is slight,washing may remove all evidence of injury eliminating the need for painting.

No painting should ever be done on wet wood or wet plaster surfaces. Nail holes, scratches, raised grain, and other surface defects and blemishes should be filled and sanded smooth. Cracks in plaster should be cut out and filled.

Many times steel members, masonry, metal ceilings, and so forth, must be wire brushed to remove rust, blisters or other loose material before painting. Varnished, enameled, and other glossy surfaces should be lightly sanded before applying new material.

Only a careful examination of the damaged property will make known the extent of necessary surface preparation. It most cases it will consist of washing or cleaning of the old surface, with possibly minor sanding or spackling. When there is a doubt that a surface requires, or can be, washed, brushed, sanded or scraped in order to prepare it, sample tests should be made. Tables 14-3 and 14-4 show the approximate number of square feet per hour required to prepare surfaces. The labor cost per square foot is obtained by dividing the local hourly wage rate by the number of square feet covered per hour. The cost of any materials such as washing powders, detergents, sponges, varnish remover and acids should be added.

NUMBER OF COATS TO BE APPLIED

Exterior. New exterior woodwork should be given three coats of paint. On old painted surfaces, depending on their condition, one or two coats of the existing color are adequate. If the old paint is in good condition and the damage is slight, one coat of paint will generally give a satisfactory finish. If the old paint is in poor condition or has been seriously injured, it may require two coats.

Interior. Ordinarily new interior woodwork receives a prime coat and a finish coat of paint. Newly plastered walls should be given a coat of size, followed by two coats of paint. Calcimine or cold water paints are applied over new plaster after applying a thin coat of shellac or sizing.

Old woodwork and plastered or dry walls may require one coat only if the soilage is slight, or if pre-washing will remove most of the stain. Otherwise two coats are required to restore the surface.

PAPERHANGING

Wallpaper is sold by either the single or the double roll. A single roll is a strip 18 inches wide, 24 feet long, and contains 36 square feet. A double roll is 48 feet long and contains 72 square feet. Wallpaper borders are sold by the lineal yard. The quality and the price of wallpaper has a wide range. It is, therefore, important to determine as closely as possible, the quality of paper that was on the walls at the time of a loss.

Preparation of the Surface

Before walls are papered, they should be given a coat of glue size, the cost of which has been discussed under "Painting."

Old wallpaper should be removed, the walls washed, cracks patched and a coat of sizing applied. Wallpaper can be taken off by wetting it down with water, by brush or spray, and then scraping it off with a wide putty knife. An electric-steam appliance is available which speeds up the work of loosening the paper.

When permissible, new wallpaper can be applied directly over old paper provided the surface is smooth (not embossed) and the old paper is butt-jointed and tight on the wall. Under these conditions papering over old paper is entirely satisfactory.

TABLE 14-5

ESTIMATING QUANTITIES FOR PAPERHANGING WALLS AND CEILINGS: Measure straight area to secure number of square feet and divide by 30. (This allows 6 sq. ft. for waste.) Deduct ½ roll for every opening. Or, for the following standard-size rooms use this handy chart:

Size of Room	Single Rolls of Side Wall Height of Ceiling			Yards of Border	Rolls of Ceiling
	8 feet	9 feet	10 feet		
4 x 8	6	7	8	9	2
4 x 10	7	8	9	11	2
4 x 12	8	9	10	12	2
6 x 10	8	9	10	12	2
6 x 12	9	10	11	13	3
8 x 12	10	11	13	15	4
8 x 14	11	12	14	16	4
10 x 14	12	14	15	18	5
10 x 16	13	15	16	19	6
12 x 16	14	16	17	20	7
12 x 18	15	17	19	22	8
14 x 18	16	18	20	23	8
14 x 22	18	20	22	26	10
15 x 16	15	17	19	23	8
15 x 18	16	18	20	24	9
15 x 20	17	20	22	25	10
15 x 23	19	21	23	28	11
16 x 18	17	19	21	25	10
16 x 20	18	20	22	26	10
16 x 22	19	21	23	28	11
16 x 24	20	22	25	29	12
16 x 26	21	23	26	31	13
17 x 22	19	22	24	28	12
17 x 25	21	23	26	31	13
17 x 28	22	25	28	32	15
17 x 32	24	27	30	35	17
17 x 35	26	29	32	37	18
18 x 22	20	22	25	29	12
18 x 25	21	24	27	31	14
18 x 28	23	26	28	33	16
20 x 26	23	28	28	33	17
20 x 28	24	27	30	34	18
20 x 34	27	30	33	39	21

Deduct one single roll of side wall for every two ordinary sized doors or windows or every 36 square feet of opening. Yard goods with no match such as wide vinyls, measure area, take out for openings and allow 10" for waste.

Courtesy of Painting and Decorating Contractors of America, Falls Church, Va. 22046

Estimating Materials

When hanging wallpaper, some allowance must be made for the waste which results from cutting, fitting around openings and built-in units, and for matching patterns. As a general rule the waste averages between 15 and 20 percent. A single roll of paper contains 36 square feet. A practical and acceptable method for allowing waste is to figure the rolls at 30 square feet. Double rolls are figured at 60 square feet to take care of waste.

The cost of material and labor for papering is estimated by the roll. If the cost per square foot or square yard is desired, it may be obtained in two ways: by dividing the cost of papering a room by the wall area, and by computing the cost of hanging one roll of paper and dividing by 30 square feet. (See "Methods of Estimating.")

To obtain the number of rolls of paper required, the distance around the room (perimeter) is multiplied by the height from the top of the baseboard to the ceiling. *All window and door openings are deducted.* The result is the wall area to be papered. That area is divided by 30 to obtain the number of single rolls of paper. Paper is sold in full rolls and the number of rolls needed must be carried out to the next whole roll. If double rolls are used, the area should be divided by 60.

Where walls are papered above a chair rail only, or above a wainscoting, the height measurement should be taken from where the paper starts to the ceiling. When a single wall is to be papered the length of that wall is measured and multiplied by the height.

Where ceilings are to be papered, the area is obtained by multiplying the length of the ceiling by the width.

If a border is required, the distance around the room divided by 3 will give the number of yards required.

Illustrative Example

Find the number of single rolls of paper and the number of yards of border required for the walls of a room 14' x 18' x 8' 6". The baseboard is 6" high. Openings consist of two windows 3' x 5' and two doors 3' 6" x 7'.

Area of Walls (above baseboard)	
2(14' + 18') x 8'	= 512 sq ft
Deduct openings	
2 windows 3' x 5' = 30 sq ft	
2 doors 3½' x 7' = 49 sq ft	
	-79 sq ft
Area to be papered	= 433 sq ft

$$Rolls\ of\ paper = \frac{433}{30} = 14.4 = 15\ rolls$$

$$Yards\ of\ border = \frac{2(14' + 18')}{3} = \frac{64}{3} \qquad = 22\ yards$$

The amount of paste needed to hang paper varies with the type and weight of the paper. Heavy or rough textured papers require more than lightweight papers. For average estimating purposes a pound of paste makes 2 gallons, which will hang approximately 10 single rolls of paper.

Estimating Labor

A paperhanger will hang paper at the rate of 3 single rolls per hour under normal conditions. This rate is for butt-jointed work. If the joints are lapped, the rate should be increased to about 3 1/2 or 4 rolls per hour.

Where workmanship and materials are of the highest grade, a paperhanger will hang paper at the rate of 2 rolls per hour.

Paper borders can be hung at the rate of approximately 30 yards per hour.

Table 14-6 shows the approximate labor hours required for various paperhanging operations.

SANITAS AND WALLTEX

Sanitas and Walltex come in rolls containing 9 square yards. They are frequently used on the walls of bathrooms and kitchens because they are water-resistant and washable. The area is obtained in the same manner as for wallpaper. The quantity of Sanitas or Walltex required is equal to the area to be covered plus 10 percent for waste. Because they are expensive, measurements for Sanitas and Walltex should be taken with care.

A paperhanger can hang approximately 2 rolls of Sanitas or Walltex per hour.

Unit Cost per Square Foot

A unit cost per square foot may be obtained in two ways. The cost of the material and labor for a particular room may be computed and divided by the number of square feet to be papered. In the previous Illustrative Example the total cost of labor and material was $49.44. The area papered was 421 square feet. By dividing $49.44 by 421 a unit cost of $.117 is obtained. This method is useful where several rooms are to be papered. Once a unit cost is obtained for a typical room, it can be applied to other similar rooms.

TABLE 14-6
LABOR TO HANG PAPER

A. Ceilings (residential or commercial) rolls per 8 hr. day.

	Occupied	Vacant
a. Mica	18 to 24 rolls	26 to 30 rolls
b. Brush-tint	16 to 18 rolls	22 to 24 rolls
c. Lining (blank stock)	16 to 18 rolls	22 to 24 rolls

B. Sidewalls (residential or commerical) rolls per 8 hr. day.

	Occupied	Vacant
a. Lining (blank stock)	18 to 20 rolls	22 to 24 rolls
b. Ordinary pre-trim.	16 to 18 rolls	22 to 24 rolls
c. Hand-knifed	12 to 14 rolls	14 to 16 rolls
d. Cork	12 to 14 rolls	14 to 16 rolls
e. Grasscloth	14 to 16 rolls	18 to 20 rolls
f. Flock	8 to 10 rolls	. . . .
g. Burlap	12 rolls	. . . .
h. Silk	8 to 10 rolls	. . . .
i. Foil	8 rolls	. . . .
j. Scenics	10 panels per day	
k. Coated fabrics	14 to 16 rolls	18 to 20 rolls
l. Lining canvas	14 to 16 rolls	18 to 20 rolls

Vinyl or Coated Fabrics (commercial) rollage & yardage per 8 hr. day:
A. Light weight, 24″ to 27″, 18 to 22 rolls.
B. Heavy weight, 54″, 30 lineal yards average; 45-65 lin. yds. corridor & volume.
Wood Veneer, Flexwood etc. sq. ft. basis
A. 105 to 145 sq. ft. per day.
Courtesy, Painting and Decorating Contractors of America

A much simpler procedure is to develop a unit cost from a single roll of paper. The following basic formula may be used and local wage rates and material prices inserted. The illustration contemplates that a paperhanger can hang an average of 3 single rolls of paper an hour. When the rate of hanging is changed because of better grade work or an inferior grade, the formula may be adjusted accordingly. If 2 rolls are to be hung each hour, an allowance of 1/2 hour for labor is made in the formula. If 4 rolls per hour are contemplated, then the labor figure should be 1/4 hour. Paste is a relatively insignificant item and may be omitted if desired.

METHODS OF ESTIMATING COST

Material and Labor Per Room

Some estimators prefer to obtain the cost of wall papering by estimating the material and labor required for each room rather than using a unit cost per square foot.

Illustrative Example

To find the cost of papering a room 12' x 16' x 8½'. It has two windows 3' x 5' and one door 3'6" x 7". Assume a baseboard 6" wide. The cost of the paper is $1.50 per roll, paste is $.30 per lb and paperhanger's wages are $6.00 per hour.

Area of walls = 2(12' + 16') x 8½'   = 476 sq ft
Deduct openings
2 windows 3' x 5' = 30 sq ft
1 door 3½' x 7'   = 25 sq ft

| | -55 sq ft |
| Area to be papered | 421 sq ft |

Material cost

Rolls of paper needed $= \dfrac{421}{30} = 14$@$1.50   = $21.00

Paste (1 lb for 10 rolls)
$$\dfrac{14}{10} = 1.4 \text{ lbs.} @ \$.30 \qquad = \quad .42$$

Total cost of materials = $21.42

Labor cost (Table 14-6)

Number of rolls $= \dfrac{14}{3} = 4.67$ x $6.00   = 28.02
Rolls per hour

Total cost of papering = $49.44

Basic Formula

1 roll of wallpaper	@ $	= $
.l lb wallpaper paste	@	=
Hours labor	@	= _____
	Total cost per roll	= $

$$\text{Unit cost per sq ft} = \dfrac{\$}{30} = \$$$

Illustrative Example

Using the material prices and labor wage rate in the previous example, a unit cost may be obtained.

1 roll of wallpaper	@ $1.50 = $1.50
.1 lb wallpaper paste	@ .30 = .03
⅓ hours labor	@ 6.00 = 2.00
	Total cost per roll = $3.53

$$\text{Unit cost per sq ft} = \dfrac{\$3.53}{30} = \$.117$$

Papering and Painting

When it is required to paper and paint a room, the cost of painting the ceiling, the openings, and the trim is estimated as shown under the section on "Painting." Assume paint costs $6.00 per gallon and painter's wages are $6.00 per hour.

Illustrative Example
(Using the foregoing problem, page 387)
Cost of papering = $49.44

Considering the baseboard equivalent to one opening, the number of openings (one side) to be painted two coats would be:

2 Windows
1 Door
1 Baseboard
4 Openings

From Table 14-3 (Footnote)

1/16 gal per opening	@ $6.00 = $ .38
⅓ hours labor per opening	@ 6.00 = 2.00
	$2.38
Two coats	x 2
Cost to paint each opening	$4.76
Cost for 4 openings $4.76 x 4 =	$19.04

The cost to paint the ceiling two coats would be:

Area 12' x 16'	= 192 sq ft
Two coats	x 2
	384 sq ft
Materials = $\frac{384}{450}$ = .85 gallons @ $6.00	= $ 5.10
Labor = $\frac{384}{150}$ = 2.6 hours @ $6.00	= 15.60
Total cost to paint ceiling	$20.70

Summary

Cost to hang paper	= $49.44
Cost to paint openings	= 19.04
Cost to paint ceiling	= 20.70
Total =	$89.18

The unit cost per square foot for the entire wall and ceiling area of this room is obtained by dividing the gross area of walls and ceiling by the total cost.

Gross area of walls (including openings)	= 476 sq ft
Area of ceiling	= 192 sq ft
Total =	668 sq ft

$89.18 ÷ 668 = $.133 or $.135 per square foot

The unit cost of $.135 developed here could be applied to the square foot area of any room where the specifications and the quality of materials and workmanship were similar. A great many estimators use this method to arrive at a unit cost. It is generally considered less accurate than when the painting and papering is figured individually for each room because the number of openings, the quality of paper and paint, and the working conditions vary from one room to another.

DAMAGEABILITY

There is little to be said concerning the damageability of wallpaper that is not generally known. The problem that frequently confronts an estimator is deciding whether soiled wallpaper should be cleaned rather than replaced. Many times it is necessary to make sample tests, or call in wall cleaning firms for advice when they are available.

Most wallpapers that have been wet will have to be replaced. Unless perfect matching of the existing paper is possible, the entire room will have to be repapered even when small areas are stained.

Sanitas and Walltex that has been wet from the back side should be carefully examined to determine whether or not the glue has been injured.

Smoked or otherwise soiled walls can frequently be cleaned with commercial dough-type cleaners. A man can clean wallpaper at the approximate rate of 100 square feet per hour by this method. Water-resistant fabric type materials, when subjected to cleaning can usually be lightly washed with water and soap or detergents. Care should be taken not to remove any of the painted surface of the fabric.

GLAZING

The cost of glazing depends mainly on five conditions which should be considered in estimating:

1. Type and thickness of glass.
2. Quality of glass.
3. Size and shape of glass.
4. Type of sash.
5. Difficulties of installing.

Type and Thickness of Glass

There are many types of glass used in buildings. Clear window glass ranges in thickness from about 1/12 inch to 1/4 inch. It is available in two grades, or qualities, "A" and "B." Window glass is customarily identified in thickness as Single Strength (S.S.), Double Strength (D.S.), Semi Plate (S.P.), and Plate (Pl.) glass. The respective thicknesses are approximately 1/12 inch for Single Strength, 1/8 inch for Double Strength, 3/16 inch for Semi Plate, and 1/4 inch or 5/16 inch for Plate glass. Single Strength glass is not recommended for lights containing areas in excess of about 3 square feet. Double Strength may be used for lights up to 5 or 6 square feet. Crystal sheet glass (Semi Plate) is a type used where plate glass is too heavy and expensive, but where heavier than D.S. glass is desired. It comes in different thicknesses, but the 3/16 inch or 39 ounce per square foot is most commonly used.

In mercantile and industrial buildings wire glass is commonly used in doors and windows that are required to be fire resistive, and in areas exposed to possible fires in adjoining properties. The wire glass may be rough, ribbed, or polished. In specifying which type of wire glass is required, the thickness should also be indicated.

The ornamental types of glass include Florentine, Pressed, Ribbed, Flutex, Maze, Syenite, Frosted, Leaded and other art glass.

Size and Shape of Glass

The size of each light of glass should be measured carefully. In window glass the available sizes range in width from 6 to 12 inches by 1-inch intervals, and from 12 to 30 inches in 2-inch intervals. The sizes in length range from 8 to 40 inches in 2-inch intervals. Plate glass should be measured in even inches. Wire glass, Window glass, and Fancy glass lights should be measured exactly before obtaining prices from local dealers.

Glass is sold on discounted price lists. The amount of discount varies with the type, quality, size and quantity.

Specifications for glass in an estimate should include details of any special shape or finish of edges. Bent glass, and beveled or polished edges all add to the price of the glass.

Quality of Glass

Single strength glass, because of its thinness and slight variation in thickness, may have certain imperfections in the form of blisters or lines. There are three grades of quality in window glass. The most perfect glass is graded AA which has no objectionable imperfections. Grade A is the next highest and may contain a few flaws, none of which are prominent. Grade B is the lowest grade and generally contains noticeable imperfections.

Type of Sash

Wood sash with 1 3/8-inch frames require less putty than those with 1 3/4-inch frames. Steel sash generally require more putty than wood sash.

Difficulties of Installing

Allowances must be made for any difficulties a glazier will encounter when replacing lights of glass. If the putty is hard and adheres tightly to the frame, it may take more than normal time to cut it out. Glazing sash is easier from the floor or ground level than from exterior ladders or scaffolding. A glazier works outside faster in summer than in winter when the putty tends to stiffen with the cold.

ESTIMATING MATERIALS

In estimating glass, the number of lights of each size should be noted with a description of type, thickness and quality. The prices should be obtained from a local dealer.

The amount of putty needed for glazing varies, and in relation to the total cost of glazing it is a minor item. For estimating the following amounts are recommended:

Type of Sash	Quantity of Putty
Wood sash (1 3/8-inch frames)	1 lb per 8 lin feet of puttying
Wood sash (1¾-inch frames)	1 lb per 7 lin feet of puttying
Steel sash	1 lb per 5 lin feet of puttying

For steel sash ½ lb of putty per square foot is sometimes used.

Plate glass in doors, store fronts, transoms, and so forth should be measured to the next even inch. Individual sizes should be listed.

ESTIMATING LABOR

In estimating the time required to re-glaze sash, thoughtful consideration must be given to the physical problems that confront the person who will do the actual work. On new jobs, or those involving a large number of lights of glass in a concentrated area, the probable average number of hours for glazing can be estimated fairly closely. If, however, a windstorm has broken a sizeable number of lights in a building, which requires moving constantly from one area to another, or from one room to another, much time will be taken up by that activity. Hardened putty in older wood frames is difficult to clean out without splitting away portions of the wood.

When the wood sash can be taken out of the frames, the glazing is easier and usually results in better workmanship.

If a glazier must be hired to install one light of glass, the entire cost of coming from and returning to his shop may have to be charged against the cost of the one light.

The glazier's time when working from ladders or scaffolds is usually greater than when working from ground or floor level.

The best way to estimate the labor for reglazing is to separate the time for removing old putty and cleaning the frame, from the time required to install and putty in the new glass. When it is not necessary to remove old putty, no allowance should be made. Table 14-7 shows the approximate number of hours of labor for each operation for various sizes and kinds of glass. These hourly rates should be used as a starting basis and modified according to the conditions found.

While glaziers or painters do most glazing, it is not uncommon in some localities to have carpenters or other mechanics install glass. If repairs in addition to glazing are being made, and one or even a few lights of glass are to be put in, the estimator should consider the possibility of having it done by workmen on the job.

Table 14-7
Approximate Number of Hours to Remove Putty and
Glaze Sash Per Light Under Favorable Working Conditions

Size of Glass (Inches) and Kind of Work	Hours to Clean, Remove Old Putty	Hours to Glaze	Total Hours
Glazing windows in wood sash:			
(S.S. and D.S.):			
8 x 12*	.21	.14	.35
10 x 16	.26	.17	.43
12 x 14	.26	.17	.43
14 x 20	.30	.20	.50
16 x 24	.40	.27	.67
20 x 28	.51	.34	.85
24 x 30	.55	.37	.92
Glazing windows in steel sash:			
12 x 16	.18	.12	.30
12 x 18	.25	.17	.42
16 x 20	.30	.20	.50
16 x 30	.37	.25	.62
16 x 48	.40	.33	.73
Monitor and skylight sash:			
24 x 36	.67	.50	1.17
24 x 48	.75	.67	1.42
24 x 60	.90	.75	1.65

*For glazing plate glass in wood sash, double the time shown for S.S. and D.S. For glass sizes other than those shown, interpolate between the size below and that above it.

METHODS OF ESTIMATING

Plate glass is estimated by the square foot, and local dealers should be consulted for guidance on prices. Every estimator should avail himself of the book of standard list prices, and current discounts on glass, so that he can approximate prices when necessary. There is frequently salvage in the larger plate glass windows. The amount of salvage will depend on the size of the remaining section. Glass companies customarily have standard practices in allowing for salvage.

Other types of glass are estimated by the individual kind, size and quality of each light. Material and labor costs should be figured for the locality. When a large number of lights other than wire glass are to be installed it is customary to allow approximately 3 percent for breakage. The following is an example of an estimate for glazing:

Illustrative Example No. 1

To estimate the cost of removing 150 lights of 1/4" x 16" x 24" cracked polished wire glass in steel sash, and replacing with new lights.

Assume wages to be $6.00 per hr., glass $2.10 per sq ft and putty $.20 per lb.

First step—Develop a unit cost:

1 16" x 24" light = 2⅔ sq ft	@ $2.10 = $	5.60
(½ lb putty per sq ft page 391)		
½ x 2⅔ = 4/3 lbs x $.20	=	.27
(Labor from Table 14-7) = .55 hrs	@ 6.00 =	$3.30

Unit cost per light = $ 9.17

Cost of 150 lights = 150 x $9.17 = $1,375.50

In this example had it not been necessary to remove old glass, as in the case of new work, the unit cost would be:

1 16" x 24" light	=	$5.60
4/3 lbs putty	@ $ .20 =	.27
Labor .22 hours	@ 6.00 =	1.32

Unit cost per light = $7.19

Cost of 150 lights = 150 x $7.19 = $1,078

Illustrative Example No. 2

To estimate the cost to remove old putty and glass from 1 3/8 inch wood sash and reglazing with double thick 14" x 20" lights.

Assume wages to be $6.00 per hour, glass $.50 each and putty $.20 per pound.

Unit cost developed:

1 light 14" x 20" d.t. glass	=	$ .50
Putty (8 lin ft per lb, page 391)		
$\frac{2(14" + 20")}{12} = \frac{6}{8}$ = ¾ lbs	@ $.20 =	.15

Labor (Table 14-7) .50 hrs   @ 6.00 =   3.00

Cost per light = $3.65

Chapter 15.

Lathing and Plastering

LATHING

Originally *plaster lath* denoted narrow strips of wood that were placed close together to provide a base or background for the application of plaster. In recent years, however, it has come to include the numerous substitutes for that method of construction. Perforated sheets of thin steel and gypsum board, either solid or perforated, are known in the trade as metal lath and gypsum lath respectively. The cost of gypsum lath, in most places, is slightly less than wood lath, and wood lath is slightly less than metal lath. Lath of all kinds is applied by a lather where union regulations prevail. In other areas of the country lath is applied either by a lather or a carpenter.

MEASUREMENT OF AREAS

As in measuring areas for applying numerous other building materials, there are different practices and varying opinions on the question of deducting window and door openings. One person may deduct all openings in full, while another will deduct only one half of the area of the openings. Some estimators do not deduct any openings. Those who do not deduct openings, and those who deduct half of them, feel that their method will take care of the waste and the additional labor necessary to work around them. Those who deduct openings in full add a percentage allowance for waste, giving the waste a more realistic relationship to the actual quantity of material required to cover the surface. In the labor rate of application, a further allowance in time is then made for working around openings.

Although both methods have merit, the author recommends deducting all openings in full. An allowance of about 10 percent should be added to the "net" area for waste. This method of measuring areas is especially preferred in many of the modern industrial buildings and in house construction where large windows, glass walls, and trimmed openings are frequently encountered.

Whichever method is used, good judgment should be exercised to make certain that an adequate amount of material is allowed, and that sufficient labor is included for the physical layout of the surface to be covered.

WOOD LATH

Wood lath are rough wood strips that measure 1 1/2" x 3/8" x 4' in length. They will reach from center to center on four studs, joists, or furring placed 16 inches apart. They are sold in bundles containing 100 lath, and are priced either by the bundle or per 1,000 lath. They are applied with 3d fine, No. 16 gauge, blued nails, and are spaced 3/8 inch apart for lime plaster and 1/4 inch for light gypsum plaster. This allows the plaster to be forced between the lath to form a hardened blob or "key" on the backside.

Estimating Quantities

The customary unit for computing quantities of wood lath is the square yard. It takes approximately 1,500 lath, 1 1/2" x 3/8", spaced 3/8" apart, to cover an area of 100 square yards. Approximately 10 pounds of nails are required for 100 square yards.

The *gross* area of the surface to be lathed is computed. *All openings are deducted,* and an allowance of 10 percent is added for waste (See Measurement of Areas).

Estimating Labor

Under average conditions such as might be found in dwellings with rooms that have one or two windows and doors, a lather can apply wood lath at the rate of about 10 square yards an hour, or 1,500 lath in 10 hours. On larger surfaces, with few or no openings, the rate of application can be increased to about 12.5 square yards an hour, or 1,500 lath in 8 hours.

Unit Costs

Illustrative Example
(Material Prices and Wages Are Assumed)

1500 spruce lath	@ $28.00 per M =	$ 42.00
10 lbs 3d lath nails	@ .20 per lb =	2.00
10 hours labor	6.00 per hr =	60.00
	Unit cost for 100 square yards =	$104.00

$$\text{Unit cost per sq yd} = \frac{\$104.00}{100} = \$1.04$$

GYPSUM LATH

One of the most popular plaster bases is gypsum board. This is a core of gypsum between two layers of paper. It is relatively easy to apply, and the cutting and fitting around openings is quickly done by scoring the material with a knife or a hatchet. It is fire-resistive, offers desirable insulating qualities, and provides a better adhesive bond that wood or steel lath.

Gypsum lath is generally 16" x 48" and is 3/8" thick. Whether applied vertically or horizontally, the sheets will reach from center to center of studding that is spaced 16 inches apart. When studding is more than 16 inches on center but not more than 24 inches, gypsum lath of 1/2-inch thickness is used. It comes in solid sheets, or in sheets perforated with 3/4-inch holes every 16 square inches. The latter permits the plaster to form a key, or mechanical bond, on the back side; this type is frequently preferred for ceilings or is specified where a better fire-resistive rating is required. The mechanical bond tends to hold the plaster in place after the adhesive bond has been weakened by heat.

Estimating Quantities

The unit for computing quantities of gypsum lath is the square yard. Special flathead blued nails, 1 1/8-inches, long No. 13 gauge, are used in nailing gypsum lath. The nails are spaced about 5 inches apart. Approximately 15 pounds are required for 100 square yards. For gypsum lath 1/2-inch thick, nails 1 1/4 inches long are used and spaced 4 inches apart.

The gross area of the surface to be lathed is measured, and *all openings are deducted* (See Measurement of Areas). An allowance of about 10 percent is added for cutting and fitting.

Estimating Labor

A lather should be able to apply gypsum lath on frame construction at an average rate of 10 square yards an hour, or 100 square yards in 10 hours.

When it is applied with staples, a lather can put on 100 square yards in about 8 hours. The difference in cost between nails and staples is not significant.

Unit Costs

Illustrative Example
(Material Prices and Wages Are Assumed)

100 sq yds gypsum lath	@ $37.00 =	$ 37.00
15 lbs gypsum board nails	@ .20 =	3.00
10 hours labor	@ 6.00 =	60.00
	Unit cost for 100 sq yd =	$100.00

$$\text{Unit cost per sq yd} = \frac{\$100.00}{100} = \$1.00$$

METAL LATH

Metal lath is made of steel. It comes either painted or galvanized to protect it against rusting. Metal lath gives the best mechanical bond of all laths. Portland cement mortar does not adhere well to gypsum lath; where that type mortar is required, metal lath is used.

Metal lath is made in a variety of patterns that may be generally classified as "expanded," "ribbed," "sheet," or "wire" laths. It comes in sheets measuring 27" x 96" with 10 sheets to the bundle. One bundle will cover 20 square yards. Metal lath is fastened to wood framing and furring by blued or galvanized nails. Where sheets are joined in between studding and joists, the metal lath is wired together with annealed wire.

Estimating Quantities

The unit for computing quantities of metal lath is also the square yard. Approximately 8 pounds of 1 1/2-inch No. 11 gauge, flat-head, blued nails are required for 100 square yards of lath. *In measuring areas for wire lath all openings are deducted in full.* (See "Measurement of Areas"). An allowance of 10 percent should be added for waste in cutting, lapping, and fitting.

Estimating Labor

Under average working conditions a lather can apply 100 square yards of metal lath to frame construction in 10 hours, or 10 square yards per hour.

Unit Costs

Illustrative Example
(*Material Prices and Wages Assumed*)

100 sq yds metal lath	@ $54.00 =	$ 54.00
8 lbs metal lath nails	@ .20 =	1.60
10 hours labor	@ 6.00 =	60.00
	Unit cost for 100 sq yds =	$115.60

$$\text{Unit cost per sq yd} = \frac{\$115.60}{100} = \$1.16$$

GROUNDS AND SCREEDS

Grounds are wooden or metal strips that are nailed around rough interior door openings and some window openings to establish the thickness of the plaster when it is applied. They also act as a leveling guide for the mason. Grounds are also nailed along the base where the plaster is to stop. These may be nominal 1" x 2" furring. Usually they are left in place when they are flush with the plaster, and the baseboard is nailed over them.

Screeds are plaster strips placed by the plasterer along the outer edge of the ceiling to serve as a thickness and a leveling guide. Sometimes, they are also placed along the base of the wall.

The cost of screeds is included in the cost of the plastering labor. The cost of grounds at door or window openings may be estimated at approximately 1/2 hour per average opening. The cost of material is not significant. Baseboard grounds may be estimated at the rate of 100 lineal feet in 2 hours, and the material cost should be added.

CORNER BEAD

Galvanized steel corner bead is used at all corners for protection of the finished plaster edge. It also serves as a ground and leveling guide. Corner bead can be used with all types of lath.

The cost of corner bead is estimated on the basis of a lather's installation at the rate of 100 lineal feet in 2 hours. The cost of the corner bead itself should be added.

Illustrative Example
(Prices and Wages Assumed)

100 lin ft corner bead	@ $	.05 =	$	5.00
2 hours lather	@	6.00 =		12.00
	Unit cost per 100 lin ft =		$	17.00

$$\text{Unit cost per lin ft} = \frac{\$17.00}{100} = \$.17$$

PLASTERING

Plaster is actually mortar that is applied in successive coats to wall and ceiling surfaces. It is composed of a fine aggregate held together by either gypsum or portland cement as a cementatious material. Water is added to form a workable plastic mass that can be spread over the surface and smoothed out with a trowel or float. Gypsum is used for interior plaster, and portland cement is used for exterior plaster because it is weather resistant. Portland cement is also used for certain kinds of interior plastering work as basement walls or backing for ceramic tile.

A mechanic who applies plaster is known in the building trades as a *plasterer.* Because their work is somewhat related to masonry some plasterers are also masons and *vice versa.*

Plaster is applied in one, two, or three coats. Three-coat plaster consists of a *scratch coat,* a *brown coat,* and a *finish coat.* Three-coat work is considered to be the highest grade of plastering, but two-coat work is becoming increasingly common, particularly in house building. One reason for the popularity of the latter is the lower cost in labor. Another reason is that the solid gypsum board lath enables a plasterer to obtain a true surface with two coats rather than three, which is generally necessary on wood and metal lath. One-coat plaster is generally limited to secondary structures and out-buildings, where a rough or uneven surface offers no objection and where the plaster is basically functional.

Both gypsum and portland cement plaster may be applied to masonry walls. In such cases two-coat work is considered adequate. Local building codes may, in certain instances, require three coats.

INTERIOR PLASTERING

Three-coat Plastering

Three-coat plaster is applied over metal or wood lath. It consists of two base coats and one finish coat. The first base coat in three-coat work is called a *scratch coat* because it is scored, or scratched, while still wet to provide a better bond for the second base coat. Much of the plaster applied on the first coat is forced through the spaces in wood and metal lath to provide a *key* effect behind the lath, giving the plaster a mechanical bond. Because first-coat plaster is applied about 1/4 inch thick, it leaves the ridge marks of the lath. A second, or *brown coat,* is applied after the scratch coat has set firm but not dried out. This coat is troweled level and smooth in preparation for the finish coat. The finish coat usually is a mortar of gypsum and hydrated lime. It is called a putty or *white coat.* It is applied approximately 1/16 inch thick.

Two-coat Plastering

Two-coat plaster is applied over a solid plaster base such as gypsum board or masonry. The first coat is put on either in one operation, or by the *double back* method. In the latter method a thin coat is first applied. Before it has set, the plasterer goes back over the area filling in and leveling the surface to the full thickness of the grounds. A finish coat is then applied in the same manner as in three-coat work after the base coat has set.

White Putty Finish

The plaster finish in most common use is the lime-putty finish. It is applied over the base plaster in a thickness of about 1/16 of an inch. It produces a smooth hard white surface, which may be painted or papered.

Lime that has been partially hydrated by the manufacturer is known as "quicklime." It must be further hydrated on the job by soaking (slaking) from 12 to 24 hours before it can be used. Double hydrated finishing lime such as USG Ivory Finishing Lime, requires no slaking on the job because it is 92 percent hydrated during manufacture. When mixed with water, the lime immediately develops high plasticity, eliminating overnight soaking.

Neat gypsum plaster is added to the lime putty as a setting agent to produce hardness, and also to resist shrinking which is a characteristic of hydrated lime. The proportions specified by the ASA are:

1 (100-lb) bag of gypsum gauging plaster to not more than
4 (50-lb) bags of hydrated lime

Keene's Cement Finish

During the calcination of natural gypsum, if all of the chemically combined water is removed, instead of 75 percent as in the case of plaster of Paris, the resultant product is called Keene's cement. It is a form of gauging plaster which, when mixed with lime putty, produces a very hard and durable surface.

The ASA (American Standards Association) specifications for mixing proportions of *medium hard finish* Keene's cement are:

1 (100-lb) bag of Keene's cement to not more than
1 (50-lb) bag of hydrated lime.

The ASA specifications for *hard finish* Keene's cement are:
1 (100-lb) bag of Keene's cement to not more than
1/2 (50-lb) bag of hydrated lime

Skimcoat Plastering

Where economy is a factor in less expensive dwelling construction, a thin coat of lime putty plaster finish is applied *directly* to gypsum board after the joints have been taped. It is found in many of the low-cost homes in the southern states where extreme changes in temperature are rare.

The thickness of this skimcoat is similar to the 1/16-inch finish coat applied over regular two- and three-coat plastering. The cost may be obtained as follows:

Illustrative Example
(Material Prices and Wages Assumed)

2 (100-lb) sacks gypsum plaster	@ $1.40 =	$ 2.80
8 (50-lb) sacks hydrated fin. lime	@ 1.00 =	8.00
12 hours plasterer	@ 6.00 =	72.00
12 hours laborer	@ 5.00 =	60.00
	Unit cost for 100 sq yd =	$142.80

$$\text{Unit cost per sq yd} = \frac{\$142.80}{100} = \$1.43$$

Sand Finish

A *sand finish* employs mortar similar to the scratch and brown coats except that a very fine sand (sometimes white sand) is used. A cork float is employed to obtain a sandpaper-like surface. Occasionally colors are used to tint the sand-finish making it unnecessary to paint or decorate.

MATERIALS

Gypsum is a grey or white, calcium, rocklike mineral that is found abundantly in nature in most parts of the world. It contains, by weight, approximately 21 percent of chemically combined water. When the gypsum rock is pulverized, and subjected to heat, water is driven off. This process is called *calcination*. If 75 percent of the water is removed the resulting product is *plaster of Paris*. This calcined gypsum when remixed with water, sets and returns to gypsum rock *with its original chemical composition.*

Calcined gypsum is the cementatious base material for most of the interior plaster mortars. Mixed with sand, perlite, or vermiculite, it is used as a rough or base plaster for scratch, brown, or sand-finish coats. When mixed with hydrated lime, it is used as a putty or white finish coat.

PROPORTIONS

Job-mixed Plaster

Gypsum plaster for base coats may be mixed on the job by machine or by hand in a mortar box. It may also be purchased ready-mixed with mineral aggregate already added.

When job-mixed, the following proportions of calcined gypsum to aggregate are recommended by the Gypsum Association:

For two coat or "double back" plastering: 100 pounds of calcined gyspum plaster to 250 pounds of damp, loose sand, or 2½ cubic feet of perlite* or vermiculite.

For three coat application of plasters: The first (scratch) coat is 100 pounds of calcined gypsum plaster to 200 pounds of damp, loose sand, or 2 cubic feet of perlite or vermiculite.

The second (brown) coat is 100 pounds of gypsum to 300 pounds of damp, loose sand, or 3 cubic feet of perlite or vermiculite.

In lieu of the above, the scratch coat may consist of 100 pounds of gypsum to 250 pounds of sand, or 2½ cubic feet of perlite or vermiculite providing the brown coat is of similar proportioning.

It is common practice to designate proportioning by symbols such as 1:2 meaning 100 pounds of gypsum to 200 pounds of sand, or 2 cubic feet of perlite or vermiculite depending upon the aggregate indicated. Similarly, 1:2½ means 100 pounds of gypsum to 250 pounds of sand, or 2½ cubic feet of perlite or vermiculite.

Where double symbols occur as "1:2, 1:3," the first symbol refers to the scratch coat proportioning and the second symbol to the brown coat. A single designation "1:2½" indicates that proportioning is the same for both coats.

The use of sand, perlite, or vermiculite, as an aggregate in job-mixed plaster is a matter of choice and also dependent upon several conditions. For ordinary plastering, sand is generally used where an ample supply of clean, sharp sand is available at low cost. Because of the better insulating and fire-resistive qualities of the lightweight aggregates, they may be preferred over sand. Where it is necessary to comply with specific regulations on fire ratings, perlite and vermiculite are used. When plaster thicknesses are specified in excess of those normally used, lightweight aggregates may be preferred.

Neat gypsum is the term applied to gypsum with no aggregate added. It comes in 100-pound sacks. Perlite and vermiculite aggregate are sold in

---

* *Perlite* is a siliceous volcanic glass expanded by heat.

Vermiculite is a micaceous mineral expanded by heat. Both are lightweight aggregates with fire resistive properites superior to sanded plasters.

sacks containing 3 to 4 cubic feet. Their weight varies between 7 1/2 and 10 pounds per cubic foot. Sand is sold by the ton or cubic yard. A cubic yard contains 27 cubic feet and weighs between 2,600 and 2,800 pounds.

Ready-mixed Plasters

Much of the gypsum for scratch and brown coats is mill-mixed, and delivered to the job in sacks. It requires only the addition of water for use.

While at one time sand was the principal aggregate used in mill-mixed plasters, it has almost been superseded by the lightweight perlite and vermiculite aggregates. These lightweight aggregates weigh about 10 percent as much as sand. They have superior sound and heat insulating qualities. They also have a better fire-resistive rating than sand. Plasters with the lightweight aggregates handle with greater ease and, to some extent, can be applied more rapidly by a plasterer. An important advantage of mill-mixed plasters is their uniformity of mixture, which is not always easy to control with job mixing.

Ready-mixed plasters are favored by many builders for winter plastering. Mixing gypsum and sand on the job in wet and freezing weather can be awkward and costly.

ESTIMATING QUANTITIES

The unit of measurement in plastering is 1 square yard. The gross area of a room is computed and *all window and door openings are deducted in full as in lathing.* (See Measurement of Quantities under Lathing.)

When estimating the quantity of materials required for a square yard of one-, two-, or three-coat work, it is important to determine the thickness of the plaster that existed in the damaged building. This can be done usually by measuring with a rule *from the face of the lath* to the outer surface of the finished plaster. It is general practice in most parts of the country to apply plaster over the face of the plaster base in approximately the following thicknesses including the finish coat:

Wood lath . . . . .5/8"
Metal lath. . . . . .5/8"
Gypsum lath . . . . .3/8" and 1/2"
Masonry Walls. . . . . .5/8"

Since wood and metal lath have openings through which the plaster is forced to provide a key, or mechanical bond, much more plaster is required than is indicated for either gypsum or masonry bases. It is more difficult to control the amount of plaster on wood and metal lath. Too much pressure on the trowel forces more plaster than is necessary through the openings and it drops behind the lath. Table 15-1 shows the approximate number of sacks of prepared perlite gypsum plaster required for 100 square yards, applied on different kinds of plaster base. The thickness shown includes a 1/16-inch white-coat finish.

Table 15-1
Approximate Number of Sacks of Prepared Perlite Gypsum
Plaster Required to Cover 100 Square Yards

Kind of Base	Total Plaster Thickness (Inches)	Number of Sacks
Wood lath	5/8	22 (80 lbs per sack)
Metal lath	5/8	35 (80 lbs per sack)
Gypsum lath	1/2	20 (80 lbs per sack)
Masonry walls	5/8	24 (67 lbs per sack)

The approximate quantities of materials for job-mixed, sanded, or perlite gypsum plaster for 100 square yards is shown in Table 15-2.

Table 15-2
Approximate Quantities of Materials Required for 100 Square Yards
of Job Mixed Plaster

Kind of Base	Total Plaster Thickness Inches	100 Lb Sacks of Neat Gypsum	Cu Yds Sand 1:2½ mix	Aggregate Perlite		
				Cu Ft	4 Cu Ft Sacks	Mix
Wood lath	5/8	11	1.1	27.5	7	1:2½
Metal lath	5/8	20	2.0	50.0	12.5	1:2½
Gypsum lath	1/2	10	1.0	25	6.5	1:2½
Masonry walls	5/8	12	1.2	30	7.5	1:3

ASA permit 250 lbs damp loose sand or 2½ cu ft of vermiculite or perlite provided this proportioning is used for both scratch and brown coats on three-coat work.

The number of 4 cu ft sacks of perlite are shown to nearest ½ sack.

Table 15-3 shows the approximate quantities of gypsum and hydrated lime required to cover 100 square yards with white coat finish 1/16 inch thick.

The quantities indicated in Tables 15-2 and 15-3 are within the coverage ranges published by the U.S. Gypsum Company.

Table 15-3
Approximate Quantities of Materials Required for 100 Square Yards
White Coat Finish 1/16″ Thick

Kind of Finish	100 Lb Sacks Neat Gypsum	100 Lb Sacks Keene's Cement	50 Lb Sacks Hydrated Lime
Lime putty	2	—	8
Keene's cement	—	4	4

Based on ASA mix of 1(100-lb) sack neat gypsum gauging plaster to 4(50-lb) sacks hydrated lime, and 1(100-lb) sack of Keene's cement to 1(50-lb) sack hydrated lime for medium hard finish.

ESTIMATING LABOR

When estimating labor costs for plastering, just as is done in the case of painting or other operations which involve the interior of a building, particular attention should be given to the physical circumstances of each job. Where patching of walls or ceilings are required, or where certain ceilings and walls require complete replacement, and the premises are furnished and occupied, the labor factor is considerably more than when the plasterer is working in a vacant and unoccupied building. Where the men are free to move about, and are unhampered by furnishings that need to be moved or protected, their rate of applying plaster approaches that for new construction.

The cost of plaster per square yard on small areas is usually much higher than that on large areas. The cost of plastering ceilings only in most instances is greater per square yard than it is for side walls. Closets and cupboards slow a plasterer down. All factors affecting labor, as discussed in Chapter 2, should be reviewed from the standpoint of their influence on plastering work.

When estimating the cost of completely plastering rooms, several rooms, or an entire building, the use of average labor rates of application are entirely in order.

It was formerly a custom to allow the same number of hours for a plasterer's helper, or laborer, as for a plasterer, when most plastering was done on wood lath, and the aggregate was mainly sand and was mixed by hand. The plaster was heavy, and much time was required to mix materials, carry them to a plasterer and arrange scaffold horses and plank. With present day lightweight aggregates, and machines for mixing plaster as it is required, the laborer's time is reduced. A rough ratio of a laborer's hours to a plasterer's is approximately 75 percent. There will be situations, however, where a laborer is required for the full time that the plasterer is working. Proper consideration should be given in each instance to the probable number of hours a laborer will be needed to serve the plasterer.

Table 15-4
Approximate Hours Required to Apply 100 Square Yards Gypsum
Plaster on Various Kinds of Plaster Base

Kind of Base	Plaster Thickness Inches	No. of Coats Excluding White Coat	Plasterer	Laborer
Wood lath	5/8	2	14	10
Metal lath	5/8	2	15	12
Gypsum lath	½	1	10	8
Gypsum lath	½	2	14	10
Masonry walls	½	1	10	8
Masonry walls	5/8	2	12	9
White Finish Coat				
Lime putty	1/16	1	10	8
Keene's cement	1/16	1	13	9

Table 15-4 shows the approximate number of hours required for a plasterer and a laborer, or plasterer's helper, to apply 100 square yards of various kinds of plaster on different types of plaster base. It is assumed that the plaster base has been prepared for the plasterer.

ESTIMATING COSTS

The cost of plaster applied to a surface depends on the type of plaster base or lath, the thickness of the plaster, and the number of coats to be applied. In most sections of the country the cost of plaster is figured on a square yard basis, as opposed to estimating the materials and labor for each room or for the area to be covered. However, the cost per square yard must be determined first, and this is done by obtaining the cost for 100 square yards, because most coverage tables for material, and also for labor, are in terms of 100 square yards. These unit costs are developed for each type of plastering on different types of bases.

Before attempting to compute a unit cost for plastering on any loss, it is important to determine by inspection the kind of plaster that existed. Insurance policies that cover property damage provide for replacing with like kind and quality. The kind of plaster that is originally applied in a building is controlled by several considerations, among which are local custom and practice, climate, economy, and building code regulations. The quality ranges from the application of *skim coat* plaster on gypsum board, two-coat plaster on gypsum lath, to three-coat work on wire lath. It is also significant to base unit costs on the exact specifications necessary to restore a plastered wall to its original condition. If only the white finish coat is affected, or if the lath does not require replacing, the unit cost should be confined to those operations. As an illustration, while wood lath is used infrequently on new construction, much of it is found in existing older buildings. Unless wood lath is burned or destroyed, the old plaster frequently can be removed and new plaster applied without replacing lath.

Unit Costs

The costs for common kinds of plastering may be computed as shown in the following examples: By referring to Tables 15-1 through 15-4 unit prices may be obtained for types of plastering bases and finishes not shown in the illustrations. Where a different finish coat is used, the cost of "lime putty finish" should be deducted and the cost of the finish coat to be applied should be added.

The following are illustrative unit costs with material prices and hourly labor rates assumed.

Prepared Perlite Gypsum Plaster-Unit Cost of Two-Coat "Double Back" Plaster on Gypsum Lath (*Grounds 7/8"-Plaster Thickness ½" Including 1/16" finish*)

Illustrative Example

Table	Base Coat		
15-1	20(80-lb) sacks prepared plaster	@ $1.70 =	$ 34.00
15-4	10 hours plasterer	@ 6.00 =	60.00
15-4	8 hours laborer	@ 5.00 =	40.00
	Lime Putty Finish Coat		
15-3	2(100-lb) sacks gauging plaster	@ 1.40 =	2.80
15-3	8 (50-lb) sacks hydrated finish lime	@ 1.00 =	8.00
15-4	10 hours plasterer	@ 6.00 =	60.00
15-4	8 hours laborer	@ 5.00 =	40.00
	Unit cost for 100 sq yds =		$244.80

$$\text{Unit cost per sq yd } \frac{\$244.80}{100} = \$\ 2.45$$

Unit cost of Three-Coat Plaster on Metal Lath (*Grounds ¾"—Plaster Thickness 5/8" Including 1/16" Finish*)

Table	Base Coat		
15-1	35 (80-lb) sacks prepared plaster	@ $1.70 =	$ 59.50
15-4	15 hours plasterer	@ 6.00 =	90.00
15-4	12 hours laborer	@ 5.00 =	60.00

Lime Putty Finish Coat

15-3	2 (100-lb) sacks gauging plaster	@ 1.40 =	2.80
15-3	8(50-lb) sacks hydrated finish lime	@ 1.00 =	8.00
15-4	10 hours plasterer	@ 6.00 =	60.00
15-4	8 hours laborer	@ 5.00 =	40.00

Unit cost for 100 sq yds = $320.30

$$\text{Unit cost per sq yd } \frac{\$320.30}{100} = \$ \; 3.20$$

Unit cost of Three-Coat Plaster on Wood Lath (Grounds 7/8"—Plaster thickness 5/8" Including 1/16" Finish)

Illustrative Example

Table	Base Coats		
15-1	22(80-lb) sacks prepared plaster	@ $1.70 = $	37.40
15-4	14 hours plasterer	@ 6.00 =	84.00
15-4	10 hours laborer	@ 5.00 =	50.00
	Lime Putty Finish Coat		
15-3	2(100-lb) sacks gauging plaster	@ $1.40 = $	2.80
15-3	8(50-lb) sacks hydrated finish lime	@ 1.00 =	8.00
15-4	10 hours plasterer	@ 6.00 =	60.00
15-4	8 hours laborer	@ 5.00 =	40.00

Unit cost for 100 sq yds = $282.20

$$\text{Unit cost per sq yd } \frac{\$282.20}{100} = \$ \; 2.82$$

Unit Cost of Two-Coat Plaster on Masonry (Grounds 5/8"—Plaster Thickness 5/8" Including 1/16" Finish)

Illustrative Example

Table	Base Coat		
15-1	24(80-lb) sacks prepared plaster	@ $1.70 = $	40.80
15-4	12 hours plasterer	@ 6.00 =	72.00
15-4	9 hours laborer	@ 5.00 =	45.00
	Lime Putty Finish Coat		
15-3	2(100-lb) sacks gauging plaster	@ 1.40 =	2.80
15-3	8(50-lb) sacks hydrated finish lime	@ 1.00 =	8.00
15-4	10 hours plasterer	@ 6.00 =	60.00
15-4	8 hours laborer	@ 5.00 =	40.00

Unit cost for 100 sq yds = $268.60

$$\text{Unit cost per sq yd } \frac{\$268.60}{100} = \$ \; 2.69$$

JOB-MIXED GYPSUM PLASTER USING SAND AGGREGATE

The following basic formulas and illustrative examples use sand as the aggregate to be mixed with gypsum plaster. Where specifications call for perlite aggregate refer to Table 15-1 for quantities required.

Unit Cost of Two-Coat "Double Back" Gypsum Plaster on Gypsum Lath (Grounds 7/8"—Plaster Thickness ½" Including 1/16" Finish)

Illustrative Example

Table	Base Coat		
15-2	10(100-lb) sacks gypsum plaster	@ $1.40 =	$ 14.00
15-2	1 cu yd sand	@ 3.75 =	3.75
15-4	10 hours plasterer	@ 6.00 =	60.00
15-4	8 hours laborer	@ 5.00 =	40.00
	Lime Putty Finish Coat		
15-3	2(100-lb) sacks gauging plaster	@ 1.40 =	2.80
15-3	8(50-lb) sacks hydrated finish lime	@ 1.00 =	8.00
15-4	10 hours plasterer	@ 6.00 =	60.00
15-4	8 hours laborer	@ 5.00 =	40.00

Unit cost for 100 sq yds = $228.55

$$\text{Unit cost per sq yd } \frac{\$228.55}{100} = \$ \ 2.29$$

Unit cost of Three-Coat Plaster on Metal Lath (Grounds ¾"—Plaster Thickness 5/8" Including 1/16" Finish)

Illustrative Example

Table	Base Coat		
15-2	20(100-lb) sacks gypsum plaster	@ $1.40 =	$ 28.00
15-2	2 cu yds sand	@ 3.75 =	7.50
15-4	15 hours plasterer	@ 6.00 =	90.00
15-4	12 hours laborer	@ 5.00 =	60.00
	Lime Putty Finish Coat		
15-3	2(100-lb) sacks gauging plaster	@ 1.40 =	2.80
15-3	8(50-lb) sacks hydrated finish lime	@ 1.00 =	8.00
15-4	10 hours plasterer	@ 6.00 =	60.00
15-4	8 hours laborer	@ 5.00 =	40.00

Unit Cost for 100 sq yds = $296.30

$$\text{Unit cost per sq yd } \frac{\$296.30}{100} = \$ \ 2.96$$

Unit Cost of Two-Coat Plaster on Masonry (Grounds 5/8"—Plaster Thickness 5/8" Including 1/16" Finish)

Illustrative Example

Table	Base Coat			
15-2	12(100-lb) sacks gypsum	@	$1.40 =	$ 16.80
15-2	1.2 cu yds sand	@	3.75 =	4.50
15-4	12 hours plasterer	@	6.00 =	72.00
15-4	9 hours laborer	@	5.00 =	45.00
	Lime Putty Finish Coat			
15-3	2(100-lb) sacks gauging plaster	@	1.40 =	2.80
15-3	8(50-lb) sacks hydrated finish lime	@	1.00 =	8.00
15-4	10 hours plasterer	@	6.00 =	60.00
15-4	8 hours laborer	@	5.00 =	40.00
		Unit cost for 100 sq yds =		$249.10

$$\text{Unit cost per sq yd } \frac{\$249.10}{100} = \$ \ \ 2.49$$

Unit Costs Including Lath

Because of the numerous combinations of the different kinds of lath or plaster base, and the various kinds of plaster applied, it is recommended that a unit cost per square yard be obtained for the particular *lath* involved and added to the cost per square yard for the type of plaster being considered.

MACHINE PLASTERING

Plastering machines have been developed for applying rough or basecoat plasters. They are being improved continually and manufacturers of plastering materials are revising their formulas to adapt them to machine application. There are some who predict that much, if not most, of the future plastering will be applied by machine. They are portable units operated by electric or gasoline motors.

These machines actually spray mortar on the walls and ceilings. Where a true and smooth surface is required, a plasterer must still darby and trowel the surface after the machine has deposited the mortar.

Machine application uses more mortar, but requires considerably less labor, which is the more expensive factor in plastering. Because it is a specialized operation and used mainly where large quantities are needed, no attempt should be made to estimate the unit cost of machine plastering without obtaining the aid of a subcontractor familiar with the work.

EXTERIOR PLASTERING

Stucco, which is the term often used for exterior plaster, is composed of a mortar made with portland cement rather than gypsum because it is exposed to the weather. The aggregate in the mortar is sand, although a fine pebble gravel, marble dust, white sand, and colors are sometimes used in the finish coat. Hydrated lime is added as a plasticiser to make the mortar workable when being applied.

Exterior plaster can be applied directly to brick, concrete, concrete block, hollow terra cotta tile, and other kinds of masonry. It can also be applied on wire lath nailed directly to studding, or over sheathing. Three-coat exterior plaster is considered the best grade of work; under many building codes it is a requirement. However, there is some two-coat work being done, both on masonry and lath bases in low-priced homes and on secondary buildings. As with interior plastering the successive coats are scratch, brown, and finish coats. The customary thickness of exterior plaster is about 3/4 inch but may be found in thicknesses that vary between 5/8 and 1 inch. Before estimating, the actual thickness of the plaster should be scaled at the loss as it affects the quantity of material required, and also the labor.

Measurement of Areas

Under *Lathing* the different customs for measuring areas are discussed, and the reader is referred to that section as the comments apply also to exterior plaster work. It is recommended that gross areas be computed and *all window and door openings be deducted in full.* An allowance of 10 percent should be added to the net area for waste.

Metal Lath Base

The metal lath applied to frame buildings as an exterior plaster base comes in many forms. They may be classified into three kinds: the expanded metal lath, the welded wire lath, and the woven wire lath. The latter, because of its similarity, is frequently called *chicken wire,* and is used extensively. Because it is not good practice to nail lath solidly to studding or sheathing, a special 1 1/2-inch self-furring nail is used which holds the lath 1/4 to 3/8 of an inch away from the surface. This allows the mortar to flow in back of the lath and form a proper key at all points.

Whether the metal lath is applied over open studding, or solid wood sheathing, a layer of 15-lb saturated felt paper should first be applied. In the case of open studding it acts as a backing behind the lath for the plaster. When applied over solid wood sheathing, it prevents the wood from soaking up the water in the plaster. When asphalt-coated composition sheathing is used, it is not necessary to apply saturated felt paper, as the coating prevents absorption of the water from the mortar.

Before applying saturated felt paper on open studding, 18-gauge steel wires are stretched horizontally, spaced about 6 or 8 inches apart. In some sections of the country the wires are drawn tightly in the horizontal position; in other sections it is the practice to staple the wires on every other stud. At the alternate studs the wires are then drawn upward until taut and stapled to the stud. The effect is the same: The wires provide ridigity back of the wire lath and paper against pressure from the plastering trowel when the mortar is being applied.

To estimate the quantity of wire lath and the labor for applying it, refer to that subject under "Lathing" in this chapter. Costs may be obtained by using the formula shown for metal lath. The labor cost of stringing 18 gauge wire every 8 inches should be estimated at an average of 6 hours per 100 square yards of surface area. The number of lineal feet of wire is obtained by multiplying the area by 1.5. It requires about 8 pounds of 18 gauge wire per 100 square yards.

Another type of metal lath used on exterior open-stud stucco work is a paper-backed product of 2" x 2" mesh 16 gauge steel wire. The trade name is *Steeltex*. When this type of metal lath base is applied over open studding, neither the wire back nor 15-lb saturated felt paper are required.

The labor required to apply 100 square yards of paper-backed stucco lath would be an average of 10 hours.

Masonry Base

All masonry bases must be completely cleaned to remove dirt and grease before plastering; otherwise, the mortar may not adhere. In some cases chipping, scratching or roughing the surface is necessary to obtain a good mechanical bond.

Exterior Plaster Mortar

For estimating the cost of portland cement mortar, the proportions of materials for a cubic yard are identical to those used for masonry cement mortar (Chapter 8) employing a 1:3 mix with hydrated lime equal to 10 percent of the weight of cement added.

Cement	10 sacks
Sand	1 cu yd
Hydrated lime	2 (50-lb) sacks

Estimating Quantities

The unit used for quantities of portland cement plastering is 1 square yard. The area to be covered is obtained by measurement and converted to square yards by dividing the number of square feet by nine. Table 15-5 shows the number of cubic yards of cement mortar required for 100 square yards of plastering of the thickness indicated.

Table 15-5
Cubic Yards Portland Cement Plaster Required for 100 Square Yards
for the Various Thicknesses Shown (1:3 Mix)

Plaster Thickness (Inches)	Cubic Yards on Masonry Base	Cubic Yards on Wire Lath Base
¼	0.75	0.90
½	1.50	1.80
5/8	1.37	1.64
¾	2.25	2.70
1	3.00	3.60

Table 15-6
Approximate Hours Required to Apply 100 Square Yards Portland
Cement Plaster on Wire Lath and Masonry Bases (Float Finish)

Base	Plaster Thickness (Inches)	Number of Coats	Plasterer	Labor
Masonry	½	1	13	6
	5/8	2	21	10
	¾	3	28	20
Wire lath	5/8	2	30	20
	¾	3	40	30
Add approximately 4 hours to plasterer's time for a troweled finish.				

In computing the quantity of cement plaster mortar, no regard need be given to the number of coats as the thickness of the plaster controls the quantity. When estimating labor, the number of coats is important because each coat is separately applied.

Estimating Labor

A plasterer and a laborer work together on exterior work in the same manner as interior plastering. The laborer mixes the mortar, arranges horse and plank scaffold, and serves the plasterer with necessary materials. On small to medium-sized jobs a laborer's time will be practically equal to the plasterer's. When larger areas are being plastered and the mortar is machine mixed, the laborer's time is reduced considerably. The ratio of a laborer's time to that of the plasterer should be judged for each particular job. Table 15-6 shows the approximate number of hours required for a mason and laborer to apply 100 square yards of plaster of various thickness and number of coats on both masonry and wire lath base.

Scaffolding and Hoisting

When structural scaffolding is to be erected by carpenters, the cost should be added as a separate item. The cost of hoisting materials or equipment to floors above grade should also be added when such operations are indicated.

Estimating Costs

It is customary to compute the cost of exterior plastering for 100 square yards, and then reduce this to a cost per square yard as is done in the case of interior plastering. (Refer to "Estimating Costs" under "Interior Plastering" for considerations prior to estimating).

The following examples are given for different kinds of exterior plaster work (prices and wages assumed):

Unit Cost of 100 Sq Yds Two-Coat Stucco on Masonry Base (1:3 Mix-5/8″ Thick—Float Finish)

Illustrative Example

Table			
15-5	1.37 cu yds plaster mortar	@ $19.45 =	$ 26.65
15-6	21 hours plasterer	@ 6.00 =	126.00
15-6	10 hours laborer	@ 5.00 =	50.00

Unit cost for 100 sq yds = $202.65

$$\text{Unit cost per sq yd} = \frac{\$202.65}{100} = \$2.03$$

Unit Cost of 100 Sq Yds One-Coat Stucco on Masonry Base (1:3 Mix— ½″ Thick-Float Finish)

Illustrative Example

Table			
15-5	1.5 cu yds plaster mortar	@ $19.45 =	$ 29.18
15-6	13 hours plasterer	@ 6.00 =	78.00
15-6	6 hours laborer	@ 5.00 =	30.00

Unit cost for 100 sq yds = $137.18

$$\text{Unit cost per sq yd} \frac{\$137.18}{100} = \$1.37$$

Unit Cost of 100 Sq Yds Two-Coat Stucco on Woven Wire Mesh Lath on Open Studs (1:3 Mix—5/8" Thick-Float Finish)
Table

	8 lbs 18 gauge wire	@ $	.25 = $	2.00
	100 sq yds wire mesh lath	@	.30 =	30.00
	2½ rolls 15-lb saturated felt paper	@	3.20 =	8.00
	8 lbs self-furring lath nails	@	.30 =	2.40
	6 hours lather applying wire	@	6.00 =	36.00
	10 hours lather applying lath	@	6.00 =	60.00
15-5	1.64 cu yds plaster mortar	@	19.45 =	31.90
15-6	30 hours plasterer	@	6.00 =	180.00
15-6	24 hours laborer	@	5.00 =	120.00

Unit cost for 100 sq yds = $470.30

$$\text{Unit cost per sq yd } \frac{\$470.30}{100} = \$4.70$$

DAMAGEABILITY OF GYPSUM PLASTER

The degree of damage that gypsum plaster has suffered from fire, water, explosion, blasting, settlement, or other perils can best be determined only by careful inspection. To assume that plaster must be removed and replaced simply because it has been exposed to fire or has been wet is an unsound approach. Much will depend on the type of lath or plaster base, the extent and duration of a fire, whether it is sidewall or ceiling plaster, the age of the plaster, its composition and its physical condition at the time the damage took place.

Damage by Fire

When gypsum plaster is directly exposed to the heat of fire, the water of crystallization is slowly driven off. Long continued heat causes the surface to powder which, although increasing its insulating properties, weakens it. The thin white putty coat finish becomes chalky and may spall or scale off of the rough plaster base. In extremely hot fires of long duration the plaster may completely crumble and disintegrate.

When the back of plaster on metal lath is exposed to hot fires for any length of time, the key can be weakened seriously. Wood lath will char and burn when directly exposed to a fire so that, even though the plaster appears sound on the room side, replacement may be required. Gypsum lath base will withstand the effects of heat and fire in back of it for a longer period than either wood or metal lath without serious injury to the plaster. Where any question arises concerning the necessity of replacing plaster, proper examination and tests should be made to determine the condition and firmness of the plaster. Scraping the surface with a knife will disclose whether it has powdered or is chalky. Tapping firmly with a stout rod, broom handle, or any other convenient object will indicate by the sound, and by any movement, whether or not the plaster has been loosened.

Damage by Water

One of the characteristics of gypsum is that it can be wet down to the point of being plastic, but upon drying out, it returns to its original hardness. This is the reason it makes such an excellent cementatious material for plaster.

Plaster that has been wet by water from a leaky roof, a fire hose, or even inundation will show no harmful effects, provided it is not continuously or periodically exposed to water and is allowed to dry out. Probably there is no better example of the ability of gypsum plaster to withstand water effects than that contained in the Gypsum Association's report on the Topeka flood of 1951. Two motels were submerged under 8 feet of water for five days. Less than six weeks after the waters receded, they both were back in business with no repairs being needed to the walls.

There are innumerable instances of sidewall plaster being so saturated with water used to extinguish a fire that has occurred above, that a lead pencil could be pushed through the plaster. After several days the plaster returns to its original hardness with no evidence of damage. The drying out process should, however, be controlled. Usually a moderate amount of heat, with the windows opened to permit air circulation, will be satisfactory. It is important, of course, in winter months, to protect the wet plaster from freezing.

When ceilings have been seriously wet down and the plaster is on a wood lath base, there is always a strong possibility that the wood has swelled and this, together with the added weight, will break the key. This same thing may also happen in the case of certain types of metal lath. The adhesive bond of gypsum lath will generally hold, but frequently the weight of the wet plaster and gypsum board may pull the nails through.

It is always better, if possible, to allow wet plaster to dry out for several days before estimating. At that time appropriate tests may be made to ascertain its true condition.

Repeated wetting of plaster such as might take place when a roof continues to leak over a period of time, will eventually rot the plaster.

Damage by Cracking

The causes of cracks in plaster are numerous aside from inherent cracking due to poor materials, improper proportioning, and inferior workmanship. Cracks most commonly found within insurance coverage are caused by settlement of the ground, blasting, jarring or twisting of a structure during strong or violent winds, land or mud slides, earthquake, lightning, explosion, and heat from a fire. They may occur in both ceiling and sidewalls. One of the difficulties encountered is that of trying to distinguish old cracks in plaster from those attributed to the immediate casualty. This subject is more fully discussed in Chapter 18, "Analyzing Building Loss Estimates." When the plaster immediately adjacent to the cracks is sound, an experienced plasterer can perform repairs which cannot be detected after repainting or repapering.

METHODS OF REPAIRING GYPSUM PLASTER

Estimating the cost to repair damaged plaster will be governed by the specifications taken in the field. These specifications may call for patching an area of white coat that has scaled off or they may call for patching a section of three-coat work including the lath.

Lath

When lath in a patch has to be replaced before new plaster is applied, gypsum lath, when permitted, can be inserted regardless of the type of existing lath. Where metal lath is required to comply with building regulations, however, it must be used.

Wood lath, if not burned or broken, can be renailed wherever loose after all old plaster has been knocked from between the lath. Prior to replastering on wood lath, it is wet down to prevent the water in the plaster from being absorbed before it has set.

Masonry Base

All that is required to prepare a masonry base for replastering is to clean and brush it thoroughly, then wet it so that it will not absorb the water in the plaster too rapidly.

Patch Plastering

When preparing a section to be patched, it is important that all loose plaster be removed to be sure that the remaining plaster adjacent to the patch area is attached firmly to the base. This presents a difficult problem in older buildings where over a long period the plaster has sagged or pulled away from the lath, particularly wood lath. In some cases both the lath and plaster have become loosened.

If only the white-finish coat has been damaged, this is removed wherever loose, and the old brown-coat base is wet at the edges of the patch only. The brown coat is scored or scratched to provide a proper bond.

When finish coat and rough base coat, or coats, are to be replaced, the old plaster is removed and the edges are wet down. New coats are applied successively just as in new plastering. The old plaster edges serve as grounds for the final coat.

Patching Cracks

Large cracks in plaster should be cut out removing all loose plaster along the edges. Sometimes it is advisable to widen the area around the crack as much as 4 or 6 inches, and nail a strip of wire lath in the opening. In such instances rough coats and a finish coat are applied rather than attempting to fill so wide an opening in one step.

Smaller cracks are easily repaired by a skilled plasterer. They are opened slightly with a sharp tool, wet, and then filled with plaster of Paris.

Large Area Repairs

When an entire ceiling or a whole wall has to be replastered, it may be estimated either at a unit price or on a material and labor basis. If a unit price is developed and used, it must be remembered to add all operations necessary in addition to the labor of direct application of the plaster.

Estimating Repair Costs

Except where large areas are involved, all patch jobs in plastering should be estimated on a material and time basis. Consideration should be given to all factors that the workman will be confronted with on the particular job. It should be remembered that in many areas there is a minimum charge by plasterers for patching regardless of whether it takes six hours, a whole day, or thirty minutes.

DAMAGEABILITY OF STUCCO PLASTER AND REPAIR METHODS

Portland cement plaster is used on exterior work because it is water-resistant. Therefore, its damageability by the effects of water, even submersion is negligible.

In the presence of fire it is less damageable than gypsum plaster but, in severe fires of long duration, it will suffer some injury depending upon its thickness. Only careful examination will determine the extent of damage.

Patching Stucco

The procedures followed in patching gypsum plaster apply generally to portland cement plaster. Areas to be repaired are cut out back to a firm base, and new plaster is applied in successive coats. It is then finished to conform with the original surface.

Cracks are cut out and filled with mortar.

Large Areas

When large areas of cement plaster have been affected, it is frequently more economical, when permitted, to prepare the old surface by scratching and roughing, then apply new stucco over the entire area.

Chapter 16.

Sheet Metal Work

Sheet metal work is a specialized building trade encompassing the fabricating and installation of building units made of sheet metal including but not limited to such things as gutters and downspouts, roofing, flashing and counter flashing, ventilators, skylights, and metal doors.

Duct systems for heating, air conditioning, and manufacturing processes, while falling into the classification of sheet metal work, will not be taken up here. Estimating these items should be referred to those in the business who are experts in the field.

FLASHING AND COUNTER FLASHING

Flashing is used mainly in building construction to prevent water from entering where there are joints created by intersecting roofs (valleys); chimneys, stack pipes, ventilators, etc., which project through a roof; over windows and doors; at skylight curbings; and where roofs (flat or pitched) join a vertical wall. Another and more limited use for flashing is as a shield at foundation sills and post bases against insect damage, particularly by termites.

The metals used most commonly for flashing are galvanized sheet metal, copper, tin, and aluminum. In areas where the atmosphere tends to more rapidly corrode (near oceans, paper mills or certain chemical plants),

421

copper is preferable to galvanized metal. Occasionally lead or zinc flashing is used.

Copper flashing is commonly 16-ounce but is also available in 20 and 24-ounce rolls and sheets. *Galvanized sheet metal* is usually recommended to be 26-gauge. It comes in 8", 14" and 20" wide. A common thickness of *aluminum flashing* is .019 inches and is furnished in rolls 6" to 28" wide and 50 lineal feet long.

ESTIMATING COSTS

Flashing materials should be estimated by the lineal foot specifying the kind of material, thickness and width. Installation costs including material should be estimated by the square foot.

Labor should be estimated by the individual job considering the location of the work, the type, and working conditions. As a guide, a man can place roll flashing in a roof valley at the average rate of 10-20 lineal feet per hour depending on the slope. Placing flashing over windows and doors where narrow widths are used, a man should average 15 lineal feet per hour. Ridge and hip flashing can be applied at the average rate of 15-20 lineal feet per hour.

Illustrative Example

Assuming prices and wages shown, the unit cost to flash valleys on a medium pitched roof would be:

100 sq ft. .019 aluminum flashing	@ $ .35 =	$ 35.00
12 hours labor	@ 6.00 =	72.00
	Unit cost per 100 sq ft =	$107.00
	Unit cost per sq ft =	$ 1.07

GUTTERS AND DOWNSPOUTS

Gutters, installed at the cornice line of a roof to collect water runoff, are sometimes called an eave trough. The vertical downspout is sometimes called a leader pipe or conductor. Fittings include hangers, end caps, slip joints, funnels, elbows, shoes, leader straps or hooks, and wire strainers. Most of these are self descriptive. Elbows are used to connect the gutter outlet to the downspout which is located against the building. Shoes attach to the bottom of the downspout to divert the water away from the foundation.

Gutters and leaders are made in several styles such as half-round and box type. Downspouts are shaped round and rectangular, both of which may be smooth or corrugated.

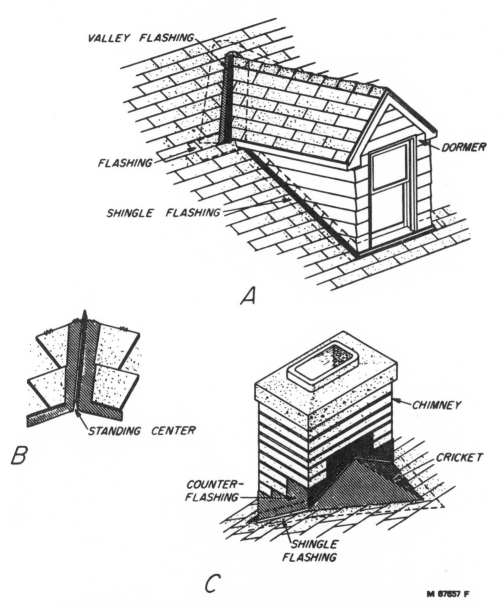

VALLEY FLASHING

DORMER

FLASHING

SHINGLE FLASHING

A

B

STANDING CENTER

CHIMNEY

CRICKET

COUNTER-
FLASHING

SHINGLE
FLASHING

C

M 87657 F

Flashing at center of rooflines. *A*, Valley and shingle flashing;
B, standing-center valley flashing; *C*, chimney and shingle flashing.

Figure 16-1

Wood-Frame House Construction . . . Agriculture Handbook No. 73

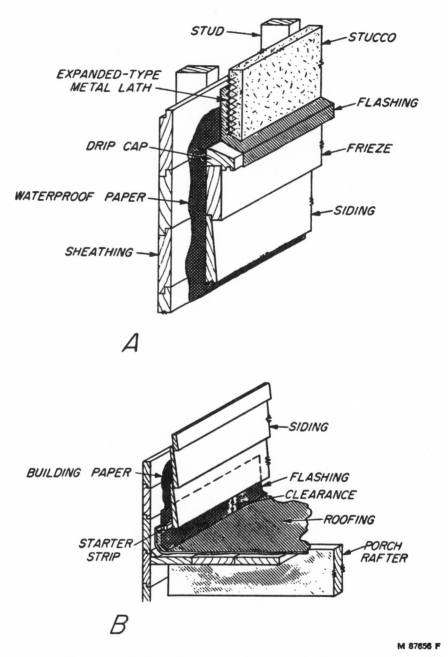

Location of flashing. *A*, at material change; *B*, at roof deck.

Figure 16-2

Wood-Frame House Construction...Agriculture Handbook No. 73

The materials may be galvanized steel, copper, vinyl, fiberglass and, less frequently, wood.

In estimating material costs, the shape of the gutter, the material and its thickness or gauge, are important considerations.

Illustrative Example No. 1

A unit cost for "Ogee" shape 26-gauge white enamel finish aluminum gutters would be as follows assuming prices and wages shown:

100 lin ft 5" enamel "ogee" 26-gauge aluminum gutters	@ $ .30 =	$30.00
Add for fittings (as required)		
4 hours labor for journeyman	@ 6.00 =	24.00
4 hours labor for helper	@ 5.00 =	20.00
	Unit cost per 100 lin ft =	$74.00
(excludes fittings)	Unit cost per sq ft =	$ .74

Illustrative Example No. 2

A unit cost for 2¼" x 3¼" aluminum downspout with enamel finish including straps would be as follows assuming prices and wages shown:

100 lin ft downspout and hangers	@ $ .25 =	$25.00
4 hours labor for journeyman	@ 6.00 =	24.00
4 hours labor for helper	@ 5.00 =	20.00
	Unit cost per 100 lin ft =	$69.00
	Unit cost per lin ft =	$ .69

Illustrative Example No. 3

A unit cost for copper, half round, 5" gutter assuming the following price and wages shown:

100 lin ft 5" copper gutter	@ $ .90 =	$ 90.00
4 hours labor for journeyman	@ 6.00 =	24.00
4 hours labor for helper	@ 5.00 =	20.00
	Unit cost per 100 lin ft =	$134.00
	Unit cost per lin ft =	$ 1.34

Illustrative Example No. 4

A unit cost for vinyl 5" gutters assuming the following price and wages would be:

100 lin ft 5" vinyl gutter	@ $ .50 =	$50.00
4 hours labor for journeyman	@ 6.00 =	24.00
4 hours labor for helper	@ 5.00 =	20.00
	Unit cost per 100 lin ft =	$94.00
	Unit cost per lin ft =	$ .94

METAL SKYLIGHTS

Prices of skylights installed should be obtained from subcontractors. Approximate estimates may be used based on the cost per square foot. The cost is dependent on whether it is a hip skylight or a gable skylight. Also, the cost varies according to whether the unit has ventilators. When estimating, all of these features should be determined as well as the size. Curbing is figured separately.

METAL VENTILATORS

Mushroom, spinner, gable, and other types of ventilators should be measured, fully described and priced by a metal worker.

RIDGE ROLLS

Ridge rolls can be placed on ridges and hips at the average rate of 100 lineal feet per 8 hour day.

Chapter 17.

Kitchen Cabinets & Built-Ins

Kitchen cabinets have been generally standardized as to dimensions and to a large extent as to construction. There are many designs and sizes to adapt them to particular areas into which they will be installed.

Basically, there are three classifications of kitchen cabinets, *base, wall* and *tall* or *utility* cabinets.

Base cabinets are those fitted with counter-tops to provide working space or into which such things as sinks and table-top ranges are recessed. They have shelves and drawers. Base cabinets may be installed directly against the walls of the kitchen or out in the open in which event they are *island counters.*

Wall cabinets, half as deep as base cabinets, are attached to the wall above the counter tops, about 18 inches, unless they are above a range, refrigerator, wall oven, or similar built-in appliance. They have shelves but no drawers. Figure 17-1 shows a profile view of base and wall cabinets with typical measurements indicated. More recent kitchen designs provide for a drop ceiling over the wall cabinets. Where there is none, the wall cabinets frequently extend to the ceiling or have a useless space of several inches from the top of them to the ceiling.

Tall cabinets, sometimes called utility cabinets, are 12" deep or the same depth (24 inches) as the base cabinet. A popular height is 7 feet with a shelf or two at the top, and open below for storage of brooms, mops and

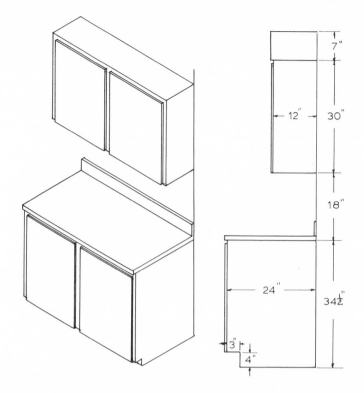

Figure 17-1

Base and wall cabinets in profile and in perspective with typical measurements shown.

similar paraphernalia. In some instances the tall cabinets have only shelves which may be used as storage for foodstuffs and cleaning materials. They have one or two doors.

MATERIALS OF CONSTRUCTION

Plywood and particleboard (flake board) are used for sides, backs, bottoms, doors and partitions. Particleboard is becoming very popular even in high grade cabinets as it does not warp. It is less costly than plywood or solid wood boards.

Framing members for fronts, sides, backs, and so forth, may also be plywood or particleboard, although pine is also used. Drawers are made of similar materials.

Finish of the cabinet fronts may be wood or vinyl veneer; stained, or painted.

The usual hardware consists of door hinges, handles or drawer pulls. Also, there are patented drawer guides and glides to take the place of wood

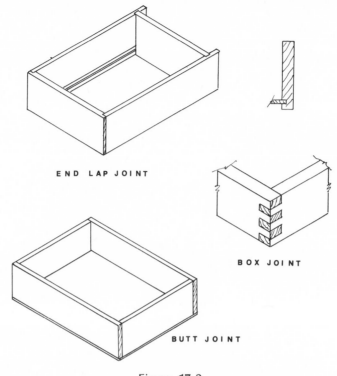

 END LAP JOINT

BOX JOINT

BUTT JOINT

Figure 17-2
Drawer joint construction.

guides. Metal, adjustable (slotted) shelf standards have a distinct ad·
vantage over fixed wood shelf supports.

Counter tops will be discussed in a later section.

METHODS OF CONSTRUCTION

There are various ways and means of cutting and joining the members
in cabinet construction. Much depends on the tools available, the quality of
workmanship and the cabinetmaker's preference, experience and
background.

Some frames are doweled or mortise and tenon, others may employ
corrugated nails or similar fasteners. Many quality made cabinets have no
backs in them as the walls serve as backs. Where cabinet backs are
necessary, 1/4 inch plywood or hardboard is used.

Drawer construction varies. Sometimes the sides are butt-jointed but
a better method is an end lap or rabbet joint, or better yet make a dado box
corner. A more expensive joint found in only the highest quality cabinets is
a box joint or dovetail. See Figure 17-2 for details.

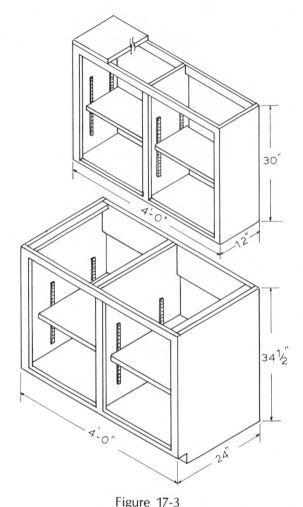

Figure 17-3

Base cabinet and wall cabinet.

In less expensive drawer construction the bottom is nailed flush to the sides. In better cabinets the sides and front of the drawers are rabbetted just above the bottom and the 1/4 inch panel is inserted into the rabbet to form the drawer bottom.

Figure 17-3 shows the dimensions and construction of an actual base and wall cabinet that was made in a cabinetmakers shop. Particleboard and plywood were used and a facing of vinyl sheet was applied to the front edges and back of the doors, and also to the face only of the front frames and side panels of the cabinets.

The materials, excluding the counter-top, are listed below. It is important to note the order in which materials are listed and, also, that the method of listing would be the same in these cabinets regardless of whether they were made with plywood veneer, plywood to be painted or solid wood to be stained and varnished. The difference would be in the cost of the material, type of hardware and perhaps some minor variations in construction details.

Base Cabinet

Platform frame	14 lin ft 1″ x 3″ = 14 lin ft		$ 1.40
Front frame—top	1-1″ x 2″ 4′-0″ = 4 ″ ″		.32
bottom	1-1″ x 1″ x 4′-0″ = 4 ″ ″		.20
stiles	3-1″ x 2″ x 3′-0″ = 9 ″ ″		.72
Counter supports	3-1″ x 2″ x 2′-0″ = 6 ″ ″		.48
Rear top ledger	1-1″ x 8″ x 4′-0″ = 4 ″ ″		1.20
			$ 4.32

Panels, ¾ ″ particle board			
bottom	1-2′ x 4′-0″ = 8 sq ft		
ends and partition	3-2′ x 3′-0″ = 18 ″ ″		
shelves	2-2′ x 1′-6″ = 6 ″ ″		
	Total panels 32 sq ft		$ 9.60
Doors, 5/8″ particle board	2-2′ x 2′-6″ = 10 sq ft		2.50
Hardware—2 pr door hinges (self closing)			4.00
2 door handles			1.90
8 metal shelf standards 2′@$1.25			12.00
		Total material	$ 34.42

Wall Cabinet

Front frame—top	1-1″ x 2″ x 4′-0″ = 4 lin ft		$ .32
bottom	1-1″ x 1″ x 4′-0″ = 4 ″ ″		.20
stiles	3-1″ x 2″ x 2′-6″ = 8 ″ ″		.64
Rear top ledger	1-1″ x 2″ x 4′-0″ = 4 ″ ″		.32
			$ 1.48

Panels, ¾ ″ particle board			
bottom	1-1′ x 4′-0″ = 4—sq ft		
sides and partition	3-1′ x 2′-6″ = 7½ sq ft		
shelves	2-1′ x 2′-0″ = 4—sq ft		
	Total panels 15-½ sq ft		4.65
Doors, 5/8″ particle board	2-2′ x 2′-6″ = 10 sq ft		2.50
Hardware—2 pr door hinges			4.00
2 door handles			1.80
8 metal shelf standards 2′@$1.25			12.00
		Total material	$ 26.43
		Total material cost, both cabinets	$ 60.85

Vinyl veneer

Vinyl sheet veneer, 1/32" thick was glued to the units in shop. It required 63 square feet of material and epoxy glue as shown below.

Base cabinet

Glue			$ 1.00
2-end panels	2' x 3'-0" =	12 sq ft	
2 doors (sides and edges)	2' x 2'-6" =	20 " "	
face of front frame and base		4 " "	
		36 sq ft	21.60
			$22.60

Wall cabinet

Glue			1.00
2-end panels (sides and edges)	1' x 2'-6" =	5 sq ft	
2 doors	2' x 2'-6" =	20 " "	
face of front frame		2 " "	16.30
		27 sq ft	17.30
Total veneer material			$39.90

Labor

It took an experienced cabinet maker 8 hours to cut out and assemble the two cabinets and another 6 hours to apply the vinyl veneer or:

14 hours @ $5 per hour	$ 70
When the cabinets were delivered it took	
4 hours to install them in the kitchen	20
Total labor cost	$ 90
Summary of cost excluding countertop	
Material for cabinets	$ 60.85
Vinyl veneer material	39.90
Labor to make	70.00
Labor to install	20.00
Subtotal	$190.75
Overhead	38.15
	228.90
Profit 10%	22.89
Grand total	$251.79

COUNTERTOPS

The materials for countertops on base cabinets range from the inexpensive linoleum type with metal edging, to the more costly molded melamine (Formica) with rounded edges and splashback or stainless steel.

Most tops are laid over 3/4" particleboard or plywood. Occasionally ceramic tile is used.

The present trend is toward laminated plastics that are heat and stain-proof. They come in widths 25 inches which permits an overhang of 1 inch, and they have a cove-molded front and a backsplash, so that it is actually a one piece top. Lengths are available from 4 to 12 feet in the straight sections, and 6 to 12 feet in L-shape corner sections. Prices range from $5 to $6 a running foot.

There is also a popular kitchen countertop of laminated maple 1 1/2" thick and 25 inches wide with a 4 inch splashback. These cost $12 to $15 a running foot. This type top is sometimes referred to as a chopping-block counter.

COST OF CABINET INCLUDING COUNTERTOP

The previously estimated cost of a 4 foot base and wall cabinet with a straight Formica countertop would be increased from $251.79 to

Cost of cabinets		251.79
4' Formica countertop @ $5. =	$20.	
Installation 1 hr. @ 5 =	5.	
	$25.	
Overhead 20%	5	
	$30.	
Profit 10%	3	
		33.00
		$284.79

Developing Unit Costs

An accepted method of estimating kitchen cabinets on an approximate basis is to use a cost-per-lineal-foot. The unit is applied to the base cabinet and the wall cabinet separately inasmuch as the construction cost for each is considerably different.

In the foregoing example the cost of the cabinets was estimated as follows assuming that the assembly and installation cost is divided equally.

Base cabinet material	$ 34.42	
vinyl veneer	22.60	
Countertop	25.00	
Labor assembling and installing	45.00	
	$127.02	
Overhead 20%	25.40	
	$152.42	
Profit	15.24	
		$167.66

Wall cabinet material	$ 26.43
vinyl veneer	17.30
Labor assembling and installing	45.00
	$ 88.73
Overhead 20%	17.75
	$106.48
Profit 10%	10.65
	$117.13
	$284.79

The unit cost for the base cabinet is $\dfrac{\$167.66}{4}$ = \$41.92 per lin ft

The unit cost for the wall cabinet is $\dfrac{\$117.13}{4}$ = \$29. 28 " " "

Modification of Unit Costs

The unit price developed for the cabinets on the previous pages are for a specific type of construction, finish and design. The cost-per-lineal-foot will vary up or down as the design and construction changes. If the vinyl finish is not put on, and a less expensive countertop is laid, then the unit cost will be much less. But if drawers are added the cost will be a little more.

It is also true that in identical type cabinets the greater the length, the lower the cost-per-lineal-foot, and conversely the shorter the cabinet, the higher the unit cost-per-lineal foot. This can be readily demonstrated by taking the catalogue of a lumber supply company, or home furnishing company, or a mail order company and check unit costs-per-lineal-foot for identical kitchen cabinets. These sources are also excellent to develop a sense of values for cabinets of various types and quality.

BASIC CONSTRUCTION TYPES

Figure 17-4 shows exploded views of the two basic types of cabinet construction. At the top is the *box or panel type* which has no frame. Plywood, particleboard or flake board 3/4-inch thick, are used for sides, bottom and back. The top may be of the same material or one of the laminated or molded types.

The panels are assembled by gluing and nailing with finishing nails. Doors are usually 1/2 or 5/8-inch stock and may have a lip or may close flush on the face. Drawer and shelving arrangement is optional. This type cabinet is used for base, wall and tall utility cabinets. The finish may be paint, stain or laminated.

BASE CABINETS

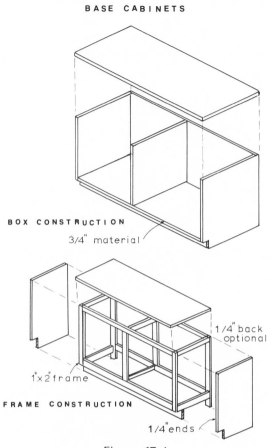

BOX CONSTRUCTION
3/4" material

1/4" back optional

1"x2" frame

FRAME CONSTRUCTION

1/4" ends

Figure 17-4

Exploded view of both "box" and "frame" construction of cabinets.

At bottom of Figure 17-4 is shown an exploded view of the *frame type* of cabinet. As its name indicates a frame for the cabinet is made using nominal 1" x 2" stock which is actually 3/4" x 1 1/2". The front or face of the frame should be clear stock unless it is to be laminated or otherwise concealed. Other members can be most any stock since it will be covered.

Jointing of the frame varies from doweling to mortise and tenon construction depending on the quality of the product.

The ends are usually a good grade of plywood and may be prefinished. A back is optional and if used is either 1/4 inch plywood or hardboard. Tops are also optional as are drawer and shelving arrangements. Doors are generally the same as in the box or panel type cabinets.

While we have illustrated the two basic types of construction in Figure 17-4 there are minor variations. Sometimes a certain amount of miscellaneous framing is used in the box or panel type as shown earlier in Figure 17-1. The important thing is to establish exactly how the cabinet you are estimating is constructed, then proceed.

ESTIMATING A BOX TYPE CABINET

If we assume that the box type base cabinet in Figure 17-4 is 48 inches long, 24 inches wide and 34-1/2 inches high, the material including doors but excluding the countertop would be as follows, keeping in mind the *nearest practical size* the material will have to be cut from. For example the height while 34-1/2 inches requires end panels cut from a 36-inch board.

¾″ Plywood

Ends	2 pcs	24″ x 36″ = 12	sq ft		
Partition	1 ″	24″ x 30″ = 5	″ ″		
Bottom	1 ″	24″ x 48″ = 8	″ ″		
Kick board	1 ″	4″ x 48″ = ⅔	″ ″		
		25⅔ sq ft		$ 7.70	
¼″ Plywood back panel		36″ x 48″ = 12	sq ft	1.44	
2½″ ″ doors		24″ x 30″ = 10	″ ″	2.20	
Glue and nails				1.00	
2 pr hinges				4.00	
2 magnetic door catches and knobs				1.10	
				$17.44	

If we added a plain particleboard countertop and splash board we include the following:

Splash board	¾″ x 6″ x 48″ = 4 lin ft	$ .80	
Countertop ¾″ particleboard	24″ x 48″ = 8 sq ft	$ 3.20	
		$ 4.00	

If we add 2 shelves we include the following:

2 Plywood shelves	¾″ x 24″ x 24″ = 8 sq ft	$ 2.40	
8 Adjustable shelf standards	24″	12.00	
		$14.40	
	Total material $35.84		

If we assume the labor to build this cabinet would be 8 hours, then in an area where carpenters are paid $6 per hour, the labor charge would be $48. The total cost to make the cabinet would be $83.84 or about $21 a running foot. This does not include finishing or hanging and only has a particleboard top; also, this cabinet has no drawers and is basically quite plain.

As the quality increases, so does the cost.

ESTIMATING A FRAME CONSTRUCTED CABINET

If we assume the same size cabinet but follow the frame construction shown in Figure 17-4 we develop the following cost assuming the prices shown.

Front frame			
2-1″ x 2″ x 4′ - 0″ bottom and top rails		= 8	lin ft
3-1″ x 2″ x 2′ - 6″ stiles		= 7½	″ ″
		15½	″ ″
Back frame identical		x 2	
		31	″ ″
6 Stretchers 1″ x 2″ x 2′ - 0′		12	″ ″
Total traming (1″ x 2″)		43 lin ft	$ 3.44
2 - 5/16″ dowels			1.00
- Glue			.50
Kickboard 1-pc ¾″ x 4″ x 4′ - 0″ = ⅔ lin ft			.20

Plywood		
Ends	2 - 24″ x 36″ x ¼″ = 12 sq ft	1.44
Back	1 - 36″ x 48″ x ¼″ = 12 ″ ″	1.44
Bottom	1 - 24″ x 48″ x 3/8″ = 8 ″ ″	1.60
Doors	2 - 24″ x 20″ x ½″ = 10 ″ ″	2.20
2 pr hinges		4.00
2 magnetic door catches and knobs		1.10
		$16.92

As in the example of the box type cabinet. if we add a plain particleboard countertop we include the following:

Splashboard	¾″ x 6″ x 48″	= 4 lin tt	$ .80
Countertop ¾″ particle board 24″ x 48″		= 8 sq ft	3.20
			$ 4.00
If we add 2 shelves we include the following:			
2 - Plywood shelves ¾″ x 24″ x 24″		= 8 sq ft	2:40
8 Adjustable shelf standards 24″			12.00
			$14.40

$16.92
4.00
14.40
Total material $35.32

The labor to make the frame by doweling will be more than nailing the box type cabinet together depending on the tools used. If we assume it will take 3 hours longer or 11 hours at the rate of $6 per hour, the cost to assemble the cabinet will be $66. The total cost, not including any finish or surfacing of the countertop is $101.32 or $25 per running foot.

ESTIMATING TALL CABINETS

Figure 17-5 shows an exploded drawing of a *tall cabinet* of box or panel construction. Using the measurements shown we can estimate the cost, assuming prices and wages, as shown below.

¾″ Plywood
2 sides	12″ x 7′ - 0″ =	14	sq ft
1 - bottom	12″ x 3′ - 0″ =	3	″ ″
1 - top	12″ x 3′ - 0″ =	3	″ ″
2 - doors	18″ x 7′ - 0″ =	21	″ ″
3 - shelves	10″ x 3′ - 0″ =	7½ sq ft	
		48.5″ ″	$14.55

¼″ Plywood back 3′ x 7′ = 21.0 sq ft	2.52
Kickboard 1-pc ¾″ x 4″ x 4′ = 2/3 ″ ″	.20
2 pr hinges	4.00
2 door catches and knobs	1.10
Glue and nails	1.00
4 adjustable shelf standards 4′ - 0″	12.00
	$35.37

The labor to build this box type tall cabinet would take 6-8 hours. If we assume 6 hours at $6 per day the labor cost would be $36 and the total labor and material cost would be $35.37 plus $36 or $71.37.

The cost per lineal foot would be $23.80

POINTS TO REMEMBER WHEN ESTIMATING CABINETS

The design, construction and finish of cabinets varies greatly.

1. Draw a free-hand sketch of the cabinet. If part or all is still in existence, look inside and see how it is constructed.
2. Note the kind of material, its thickness and quality. Take all measurements necessary to develop the cost.
3. Note the type and quality of hardware, drawer glides and shelf brackets. *This is important, note type of countertop.*
4. Note how any drawers are constructed, look at the corners.

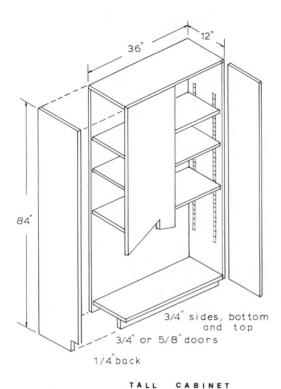

36"   12"

84"

3/4" sides, bottom
and top
3/4" or 5/8" doors

1/4" back

TALL   CABINET

Figure 17-5

Exploded view of a "tall" cabinet.

5. Carefully list the bill of materials on the premises or if your notes are complete do it in your office or home.
6. Consider very carefully the hours of labor you think it will take to build or repair the cabinet.
7. Add the cost of trucking if the cabinet is to be made away from the premises.
8. Add the cost of installation.
9. Add any subcontractor's overhead and profit if warranted.
10. Go back over your estimate and double-check the details.

Chapter 18.

Electrical Wiring

Electrical power reaches a building through service wires extending from the power company's distribution system to the building, passing through a meter to the service equipment. This generally consists of a meter, a main entrance switch, followed by a distribution panel with fuses, circuit breakers or other protective devices for each *branch* circuit. Branch circuits bring current to points within the premises by way of junction boxes, to switches, outlet receptacles, lights, motors and other equipment.

Adequate Wiring Contemplates:

1. Service entrance of sufficient capacity.
2. Wires of sufficient capacity throughout.
3. Sufficient number of circuits.
4. Adequate number of outlets, switches, etc., for livable flexibility.

ELECTRICAL TERMS

To understand the language of electrical wiring, whether in manuals, on plans, or in a discussion with electrical contractors one should be familiar with the various terms used. Those terms most frequently encountered, and their definitions, are the following:

Alternating current (A.C.). Electrical current that reverses direction in a circuit at regular intervals.

Ampacity. The maximum number of amperes a wire of specified diameter can safely carry continuously is called the ampacity of that wire.

Ampere. The rate of flow of electric current through a conductor. More technically, the amount of current that will flow through one ohm under pressure of one volt.

Circuit. A continuous path for current to travel over two or more conductors (wires) from source to point of use and back to source.

Circuit breaker. A device designed to open automatically on a predetermined overload of current, and close, either automatically or manually, all without injury to the device itself.

Conductor. Any material that will permit current to flow through it. Wire in a circuit is a conductor.

Current. A flow of electrical charge.

Cycle. The flow of alternating current in one direction, then in the other direction, is one cycle. In a 60-cycle circuit this occurs 60 times a second.

Direct current (D.C.) Current in which electricity flows in only one direction. Current from any type of battery is *always* direct current.

Fuse. A protective device inserted in circuit and containing a short length of alloy wire with a low melting point and which will carry a given amperage indefinitely. Any overload of current melts the alloy wire and opens the circuit.

(Plug fuses, when burned out, cannot be re-used.)

Fusetron. A time-delay fuse designed to tolerate temporary overloads for several seconds without blowing out.

Fustat. A nontamperable fuse designed to prevent changing fuses to a higher amperage. For example, No. 14 wire has an ampacity of 15 amps. If larger amp fuses are used there is danger of the wires in the circuit heating sufficiently to cause a fire on the premises before the fuse blows.

Grounding. A connection from the wiring system to the ground.

Horsepower. Watts measure the total energy flowing in a circuit at a given moment; 746 watts equal one horsepower.

Insulation. Any material which will *not* permit current to flow, is an insulator. When such material is used to isolate a charged conductor, it is called *insulation.*

Kilowatt. One thousand watts.

Kilowatt hours (k.w.h.). A kilowatt of power used for one hour.

National Electrical Code (N.E.C.). A set of standards for the design and manufacture of electrical materials and devices and the manner of their installation.

Ohm. The unit of electrical resistance. It is the resistance through which one volt will force one ampere.

Outlets. Any point in a circuit where electrical power can be taken and consumed.

Parallel wiring. This type of circuit wiring is commonly used in electrical construction wiring. Outlets and lights are wired across the two source wires rather than in series along one of the source wires.

Phase. The type of power available, single-phase or three-phase. The former is used in residence wiring, while three-phase is used in commerical and manufacturing properties where heavy power is required.

Resistance. Different materials offer resistance to the flow of current in different degrees. Aluminum wire, the same diameter as copper wire, has a greater resistance than the copper wire. Iron wire has a greater resistance to the flow of current than aluminum wire. Resistance depends on the kind of material, the size of the conductor and its length.

Receptacle. An outlet into which appliances can be plugged.

Series wiring. A scheme of wiring, seldom used, where the current path is along one conductor or wire. Example: Christmas tree lights, when one lamp is removed all others go out because the circuit is broken.

Source. Wherever the current comes from; may be a generator, battery, or where it enters the premises at the meter.

Switch. A device to open and close an electrical circuit or divert current from one conductor to another.

Service drop. The overhead service conductors (wires) at the building.

Transformer. An apparatus to increase or decrease voltage. The former are *step-up* and the latter *step-down* transformers.

Underwriters Laboratories. (UL) A "not-for-profit" organization whose cost of operation is supported by manufacturers who submit merchandise for testing and labels. Many cities require UL listing on all electrical materials used or sold in their jurisdiction. Underwriters Laboratories, Inc. was created in 1894 by the insurance industry but today is independent and self-supporting.

Volt. The pressure to force one ampere through a resistance of one ohm. Voltage is electrical pressure.

Voltage drop. The voltage loss due to overcoming resistance in wires and devices. The Code recommends wire sizes be used so that voltage drop will not exceed three percent in any branch circuit, measured at the furthest outlet.

Watt. The unit of electrical power. Amperes x Volts = Watts.

ELECTRICAL SYMBOLS

Electrical installation plans, whether for new work or repairs, consist of sketches or outline drawings of the room, or area involved. On these

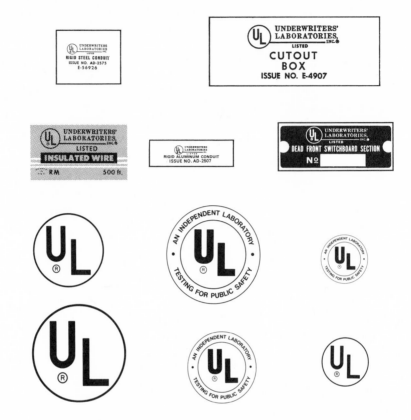

A few of the labels of the Underwriters' Laboratories, Inc. (UL). The labels are affixed to electrical wires, conduit, devices etc. when they have UL approval.

drawings are shown, by *symbols,* the location of equipment, fixtures, outlets, etc. It is as important to understand these symbols as it is to understand the language or terms used in electrical construction wiring. In Figure 18-1 are shown some of the most commonly used symbols.

Figure 18-2 is a floor plan of a residence showing how symbols are used in actual practice. The lines, dotted and solid, show the location of the wiring system.

SIZES OF WIRE

Electrical wire sizes are referred to by *number* in accord with the gauge of the American wire gauge (AWG) which is the same as that of Brown and Sharpe (B & S gauge) for non-ferrous metals. For example, No. 14 wire, a minimum size for general building wiring, is 0.064 inches (64

LIGHTING OUTLETS

CEILING	WALL	
○	—○	SURFACE OR PENDANT FIXTURE
ⓡ	—ⓡ	RECESSED FIXTURE
ⓧ	—ⓧ	SURFACE OR PENDANT EXIT LIGHT
ⓑ	—ⓑ	BLANKED OUTLET
ⓙ	—ⓙ	JUNCTION BOX

RECEPTACLE OUTLETS

GROUNDED	NOT GR'D'D	
⊖	⊖UNG	DUPLEX RECEPTACLE OUTLET
⊜	⊜UNG R	RANGE OUTLET
—△	—△UNG	SINGLE SPECIAL PURPOSE RECEPTACLE
—▲DW	—▲UNG DW	SPECIAL PURPOSE CONNECTION USE SUBSCRIPT LETTERS TO INDICATE FUNCTION*

SWITCH OUTLETS

S   SINGLE POLE SWITCH

S2   DOUBLE   "   "

S3   THREE WAY   "

—○S SWITCH AND SINGLE RECEPTACLE

Ⓢ   CEILING PULL SWITCH

CIRCUITING

_____ WIRING CONCEALED IN CEILING OR WALL

___ ___ ___ ___ _____ WIRING CONCEALED IN FLOOR ____ _____

_ _ _ _ _ _ _ _ _ _ _ WIRING EXPOSED _ _ _ _ _ _ _ _ _

*DW – dishwasher

Figure 18-1

A few of the graphic symbols for electrical wiring and layout diagrams used in architecture and building construction. Credit the Institute of Electrical and Electronics Engineers

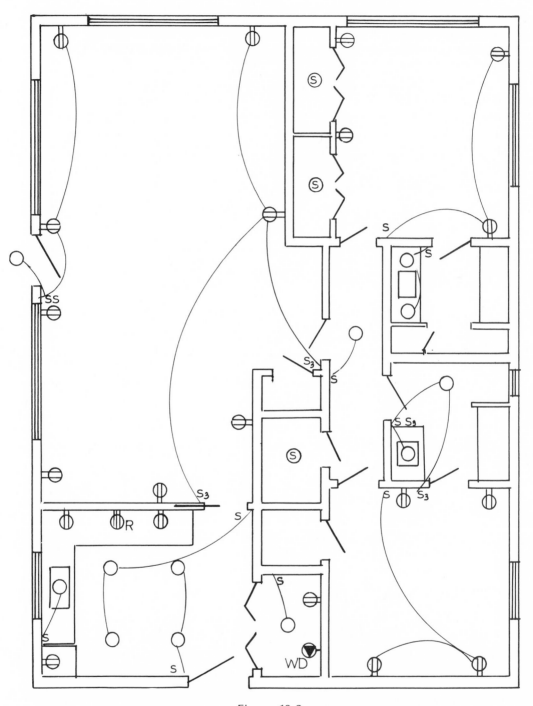

Figure 18-2
Use of symbols in the wiring layout for a residence.

mils) in diameter. Smaller wires range from No. 16 to No. 36 on the standard wire gauge. Larger wires range from No. 12 to No. 0. There are larger and also smaller wires made but we are not concerned with them. Note that as the number gets larger the wire gets smaller.

Many years ago No. 14 wire was considered adequate for wiring residential and farm buildings. With the increase in appliances that have motors and heating elements, this size wire is not approved in most cities of any size. No. 12 wire is specified and in some cases No. 10. Table 18-1 shows the ampacity (see electrical terms page 441) of various sized copper wires.

TABLE 18-1
AMPACITY OF VARIOUS SIZE WIRES

No. of Wire	Ampacity	No. of Wire	Ampacity
14	15	4	70
12	20	2	95
10	30	1/0	125
8	40	2/0	145
6	50	3/0	165

Note also that these are the maximum ratings for protective devices (fuses) for the size wire shown.

COLORS OF WIRE

The insulation on building wires is colored for a specific purpose—to assure that "hot" wires will be connected only to "hot" wires and neutral wires will run uninterrupted back to ground terminal. The white may be used as a grounding neutral wire only.

The colors of wires, established by the N.E.C. are·

2 - wire circuit . . . . . white, black
3 - wire circuit . . . . . white, black, red
4 - wire circuit . . . . . white, black, red, blue
5 - wire circuit . . . . . white, black, red, blue, yellow

Sometimes another grounding wire is used but it must be green, with a yellow stripe, or completely bare.

METHODS OF WIRING

There are many different methods of wiring but the most commonly used systems for residential, farm and non-commercial or non-manufacturing buildings are the following:

Rigid conduit
Thin-wall conduit

Flexible conduit
Non-metallic sheathed cable
Armored cable
Surface raceways

Rigid Conduit

Much like water-pipe but annealed for ease of bending and treated to prevent corrosion, this conduit comes in lengths of 10 feet with Underwriters label. It is sold in nominal inside diameter sizes from 1/2 to 6 inches although actual measurements are slightly larger. The pipe is threaded for coupling and connecting to boxes. Conduit systems are installed before the wire is pulled through.

Galvanized rigid conduit can be used inside, outside, and also underground. The *black enamel* conduit may only be used indoors away from moisture.

The N.E.C. limits the number of wires permitted in the conduit depending on the size of the pipe and the size and type of wire. Both rigid and thin-wall conduit are more costly to install than other systems.

Thin-Wall Conduit

This is the common name for *electrical metallic tubing* or EMT as designated by the N.E.C. In many respects this conduit is similar to rigid conduit; it is sold in 10 ft lengths, carries the Underwriters label, the internal diameters are the same, and the method of installation is similar. Its advantage over rigid conduit, obviously, is its ease of cutting, bending and handling. Fittings are compression types as it cannot be threaded. EMT must be supported within 3 ft of a box, fitting or cabinet, and at least every 10 ft.

Flexible Conduit

This is to be distinguished from *armored cable* (BX) which contains wires. The wires have to be pulled through *flexible conduit* in the same manner as rigid conduit or EMT. Its use is somewhat limited to situations requiring flexibility, or in conjunction with rigid conduit where freakish bends are required.

Non-metallic Sheathed Cable (originally called Romex)

This system of wiring is used where economy is a major influence due to lower material costs and ease of installation. The N.E.C. recognizes two types. In Type NM the wires are enclosed in an outer fabric braid or plastic sheath which is flame-retardant and moisture resistant. It may be used for exposed and concealed work where excessive moisture is not present. There should be a strap every 4 1/2 feet and within 12 inches of every outlet box.

Type NMC has the wires imbedded in solid plastic, and may be used in damp locations but not in water or underground.

Both Types NM and NMC are available with an uninsulated ground wire. Connections may only be made in junction boxes.

Local codes will govern the use of this system of wiring.

Armored Cable

This type of wiring was originally called BX which was a trade name. It is still referred to as BX regardless of brand. It consists of two or more insulated wires enclosed in a flexible, spiral, galvanized steel sheath to protect the wires from mechanical injury. It is not a watertight armor and nails can be driven through it. Its use should be confined to interiors, never underground, and in relatively dry surroundings. It is the second easiest system to install, non-metallic sheathed cable being first. It has a small bond wire assuring a continuous circuit through the metal covering. Only steel junction boxes, switches and outlet boxes should be used.

Surface Raceways

These are channels of metal or plastic for holding and actually concealing wires on the surface of a wall or ceiling. Special elbows, junction boxes, switches, receptacles, fittings and connectors are available. Usually the raceways and fittings are mounted and then the wires are pulled through. This system of wiring is often used in making repairs where it is undesirable to break through walls, ceilings and decorations.

NUMBER OF BRANCH CIRCUITS

In determining the number of circuits required there are four considerations to be explored:

1. Lighting circuits
2. Small appliance circuits
3. Circuits serving individual equipment
4. Special laundry circuits

Lighting Circuits

The Code recommends, in Section 220-3(a), one 20 ampere circuit for every 500 square feet of livable floor area for *lighting* (or a 15 ampere circuit for every 375 square feet). A dwelling, for example, with 1500 square feet of livable area would require 3 circuits; one with 2500 square feet would need 5 circuits. The minimum required is slightly less and is based on 3 watts for every square foot of area under Section 220-2. Practically one circuit for each 400 feet of floor area would be better. Open basements not used for living area, such as a recreation room, may be ignored in computing floor area.

Small Appliance Circuits

Section 220-3(b) of the Code requires, in addition to the previous lighting recommendation, that two or more 20-ampere *appliance branch circuits* shall be provided for the outlets servicing small appliances in the kitchen, pantry, family room, dining room, and breakfast room of dwelling occupancies. Both circuits must run to the kitchen; other rooms may be served by either of the circuits.

Special Laundry Circuit

At least one 20-ampere branch circuit shall be provided for *laundry* appliance receptacle(s).

No lighting outlets may be connected to the foregoing circuits.

A three wire 115/230-volt branch circuit is equal to two 115-volt receptacle branch circuits.

Circuits Serving Individual Equipment

Appliances or equipment that use large amounts of current are usually provided with an individual circuit. While the Code does not require that all of the following appliances have a separate circuit, it is good practice to provide one.

> Electric range
> Water heater
> Electric clothes dryer
> Automatic clothes washer
> Dishwasher
> Disposal
> Appliances rated over 1,000 watts
> Electric motors rated over 1/8 hp

These circuits may be 115 or 230-volt depending on their rating. Table 18-2 shows the watts consumed by many electrical units.

From the foregoing three considerations, the total number of circuits may be determined. Assume a dwelling with a total of 1,500 square feet of livable floor area. Also assume it has the individual equipment noted.

TABLE 18-2
WATTS CONSUMED BY ELECTRICAL UNITS

Lighter Units

Blankets (electric)	150 - 200
Blender (food)	250 - 1,000
Fans (portable)	50 - 200
Heat lamp (infrared)	250 - up
Knife	100 -
Lights, incandescent - noted on lamp	10 - up
Lights, fluorescent	20 - up
Mixer (food)	150 -
Radio (variable)	30 - 150
Refrigerator	200 - 300
Shaver	10 -
Stereo, Hi-Fi	300 -
Sump pump	300 -
Sun lamp (ultraviolet)	250 - 400
T-V	300 -
Water pump	300 -

Intermediate Units

Electric iron	800 - 1,200
Freezer	300 - 500
Power saw (depends on motor)	300 - 600
Vacuum cleaner	300 - 600
Waffle iron	500 - 1,000

Heavier Units

Air-conditioner (window type)	600 - 1,500
Coffeemaker	600 - 1,000
Deep fryer	1,300 - 1,500
Dishwasher	1,000 - 1,500
Clothes dryer	4,000 - 8,000
Fry pan	1,000 - 1,200
Garbage disposal	600 - 1,000
Grill (outdoor type)	1,300 -
Heater (portable)	1,000 - 1,600
Range - burners and oven on	8,000 - 14,000
" - oven on only	4,000 - 6,000
" - burners on only	4,000 - 8,000
Rotisserie	1,200 - 1,500
Washing machine	600 - 800
Water heater	2,000 - 5,000

Illustrative Example

1. Lighting Circuits 1,500 sq ft ÷ 500	3 circuits	
2. Small Appliances (minimum)	2	"
Laundry circuit outlets for small appliances	1	"
3. Individual Equipment		
¼ hp motor, oilburner	1	"
Range	1	"
Water heater	1	"
½ hp motor water pump	1	"
Clothes washer	1	"
Clothes dryer	1	"
Bathroom heater	1	"

Total 13 circuits

SERVICE ENTRANCE

Electrical current from the power company comes into the premises through lead-in cables called a *service drop* if it is from overhead. It passes through a watt-hour meter to measure consumption and to the main power switch or *service disconnect,* and into the *distribution panel.* This panel (*service panel*), *panel board* contains the protective devices for each branch circuit. These may be fuses or circuit breakers. The branch circuits are designed to handle, safely, subdivisions of the total anticipated electrical requirements.

60 Ampere Service. Standard for 1930-1940, but considered minimum in 1975 by NEC to "get by." Provides for lights, appliances including range, washer-dryer, *or* hot water heater. Considered adequate where gas is used to heat water and for clothes dryers. Entrance service is over three No. 6 gauge wires.

100 Ampere Service. Standard for 1950-1960 when all-electric homes became popular. Now considered *minimum* for lighting, hot water, dryer, freezer, irons, refrigeration, air conditioner, electric range and small appliances. Entrance service is over three No. 2 or No. 3 gauge wires of RH-RW insulated type.

200 Ampere Service. The coming of electric heat, plus 12,000 watt ranges, dryer, central air conditioning, roasters and rotisseries, and other appliances with heating elements, and numerous small appliances made the 100-ampere service inadequate. Entrance service is over three No. 1/0 or No. 3/0 wires of RH-RW insulated type. Some local codes now require 200-ampere minimum service.

Determining Total Service Entrance Load

The Code requires a minimum service of 100 amperes, 115/230 volts, if there are over five 2-wire branch circuits.

It is possible to fairly well approximate the size service entrance needed by totaling up the load carried. Returning to the four considerations in determining the number of branch circuits, the wattage (load) is computed.

1. The Code permits an alternate method of computing lighting and general purpose circuits. That is to figure 3 watts for each square foot area, the same area used to determine the number of circuits on pages 451-452.
2. Small appliance load under Section 220-3(b) was two 20-ampere 115-volt circuits. This works out to 20 x 115 = 2,300 watts (A x V = W); but it is recommended 3,000 watts be used.
3. The special laundry outlet required by the Code in 1968 is rounded out at 1,500 watts.
4. Individual equipment (heavy appliances) are rated individually. In Table 18-2 is a listing of the most common units and the range of wattage for each. Include each at full rating in watts.

By adding the wattage from these four considerations and dividing by 230 volts, service entrance amperage is determined.

Special Notes: The Code has provided a "demand factor" for safety. If the total for the above *items 1, 2 and 3* exceeds 3,000 watts 35 percent of the excess over 3,000 is added, *not the entire excess.*

Electric ranges get special attention. Since not all burners will be in use at once and the oven turned on, the ranges full rating in watts is not used.

If the range is rated at 12,000 watts, use an arbitrary 8,000 watts.

If the range is rated over 12,000 watts use 8,000 watts plus 400 for each one thousand watts in excess of 12,000. Add wall-mounted ovens, counter-mounted (table-top) burners together to get the equivalent range rating.

Electric heating treatment under the Code can be complicated but in general the *total* wattage consumed by all heating elements is used.

Air-conditioning requires that the rating in watts of the motor or motors be used.

Illustrative Example
(Dwelling 1,500 sq ft)

	Estimated Watts	Demand Factor %	Net Watts
1. Lighting 1,500 sq ft @ 3 watts	4,500		
2. Small appliances	3,000		
3. Special laundry outlet	1,500		
First 3,000 watts Total	9,000	100	3,000
Balance, 6,000 watts		35	2,100

Illustrative Example
(Dwelling 1,500 sq ft)

	Estimated Watts	Demand Factor %	Net Watts
4. ¼ hp motor Oil burner			300
½ hp motor Water pump			500
Range rated 15,000 watts (arbitrary)			8,000
Balance 7 x 400			2,800
Water heater			3,500
Clothes washer			600
Clothes dryer			6,000
Bathroom heater			1,000
		Total	27,800

The total watts, 27,800, divided by 230 volts equals 120.87-amp entrance service needed. A 100-amp service would not be adequate but a 150-amp service would be ideal as it would have a good safety factor for any expansion.

ESTIMATING ELECTRICAL WIRING

All of the basic principles discussed in Chapter 2, "Fundamental Considerations" are applicable to estimating the cost of electrical wiring. The major factors in forming judgment and making decisions will be *specifications, material, labor, overhead* and *profit.*

SPECIFICATIONS

In drawing up specifications for restoring electric wiring to its original condition prior to the fire, windstorm, flood or whatever peril was involved, one must be acquainted with the local ordinances, as well as the National Electrical Code. It is important that the person making the survey be knowledgeable about the damageability of wiring, appliances and devices. Frequently circuits must be tested and motors and equipment checked.

Damageability of electrical wiring. Electrical wiring, fixtures and receptacles, when exposed to heat from a fire, generally begin to deteriorate and the extent of damage is related to the temperature, its duration, and the type of material and cable. The tendency to damage by heat of the new types of cables that have a Fiberglas inner insulation and polyvinyl sheathing is somewhat less than the steel armored cable or cable in metal conduit.

Water from sprinklers in operation during a fire, and from fire hoses, which wet down the wiring system, may cause rusting of armored cable sheathing if it is not dried out quickly. Water that gets into conduit, in fixtures and in receptacles will start corrosive action or possible rust unless promptly dried out. There is also the danger of the water causing short circuits if the current is turned back on before the system is fully dried out.

The possibilities of concealed damage or the development of future trouble due to serious exposure to fire and water frequently prompts municipal authorities to require a complete rewiring of the system involved.

However, the mere exposure to fire or water does not mean that the wiring or its appurtenances have suffered permanent injury. Many electrical installations, including motors and the lighting fixtures, have been exposed to moderate heat of a fire, or have been actually submerged by flooding, yet with prompt attention, with drying out and replacing of minor units, they have continued to function satisfactorily.

A circuit that is partially damaged or destroyed can sometimes be cut, a junction box installed, and the system restored. Other times an entire circuit, back to the distribution box, is necessary. Only a thoughtful inspection will determine procedure.

Sketches are an aid in determining what work has to be done and what materials will be needed, lengths of cable or other conductors, especially where major repairs are contemplated. A line drawing of the floor plan, with room sizes, symbols and notations is usually sufficient. (See Figure 18-2.)

In some cities and towns, when the electrical wiring has been damaged a specified percentage or extent, and the wiring is substandard in the sense it is not up to the existing Code, the local ordinance requires the entire premises to be rewired to meet the new code. This may be a situation where there is knob and tube wiring or perhaps No. 14 wire is used. The Code since that installation may require No. 12 wire throughout. Outlet boxes, devices, fuses, even the distribution panel, may have to be replaced. Most insurance policies exclude any increased cost of repairs or reconstruction made necessary by the local building ordinances. Therefore the specifications for repair and replacement should be confined to what existed at the time of loss.

It is not always necessary to install the wiring in precisely the original location. Old cable may be left intact if concealed and new cable installed by lifting floor boards, boring through partition plates, or running cable down to the basement and up through a partition to the new outlet or switch. In older buildings, where cosmetic considerations are not a factor, surface wiring through raceways may be acceptable to the owners. Good judgment and common sense are necessary in writing up specifications of electrical repair in damaged buildings.

Where the property is completely destroyed and must be rebuilt, working from the new floor plan makes the job easier.

MATERIALS

It is best to make up a *branch circuit schedule* in which the location of lighting outlets, switch outlets, and convenience receptacles is shown. Figure 18-3 is a sample of a type used.

BRANCH CIRCUIT SCHEDULE

Circuit No.	Location on Premises	Light Outlets Clg	Wall	Switch Outlets S_1	S_3	S_4	Receptacles Outlets General	Special Circuit
1	Kitchen	4			2			
2	Laundry	1			1		1	
3	Kitchen						3	Oven & Range
	Dining Rm.		2		1		2	
4	Living Rm.		2	2			6	
5	Master B.R.	1		1			4	
	Master Bath	2		1			1	Heater
6	Guest Rm.	1		1			4	
	Guest Bath	1		1			1	
7	Garage	2			1			Furnace

Figure 18-3

From this schedule the number and type fixtures are listed for later pricing.

The next step for estimating materials can be quite complicated on jobs other than minor repair and replacement. It will require meticulous attention to detail. A typical *Circuit Material Schedule* is shown in Figure 18-5. There are other methods of take-off of materials less detailed and used by estimators of long experience. Some use a notebook in which they write down the materials as they inspect and examine the premises, or study a floor plan.

Since most building losses are less than total, the material listing will be rather short.

LABOR

The number of hours for an electrician to install new electrical wiring, outlets, boxes and so forth, calls for some experience and judgment. Much will depend on the type of wiring as to whether it is exposed, concealed or in raceways; also, how much cutting or boring through partitions, floors or joists is needed to "fish" wires through. Much thought and consideration must be given to those factors that affect labor as outlined in the tables.

The *labor hours* given in Table 18-3 for some of the operations encountered in run-of-the-mill electrical wiring should be used as a starting point in estimating the hours required for the particular job being con-

BRANCH CIRCUIT MATERIAL SCHEDULE

No.	Cable		Boxes			Plaster Rings			Switches			Plug Recep	Plates			Wire Nuts-Connectors Hangers-Grd. Straps			
	No. 12 2-wire	No. 12 3-wire	4" Oct.	4" Sq	SR	4" R	4" Sq	4" SS	S_1	S_3	S_4		SS	2G	PR	W.N.	C	H	G.S.
1	120'	20'	2	2	6	2	2		2			4	2		4	6	4	4	
2	160'	35'	1	3	8			1	2	4		3	2	4	3	10	4	2	2
3	40'	14'			5	1	1	1	1	3	1	8	1	1	8	4	2	6	
4	18'	70'		3	2		1			3		4	1	3	4	4	2	2	4
ETC																			
Total																			

SR Standard receptacle box   2G Two gang switch plate
SS Single pole switch plate   PR Plug receptacle plate

Figure 18-4

TABLE 18-3
APPROXIMATE HOURS TO INSTALL VARIOUS
ELECTRIC WIRING AND UNITS

Type of Work		Hours
Conduit, Cable and Wire		Hours Per 100 Lin Ft
Rigid conduit - galvanized	½" to 1"	10 - 14
" " "	1¼" to 2"	16 - 20
Rigid conduit - Thin wall	½"	6 - 10
" " " "	1"	10 - 12
" " " "	1½"	12 - 16
" " " "	2"	14 - 18
Flexible conduit (Greenfield)	½" to 1"	6 - 8
" " "	1½" to 2"	10 - 14
Non-metalic sheathed cable	#12 and #14	3 - 4
Armored cable (BX)	#10 wire	5 - 6
" "	#12 and #14 wire	4 - 5
Pulling wire through conduit	#10, #12 and #14	2 - 4
Service Entrance Installation		
Meter		1 - 2 each
Main switch		1½ - 2 "
Panel or circuit-breaker box - 200 amp		2 - 3 "
Setting Outlet Boxes		
4" x 3" and octagonal		0.5 "
2 gang		0.6 "
4 gang		1.0 "
Installing Switches and Receptacles		.3 to .5 "

sidered. In many cases, pure judgment will supplement or even supersede the hours shown. (See page 48, Estimating Labor.)

PRICING THE ESTIMATE

The principles for pricing the labor and material in an electrical wiring estimate follow those outlined in Chapter 4.

Overhead and Profit

An electrician is entitled to a reasonable charge for his overhead and profit. The same factors will apply in arriving at these charges as are outlined in detail under Fundamental Considerations (Chapter 2). As previously pointed out, the general contractor or builder adds a supervisory charge only to the electrical contractor's estimate.

Alternate Estimating Method

Frequently electrical contractors are able to estimate the cost of wiring in buildings on a unit cost for labor and material per outlet. This method is based on knowledge and experience in numerous and similar type jobs. The unit cost used will vary with the type of cable and to some extent the circumstance under which he or his men will be working.

Under this method each lighting outlet, and each convenience outlet is counted and multiplied by the unit cost. To the total must be added any special circuits to heavy duty appliances or equipment, and also any lighting fixtures, door chimes and so forth.

Chapter 19.

Thermal Insulation and Vapor Barriers

Many materials used in building construction possess some degrees of insulating quality. Even "dead air space" in the walls, floors and ceilings provide a certain amount of insulation. However, the term "dead air space" can be misleading because temperature differentials on the opposite sides of the air space can cause air movement. For example, on a very cold day one finds a current of air being drawn in at a rapid rate around loose-fitting window sash or doors due to the inside and outside temperature differences.

Table 19-1 lists various insulating materials and the type marketed for building insulation. Their relative conductivity or insulating qualities are shown. The lower the K factor, the higher the insulating value of the material based on the rate of heat transmission or heat loss.

INSULATING MATERIALS

The kinds of materials that are used for building insulation are numerous ranging from those made from minerals like glass and slag or rock, to processed vegetable fibers such as wood, cotton and bark (cork). *Polystyrene* and *urethane,* products of organic chemistry, have become popular because of their good insulating qualities and the fact they are available as foam-in-place, and in board form.

TABLE 19-1
THERMAL CONDUCTIVITY VALUES OF INSULATING MATERIALS

Type	Material	K Factor (Conductivity)
Flexible	Rock wool	.27
	Glass fiber	.25
Loose fill	Standard materials	.28 - .30
	Vermiculite	.45 - .48
Rigid	Insulating fiberboard	.35 - .36
	Sheathing ''	.42 - .55
	Corkboard	.26
Foam	Polystyrene	.25 - .29
	Urethane	.15 - .17
Wood	Low density	.60 - .65
Reflective		
(2-sides)	(vapor barrier)	(a)

(a) Insulating value is equal to slightly more than 1-inch of flexible insulation.

Mineral wool (rock wool) or silicate cotton, consists of thread-like filaments produced by the action of steam or air under pressure upon slag in a molten state.

Fiberglas is made from molten glass which is blown by machine into strands finer than human hairs. These are then spun into wool-like material for blanket and batt insulation.

Vermiculite and perlite (see pages 569 and 581).

TYPES OF INSULATION

The physical forms or types that various insulating materials are sold in, fall into five general categories.

1. Loose fill
2. Flexible (blanket or batt)
3. Rigid
4. Foamed-in-place (sprayed)
5. Vapor barriers

Loose Fill Insulation

This type of insulation is available in bulk form, in bales or bags. It can be put in place by pouring, or packing by hand in around windows and doors. The material for loose fill insulation is usually Fiberglas or mineral wool. Vermiculite, pelletized polystyrene, and ground cork sometimes are used as loose fill. When large quantities are required, loose fill can be blown into the walls or ceilings with special portable equipment. In existing buildings that were not insulated when built, this method is the most economical.

This form of insulation is best when sidewalls are filled between studs and at least 4 inches is placed between joists or rafters.

Estimating material. The quantity of loose fill needed is dependent on the type, how well it is packed down (density), and the depth. A density of 6 to 8 lbs per cu ft is recommended.

The material comes in bags weighing 40 lbs and contains 4 cubic feet. If the material is packed to a density of 6 lbs for each cubic foot, then 40 lbs would cover $\frac{40}{6}$ or 6.67 cubic feet.

This would cover an area of 80 square feet 1 inch deep (6.67 x 12 = 80). The area taken up by the joists or studs has to be added to determine the actual area of the insulation. When the area a bag will cover 1 inch deep is determined, the area a bag will cover 2, 3, 4 or 6-inches deep is obtained by dividing by the depth in inches. In other words if a bag covers 80 square feet 1 inch deep at a density of 6, it will cover 20 square feet $\frac{(80)}{4}$ 4 inches deep, and 40 square feet 2 inches deep.

The following formula (A) may be used to obtain the area any bag will cover, 1 inch deep, for a specific density.

(A) $\frac{\text{lbs per bag x 12}}{\text{density}}$ = Square foot area covered 1 inch deep

The following formula (B) may be used to obtain the area to be added which is taken up by joists (or studs).

(B) $\frac{\text{Area covered 1 inch deep x .125}}{\text{Spacing of joists or studs in feet}}$ = Square foot area of joists.

Illustrative Example

At a density of 8, determine the area, 1 inch deep, a 40 lb bag of loose fill will cover if the joists are 16 inches on center.

Area covered 1 inch deep (formula A)
$\frac{40 \times 12}{8}$ = 60 Sq ft

Area taken up by joists (formula B)
$\frac{60 \times .125}{\frac{4}{3}}$ = 5.6 Sq ft

So, the area covered 1 inch deep by the 40 lb bag of loose fill, at a density of 8, excluding that taken up by joists, is 60 sq ft. The area covered 1-inch deep including joist area is 60 + 5.6 or 65.6 sq ft.

Formulas A and B may be used for bags of different weights.

Table 19-2 gives the square foot area a 40 lb bag of loose fill will cover, including joists (or studs), for the densities and depths of insulation shown. Joists (or studs) are presumed to be 24 inches on center.

TABLE 19-2
SQUARE FEET A 40 LB BAG OF LOOSE FILL WILL COVER, INCLUDING STUDS
OR JOISTS, FOR DENSITIES AND DEPTHS OF INSULATION SHOWN.

Density-Lbs Per Cu Ft	Cu Ft of Material	Depth or Thickness of Loose Fill in Inches					
		1	2	3	4	5	6
6	6.67	85	43	28	21	17	14
8	5.00	64	32	21	16	13	11
10	4.00	51	26	17	13	10	9

Estimating labor. Pouring or packing in loose fill insulation is a relatively fast operation. Pouring in between joists takes less time than packing between studding. The physical conditions and thickness must be carefully considered.

Ordinary working conditions permit a man to pour granule insulating wool 3 to 5 inches thick at the rate of 100 to 125 sq ft per hour. Packing between studding, a man should cover 40 to 60 sq ft per hour.

Granulated vermiculite, a dry and slippery material, pours very easy and a man can cover floor areas at almost twice the rate as insulating wool. Vermiculite comes in 4 cu ft bags also. One bag will cover about 12 sq ft 4 inches thick. Density is not a concern.

Unit costs. Assume loose fill rock wool insulation costs $1.40 a bag, and labor is $6 per hour. At a rate of 100 sq ft an hour a man will use about 5 bags to pour to a depth of 4 inches and a density of 4 lbs. per cu ft. (Table 19-2 shows 21 sq ft per 40 lb. bag.)

Material 5 bags	@$1.40	$ 7.00
Labor 1 hour	@ 6.00	6.00
	Unit cost for 100 sq ft	$13.00
	Unit cost per sq ft	.13

Flexible Insulation

In new construction, the most popular type of insulation is wool batts and wool blanket or strip insulation. It is easily handled and comes in convenient widths from 15 inches to go between framing 16 inches on center, and 23 inches for framing spaced 24 inches on center.

The material consists of mineral or glass wool and is made with or without paper. Most batts and blanket wool have a form of Kraft-faced paper with a stapling flange attached to a special breather-paper layer on the side opposite the facing. This allows for inset stapling while keeping the vapor-resistant facing toward the warm side. This provides some vapor-resistance.

There is also an aluminum foil-faced type which reflects back radiant heat that comes through the walls and ceiling.

Batts and strip insulating wool are available in thicknesses of 2 to 6 inches. Batts come in lengths 6 to 8 feet.

Unfaced (no paper) batts are designed to hold in place between studs and joists until the finished surface is completed. Their main advantage is in eliminating gaps around pipes, vents and so forth.

Estimating material. The square foot area, including studs or joists, that batts will cover is readily determined. Multiply the length in feet of a 15 inch batt by 1 1/3; the length in feet of a 19 inch batt by 1 2/3; and the length in feet of a 23-inch batt by 2.

Table 19-3 shows the number of batts of different sizes needed to cover 100 sq ft including the area taken up by studs or joists. Also it shows the batts needed per sq ft of area.

Illustrative Examples:

A 15″ batt 24″ long covers 2′ x 1⅓′ = 2.67 sq ft
" 15″ " 48″ "   "   4′ x 1⅓′ = 5.33 " "
A 19″ batt 24″ long covers 2′ x 1⅔′ = 3.33 " "
" 19″ " 48″ "   "   4′ x 1⅔′ = 6.67 " "
A 23″ batt 24″ long covers 2′ x 2′   = 4.0 " "
" 23″ " 48″ "   "   4′ x 2′   = 8.0 " "

TABLE 19-3
NUMBER OF BATTS NEEDED TO COVER AN AREA OF 100 SQ FT
INCLUDING STUDS AND JOISTS

Batt Size In Inches	Square Feet Per Batt	Batts Per 100 Sq Ft	Batts Per Sq Ft
15 x 24	2.67	38	.38
15 x 48	5.33	19	.19
19 x 24	3.33	30	.30
19 x 48	6.67	15	.15
23 x 24	4.00	25	.25
23 x 48	8.00	13	.13

Estimating labor. The time to install 100 sq ft of batt or strip insulation between studding averages, on new work, 1 to 1 1/4 hours. On ceilings a little more time should be allowed.

Unit costs. If the walls of a residence measure 100 lin ft in length and are 7'-6" high, what would it cost to apply 15" x 48" batts 3" thick? Answer:

Area to be covered including that taken up by the studs is 100' x 7.5' = 750 sq ft
Table 19-3 shows that it takes .19 batts per sq ft of area.
The number of batts needed is
      .19 x 750 = 142.5 or 143 batts
If a man applies batts at the rate of 100 sq ft per hour it would take him

$$\frac{750}{100} = 7.5 \text{ hours}$$

Material (assume batts cost $.40 each)
      143 batts @ $.40   = $57.20
Labor (assume $6 per hour)
      7.5 hours @ $6   =   45.00
                                    Total cost for 750 sq ft $102.20
                                              Cost per sq ft $ .136

Rigid Insulation

The term *rigid insulation* applies to types that are made in board-form from processed wood, sugar cane, ground cork and other vegetable products. Polystyrene and urethane, the plastic foam type insulation material, are also made into board-like sheets. These are most often used to insulate the perimeter of concrete slabs on the ground.

Some familiar trade names for rigid insulating board are Celotex, Johns-Mansville Insulating Board, U.S. Gypsum Insulating Board, Beaverboard, and Gold Bond. These products are commonly referred to as *sheathing board.* They are available in sheets 4' x 6' to 12' long and range in thickness from 1/2 to 1 inch.

Installation methods and estimating quantities and costs are discussed in Chapter 9, Rough Carpentry.

REFLECTIVE INSULATION

Materials high in reflective properties are used in the enclosed stud, joist and rafter spaces to retard the transfer of heat by radiation. The reflective material used must face an air space at least 3/4 inch deep for if the reflective surface comes in contact with other material its reflective value is worthless.

The materials used for reflecting radiant heat include aluminum foil and paper products coated with a reflective oxide composition. Reflective insulation is equally effective facing the warm or cold side. Some types have aluminum foil on both sides. Some reflective insulation products include air spaces between surfaces and have nailing or stapling tabs. This type should be installed in the center of the studs to permit at least a 3/4-inch space on each side.

Estimating costs. Placing reflective insulation between studs takes more hours of labor than nailing directly on the face of the studs. The rate of application should be estimated at about 50 - 75 sq ft per hour depending on the physical working conditions and how much cutting has to be done. The aluminum foil type must be held by strips of lath nailed to the inside of the studs.

VAPOR BARRIERS

The insulating of buildings introduced a new problem that gave rise to applying *vapor barriers* in walls or floors. Warm moist air from inside is blocked from escaping through walls or floors. When the temperature inside the air space between studs or rafters is considerably lower than that within the building, the water vapor condenses and moisture collects in the wall space. If the wall or floor spaces are completely filled with insulating material the problem can be greatly reduced. However, this is not usually the case.

The condensation will eventually cause peeling and blistering of outside paint and damp spots may show up on inside walls or ceilings.

Vapor barrier materials are used to prevent passage of the warm moist air, the most effective being aluminum foil, copper coated papers, specially asphalt-treated felts and laminated Kraft papers. Also recommended where these specially treated papers cannot be used, is the application of two coats of aluminum paint underneath the interior wall and ceiling decorations.

Estimating costs. The cost of applying vapor barriers to studs should be estimated by computing the number of square feet in the area to be covered and allowing a labor rate of 250-300 square feet per hour. A unit cost may also be computed as follows using assumed material and labor costs.

$$1 - 500 \text{ sq ft roll of vapor barrier paper} \qquad \$\ 6.00$$
$$2 \text{ hours labor } \frac{(500)}{(250)} @\$6 \qquad \underline{12.00}$$

Cost per 500 sq ft $18.00

Unit cost per sq ft $ 0.036

Chapter 20.

Mechanical and
Miscellaneous Trades

There are a number of building trades that are highly specialized as to methods of installation, and the materials and labor such as plumbing, heating, electrical work, ceramic tiling, floor coverings, structural steel, sheet metal and sprinkler systems. Losses that involve these trades are best estimated by persons engaged in the particular business, especially if the cost of repairs or replacement is other than a moderate amount. There are a few general contractors who become familiar with the methods of estimating those special trades through their frequent contact with sub-contractors and by reviewing their details. They are competent to take off specifications for repairs and, by using prices and unit costs supplied by subcontractors, are able to approximate the damage with a fair degree of accuracy. However, if the amount of loss is sizable or if the general contractor's estimate is the basis of a bid to do the work, he will in most cases call in a subcontractor for a more accurate figure.

Whoever prepares the specifications and costs for these mechanical and miscellaneous trades should do so in detail, listing the materials and labor necessary. Too often the cost of repairs for specialized trades is presented in lump sum amounts.

The purpose of this chapter is to furnish the estimator with general information concerning the customary methods used to approximate repair costs and to check out estimates that are submitted. Many times a reliable subcontractor is remotely located or not available. For this reason an estimator should learn to develop a sound working knowledge of the trades so that he can take off the details, identify materials, and discuss the probable cost of repairs with a qualified subcontractor when he has the opportunity.

PLUMBING

A plumber's work is generally separated into two categories; *rough* plumbing, called *roughing-in,* and *finish* plumbing which is the setting and installation of fixtures.

The term roughing-in involves labor and materials to place piping that will be concealed in walls and under floors. This includes installing and testing the water-supply and the drainage piping. Figure 20-1 shows a schematic drawing of a typical plumbing system in a 2-story house.

WATER SUPPLY MATERIALS

Galvanized pipe, while still used to some extent, is rapidly being displaced by copper tubing because it is easier to install, takes less time and requires no threading or threaded fittings. Two types of copper tubing are available, *rigid* (hard-drawn) and *flexible* (soft-drawn). The latter can be installed with sweeping bends. For common household fixtures such as sinks, tubs, water-closets, showers and dishwashers the 1/2-inch pipe is minimum. Automatic washers should have 3/4-inch copper pipe.

Copper tubing joints and fittings are soldered (sweated) together using solder, flux and a small blow-torch.

DRAINAGE PIPING MATERIALS

A drainage system in a building includes the piping that carries off sewage and other liquid waste to the sewer. Piping materials may be cast iron; galvanized wrought iron, copper, brass or plastic. Cast iron is usually used for underground drains or those buried under concrete floors.

Plastic Piping

Carefully selected and properly installed plastic pipe offers several advantages over conventional piping materials such as galvanized steel and copper pipe or tube. There are no perfect plumbing materials and all must be installed with knowledge of their physical properties and limitations.

Today's plastic pipe and fittings are often the most economical and are nearly immune to the attack of aggressive waters. At this time PE

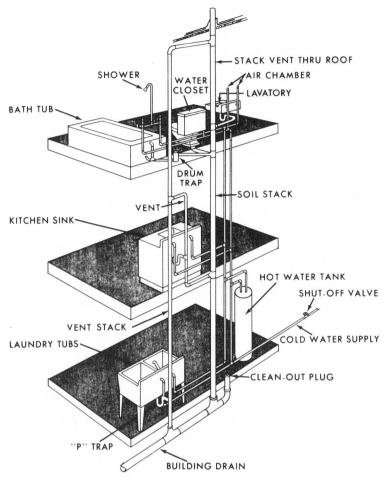

Figure 20-1

Plumbing arrangement in a two-story house with basement.

(polyethylene) pipe is used most commonly for underground service. Since it is furnished in long coils, it requires a minimum of fittings for long piping runs. For short runs, the friction loss in the insert fittings is a disadvantage. PVC (polyvinyl-chloride) pipe is available in nearly twice the pressure rating for the same cost as PE. PVC pipe is most often assembled with solvent-welded fittings. Heavy-wall PVC Schedule 80 pipe may be threaded. CPVC (chlorinated-polyvinyl-chloride) pipe is available for hot water service.

ABS (acrilonitrile-butadiene-styrene) pipe was once primarily used in potable water distribution in a size known as SWP (solvent-welded pipe).

Today ABS is used in DWV (drainage - waste - vent) systems. PVC is also used in DWV systems.

To be sure of getting quality plastic pipe and fittings, make sure that the material is marked with manufacturer's name or trademark, pipe size, the plastic material type or class code, pressure rating, standard to which the pipe is manufactured (usually an ASTM standard), and the seal of approval of an accredited testing laboratory (usually N.S.F.—the National Sanitation Foundation.)

In estimating materials, the lengths of pipe required are measured from center to center of the fittings. The nominal size of pipe is the *inside* dimension. A 3/4-inch pipe has an inside diameter of approximately 3/4 inches. Unlike combustible building materials, plumbing pipes and fixtures are seldom destroyed in a fire. They generally can be measured and identified, making it possible to write up specifications on the job in accordance with the conditions that existed at the time of loss.

Labor for cutting, threading, and installing piping is estimated according to the type of work; that is, connecting galvanized water pipe, copper pipe, waste or soil pipe, and also with the conditions under which the work is performed.

On new construction, as opposed to loss work, a contractor has his men installing pipe as the building is erected, and before the walls are plastered or floors laid. Carpenters do the cutting and notching of joists, girders and studding for the plumber.

Where a building is partially damaged, a plumber may be confronted with difficult physical working conditions. In addition he may be connecting new piping to that existing in an old structure where corrosion of the existing pipe makes it troublesome to secure tight joints. These factors should be taken into consideration when estimating labor costs on losses. A detailed breakdown of labor hours for each operation is the best method for accuracy.

In some sections of the country, Chicago and Boston being good examples, it is common practice on new construction to estimate roughing labor at a percentage of the cost of the material. The percentage may vary from 80 to 100 percent depending upon whether it is an ordinary frame dwelling (for the former) or an apartment building (for the latter). These percentages are based on experience and cost records. For industrial and fire resistive structures the labor may be figured as high as 125 percent.

While it is not practical to set fixed hourly rates that will be useful in all situations, Table 20-1 shows the approximate hours for a plumber and a helper to make certain installations. The hours shown are a guide for new construction and will have to be adjusted if applied to specific loss repairs in accordance with the conditions in each situation.

Table 20-1
Approximate Hours Labor for Roughing by a Plumber
and Helper Together

Kind of Roughing	*Plumber and Helper Team—Hours
Water closet in dwelling	8 to 10
Lavatory in dwelling	6 to 8
Bath tub and shower in dwelling	8 to 10
Kitchen sink in dwelling	5 to 6
½ to 1-inch cast or wrought iron pipe per joint	¼ to ⅓
1¼ to 2-inch cast or wrought iron pipe per joint	⅓ to ½
4-Inch soil pipe per joint	¾ to 1
Complete roughing for one-story house	50 to 60

*Add plumber and helper wage rate and multiply by hours shown.

PLUMBING FIXTURES

Plumbing fixture prices vary considerably depending on type, size, material, finish, construction and grade, all of which must be considered when estimating cost. Water closets and urinals are made of vitreous china, while bath tubs and most lavatories (sinks) for bathrooms are made of cast iron with porcelain enameled surface. Kitchen sinks are made of cast iron with porcelain enamel finish, or of stainless steel. The size, style, color and make of fixtures should be obtained before pricing.

Labor for setting fixtures after the roughing is complete is very often estimated according to the type of fixture, and the conditions under which the work is done. As with roughing, in some areas the labor to install fixtures is also based on a percentage of the cost of the fixture. Approximately 20 to 25 percent is customary.

Table 20-2
Approximate Hours Labor for a Plumber and Helper to Install
Plumbing Fixtures

Kind of Fixture	*Plumber and Helper Team—Hours
Water closet—ordinary grade	3 to 4
Lavatory—ordinary grade	3 to 4
Bath tub and shower—ordinary grade	4 to 6
Laundry tubs	4 to 6
Water heater—gas fired (45 gal)	4
Urinal—wall type	4
Urinal—floor type	6

*Add plumber and helper wage rate and multiply by hours shown.

No definite rates of labor will apply to each case, but for general estimating Table 20-2 will serve as guide.

Illustrative Example

Assuming wages and prices shown, estimate the cost of roughing in and installing three bathroom fixtures. Wages are for a plumber and a helper.

New roughing materials for 3 fixtures complete	= $ 84.00	
Labor to rough in: (Table 20-1)		
Lavatory	6 hours@($6.00 + $4.00) =	60.00
Bath tub and shower	8 hours@($6.00 + $4.00) =	80.00
Water closet	8 hours@($6.00 + $4.00) =	80.00
Price of fixtures:		
Lavatory		$ 62.00
Bath tub and shower		110.00
Water closet and seat		58.00
Labor setting fixtures: (Table 20-2)		
Lavatory	3 hours@($6.00 + $4.00) =	30.00
Bath tub and shower	4 hours@($6.00 + $4.00) =	40.00
Water closet	3 hours@($6.00 + $4.00) =	30.00
		$634.00
	20% plumber's overhead and profit =	126.80
	Total	$760.80

DAMAGEABILITY OF PLUMBING

A loss that involves plumbing, particulary when a fire or freeze-up is the cause, should be inspected carefully by an experienced person; otherwise, the damage may be overestimated or underestimated. Often the system will have to be tested for leaks. Leaded joints should be examined. Fixtures in kitchens and bathrooms frequently have to be test-cleaned to determine whether heat and smoke has caused permanent injury. Fittings such as faucets, may or may not clean up, and in some cases where they are inexpensive, the cost of cleaning may exceed the cost to replace. Medicine cabinets that have been exposed to heat and smoke often require new mirrors or, if the finish is affected, they may need new chrome or baked enamel. The cost of replacing with new should be compared to the cost of refinishing. Towel racks, shower rods and shower heads fall into the same category.

The jackets on hot water heaters and other appliances installed by plumbers such as washers, dryers, and so forth, need careful inspection to determine whether they require refinishing or replacement. Motors if submerged or wet will have to be tested, and appliances with insulation that has been damaged by water should be checked carefully.

Many times after a fire, or when a basement has been flooded, the floor drains to the sewer have to be cleaned out.

The cost of final testing and inspecting of the plumbing after repairs should be included in the estimate.

HEATING

All but minor repairs to heating systems should be estimated by a heating contractor. If such a contractor is not available, the person making the general estimate can list the items that are damaged, identify materials, and discuss the probable cost of restoration with a competent sub-contractor. Sketches and photographs will aid in the discussion, but if the loss is serious or more than a moderate amount of money is involved, the estimating should be turned over to someone in the trade. That person should be connected with a local reputable firm and preferably one that has had adequate experience in repairing damaged systems.

Where any doubt exists as to the extent of damage, or whether a fixture, control, pipe or appurtenance is injured, proper tests should be made.

An estimator should familiarize himself with the installation and parts of warm-air, steam, and hot water heating systems. He should also know the various types of floor and space heaters.

WARM-AIR HEATING

Warm-air central heating includes the *gravity, forced,* and *radiant* systems. All three types employ a furnace, fired by coal, oil or gas, to supply the warm air.

The pipeless, gravity-type, warm-air system is one of the early models in which the unit is located in the basement near the center of the house. The cast iron furnace has a metal casing that is attached to a large grille in the floor above. Cool air flows down the outer edges of the grille and the warm air rises through its center.

In a piped, gravity-type, warm-air system the warm air from the furnace flows through round pipes to registers in floors and partitions. The air is moved by the difference in the weight of the lighter warm air, and the heavier cool air at the intake. Air to this intake located at the lower part of the furnace usually is drawn through return ducts leading from registers in the baseboard of the rooms.

Forced warm-air systems have a blower at the cool air entrance to the furnace. Ducts of galvanized steel or aluminum are rectangular in shape being connected to a main trunk. The air is filtered at the blower to prevent dust and lint from passing through the system. Advantages of the forced system are that the furnace need not be located near the center of the building (as in the gravity type), the ducts may be made smaller, and the flow of air can be controlled.

Warm-air radiant heating systems (radiant panel heating) employ ducts under the floor slab through which warm air is forced. The furnace and controls are similar to those used in the forced warm-air system.

Another type of warm-air system is the floor furnace which is suspended below the floor. A combustion chamber, fired by gas or oil, permits heat to pass over radiator sections and through a floor grille. Cool air flows down the outer edges of the grille by gravity to the bottom of the furnace pushing the warm air ahead of it. Expansion of the warm air causes the circulation.

Space heaters, utilizing warm air, are used where large areas require heating such as garages, factories, and supermarkets. They are unit heaters suspended from the ceiling, and they operate by fans blowing air over steam coils, electric heating coils, or gas or oil heated radiator sections. There are many different sizes and styles of space heaters.

STEAM AND HOT WATER HEATING SYSTEMS

Steam and hot water heating systems are similar in many respects. Water is heated in a boiler which may be fired by coal, oil or gas. In a steam system the water is converted to steam and circulated through piping to radiators. The steam, after giving up its heat, condenses to water and is returned to the boiler. In a one-pipe system the water returns by gravity through the same pipes that supply the steam. In a two-pipe system the steam flows to the radiators and the water of condensation is returned by a separate system of piping connected to the opposite end of the radiators.

The hot water system provides for the circulation of water from the boiler rather than steam. The water is piped to radiators, passes through them and returns to the boiler through the return pipes. The water may circulate by gravity, in which case its movement depends on the difference between the weight of the hot water in the supply pipes and the cooler water in the returning pipes. The water may also be circulated by a pump located in the supply line near the boiler, and a one-pipe loop system.

Estimating

Steam and hot water heating systems are installed by a steam fitter and helper team. On large installations or heavy work two helpers may be used or more than one team. Much of the labor consists of pipe cutting, threading, fitting and hanging. Forced hot water systems are usually installed with copper tubing. The labor hours shown in Table 20-1, for plumbing, may be used as a guide in pipe cutting, threading and fitting. Measurements of pipe lengths are taken from center to center of the fittings. Piping for steam and hot water systems is black or galvanized wrought iron or steel. Weights and prices vary according to whether the piping is standard or extra heavy.

The specifications should include a full description of the work contemplated. Note the furnace (if involved) and its accessories by manufacturer, type, model, and any other identifying marks which will make it possible to properly price them. There are many kinds of burners and controls used in heating systems, and the price and cost of installation varies, so that adequate descriptions are very important. Insulation on piping, boilers, and duct work should be listed as to type and quantity.

ELECTRIC HEATING

There are seven principal methods of heating by electricity. Baseboard heaters, wall heaters with fans, electric heat pumps, electric furnace, ceiling cable, and wall panel heaters. There are also combinations of these systems. All electric heating systems specify particular thicknesses of insulation in walls, usually three inches; and in ceilings, usually six inches. The services of electrical contractors is recommended in estimating either repairs or replacement of electrical heating systems.

DAMAGEABILITY OF HEATING SYSTEMS

The furnaces, piping, and duct work of heating systems can withstand a considerable amount of heat during a fire without major injury. Where any question arises, tests should be made on the boilers, radiators, piping, and controls. This can be done by a competent subcontractor. Fireboxes, oil and gas burners that have been seriously wet down by water, or actually submerged, should be promptly dried out and checked for damage.

Puff-backs, or furnace box explosions, particularly in oil-fired systems, may do nothing more than blow the smoke pipe loose, crack or dislodge the asbestos covering on a boiler and deposit soot over the boiler room. If the explosion is severe, extensive damage is possible, not only to the firebox and boiler, but to the building structure. Many times a puff-back in an oil-fired furnace is preceded by leakage of oil into the firebox and in front of the furnace. A delay in ignition can ignite the oil and fire ensues. Damage may be confined to the exterior front of the boiler involving the burner, furnace and controls, or it may spread to the building itself.

Failure of a low-water cut-off may result in continued firing of a *dry* boiler. If not discovered in time, the heat can ruin the furnace, accessories, and controls.

Explosions in steam boilers are relatively infrequent due to safety devices to relieve pressure, but when they do occur the damage is usually more serious than that resulting from a firebox explosion.

STRUCTURAL STEEL

Steel beams, girders, columns and other structural members are priced per 100 pounds or per ton; therefore, each piece must be listed as to shape, size, and length in order to compute its weight. Structural steel handbooks show cross-sections for different steel shapes and list the weights per lineal foot for each type.

When measuring a steel member such as an angle iron, channel, H-beam, or I-beam, care should be taken to measure the exact width of the flanges, their thickness and, in beams and girders, the depth of the webs. Otherwise, they cannot be properly identified in the handbook.

The steel should also be listed as to its location, as different labor operations are involved. A fabricated truss, for example, should have all of its members itemized under the heading of "Truss." A girder or a column should be noted in the listing and its place in the structure identified.

It is customary to allow a percentage for rivets and details based on the weight of the steel. In built-up girders and columns, and in roof trusses, the allowance runs between 15 and 20 per cent. For plain girders and beams about 5 percent is adequate.

Small amounts of steel can be estimated by computing the weight and using a unit cost for material and labor based on a local quotation by someone familiar with the specific job. However, a steel contractor is the most qualified person to make an estimate where other than a moderate amount is involved.

DAMAGEABILITY OF STRUCTURAL STEEL

Unprotected steel members begin to fail between $1,000°$ and $1,500°$ F. Permanent injury is not always indicated by change of color, bending, or by the deforming of the shape of the steel. When it is economically sound to do so, structural steel members can be removed, straightened and replaced. This was demonstrated some years ago when structural steel framing was damaged by fire during construction of the Riverside Church in New York City. The members were removed, straightened and replaced satisfactorily.

CERAMIC TILE

Ceramic tiles are 3/8 inch thick. A popular size for floors is the 3/4 or 1 inch hexagonal, and for walls the 3" x 6" rectangular. Floor tile is laid over a concrete or cement base. The backing for wall tile may be cement board, W/R Gypsum Wall-board or portland cement plaster on wire lath. The base and backing for tile should be separately estimated.

The quantity of tile required is obtained from the square foot area to be covered to which 5 or 10 percent is added for breakage and waste in cutting. Cap, corner, cove and base tiles are individually priced for materials but the labor in setting these units is included in the over-all labor cost per square foot or for the job.

Tiles are set in portland cement mortar. A 1:3 mix is used which requires 1 sack of portland cement to 3 of sand.

A tile setter and a helper work together setting tile. The labor varies with the type and size of tile, the size and layout of the room, and the number of doors and windows. The small floor tile come attached to paper backing and can be set at about twice the rate as the individual tile. The labor for developing unit cost for an average bathroom may be figured as follows:

Kind of Tile	*Tile Setter and Helper*
3/4" or 1" paper backed floor tile	8 hrs per 100 sq ft
2" square, individual floor tile	16 hrs per 100 sq ft
4 1/4" x 4 1/4" rectangular glazed wall tile	16 hrs per 100 sq ft

Illustrative Example
Unit Costs for Ceramic Tile

Assuming the prices and wages used in the following examples, unit costs may be obtained.

Unit Cost for One-Inch Mosaic Hexagonal Floor Tile

100 sq ft 1" hexagonal tile	@ $ .70 = $ 70.00
1 sack cement	@ 1.40 = 1.40
3 (100-lb) sacks sand	@ .20 = .60
6 hours tile setter	@ 6.00 = 36.00
6 hours tile helper	@ 5.00 = 30.00
	Unit cost per 100 sq ft = $148.00

$$\text{Unit cost per sq ft} = \frac{\$148.00}{100} = \$\ 1.48$$

Unit Cost for 4¼″ x 4¼″ Glazed Tile

100 sq ft 4¼″ x 4¼″ tile	@ $ .70 =	$ 70.00
1 sack cement	@ 1.40 =	1.40
3 (100-lb) sacks sand	@ .20 =	.60
8 hours tile setter	@ 6.00 =	48.00
8 hours tile helper	@ 5.00 =	40.00
	Unit cost per 100 sq ft =	$160.00

$$\text{Unit cost per sq ft} = \frac{\$160.00}{100} = \$ \ 1.60$$

DAMAGEABILITY OF CERAMIC TILE

While ceramic tile can withstand considerable heat, it may be permanently injured by smoke stain. Glazed tile normally develops fine surface haircracks over a period of years, which are called "crazing." Frequently this is not noticed until hot smoke condenses on the surface. The condensate penetrates the cracks and deposits on the porous backing. This stain is impossible to remove, and the tile will have to be replaced, or some allowance must be made for the change in appearance.

Floor tile, which is not glazed, can be stained on the surface by excessive hot smoke. It can also be marred or scratched by broken glass, plaster and other debris being ground under foot.

In all cases the estimate of damage to stained or marred tile should be made after test-cleaning has been done.

Where sections of tile have been broken, cracked, or have come loose from the wall, repairs are made by cutting out and setting new tile, provided suitable matching is possible.

Water does not injure tile itself, but if the mortar used in the original setting is of poor grade, the tiles may come loose under conditions of submersion or excessive wetting from the back.

METAL AND PLASTIC WALL TILE

Metal and plastic wall tiles can be applied by a man at approximately 100 square feet per 8 hour day. Resin or oil type adhesives are used. These tiles can be set over plaster and composition board walls. Assuming the prices and wage shown, the unit cost would be:

Unit Cost of Plastic Wall Tile

100 sq ft 4½″ x 4½″ plastic tile	@ $ .50 =	$ 50.00
50 lbs adhesive cement	@ .24 =	12.00
8 hours labor	@ 6.00 =	48.00
	Unit cost per 100 sq ft =	$110.00

$$\text{Unit cost per sq ft} = \frac{\$110.00}{100} = \$ \ 1.10$$

The price of material for caps, corners, coves and bases are figured according to the individual job. The labor is included in the unit cost as shown for the over-all square foot area.

Metal tile can be estimated by using the same unit cost formula as shown for plastic tile and inserting local material prices and labor.

DAMAGEABILITY OF METAL AND PLASTIC TILE

Neither metal nor plastic wall tile can withstand excessive heat over a long period. The surfaces stain easily in the presence of hot smoke and they come loose from the adhesive backing. When they have been applied over plywood, plaster, or composition wall board that has been wet from the back side, a careful examination should be made to determine whether the adhesive bond has been injured.

FLOOR COVERINGS

LINOLEUM

Linoleum is a mixture of oxidized linseed oil, ground cork, other fillers and color pigments applied to a felt or burlap backing. It comes in rolls 6 feet wide and is available in three gauges; light (1/16-inch), standard (3/32-inches) and heavy (1/8-inch).

Linoleum should be laid over 15-pound saturated felt that has been pasted to a smooth surface. It is not recommended for concrete floors in basements or slabs on fill, nor directly on wood floors.

Careful planning is required in measuring the quantity of linoelum needed. Remembering that it is sold in strips 6 feet wide, the direction of joints and adequate allowance for pattern-matching is important. Frequently a pencil sketch of the room will aid in determining the number of square yards

Linoleum paste covers a little over 12 square yards per gallon. There is a waterproof type, and also one that is not waterproof which costs about one third as much.

A unit cost for linoleum may be estimated as follows, based on an average labor rate of 4 square yards per hour.

Unit Cost for 1/8" Linoleum

100 sq yds 15-lb felt	@ $ .10 =	$ 10.00
100 sq yds heavy gauge linoleum	@ 2.25 =	225.00
16 gals linoleum paste	@ 1.30 =	20.80
25 hours labor	@ 6.00 =	150.00
Unit cost per 100 sq yds =		$405.80

$$\text{Unit cost per sq yd} = \frac{\$405.80}{100} = \$ \ 4.06$$

Add cost of borders, designs, waxing and polishing as required.

FLOOR TILE

The five types of floor tile used in their order of cost are: asphalt, linoleum, rubber, vinyl and cork. Asphalt is the cheapest; cork the most expensive.

Unit costs for each type are estimated on the basis of 100 square feet of tile and lining felt, paste or adhesive, and the labor. For 6" x 6" tile approximately 4 hours are required per 100 square feet, 3 hours for 9" x 9" tile, and 2 hours for 12" x 12" tile. It takes about 50 percent more time to lay cork tile than asphalt, linoleum, rubber or vinyl.

A unit cost for vinyl tile would be developed as follows:

Unit Cost for Vinyl Tile

100 sq ft 15-lb saturated felt liner	@ $ .10 =	$ 1.00
100 sq ft 9" x 9" vinyl tile	@ .50 =	$50.00
2 gals paste	@ 2.00 =	4.00
3 hours labor	@ 6.00 =	18.00
	Unit cost per 100 sq ft =	$73.00

$$\text{Unit cost per sq ft} = \frac{\$73.00}{100} = \$ \ .73$$

The price of any floor covering varies with the type, quality, pattern or color. It is not always easy to determine the quality of tile and gauge of linoleum when it is attached to the floor, and sometimes those inexperienced with floor tile have difficulty in distinguishing asphalt, linoleum, vinyl and rubber. If a large quantity is involved, and replacement has been decided upon, a sample cut from one corner should be taken to a floor covering shop or contractor for identification.

When old floor coverings have to be removed to lay new ones, the cost of taking up the old should be figured as a separate labor operation.

DAMAGEABILITY OF FLOOR COVERINGS

All floor coverings burn and most of them can be stained by severe hot smoke, and also by dyes from contents that have been wet. Broken glass, plaster and other debris ground into the surface by people walking over them during and after a loss may cause deep scratching and marring which cannot be removed.

Some of the adhesives used to lay floor coverings are water-soluble, and if the floors are seriously wet for any length of time, the covering comes loose. This condition will not always show up immediately, therefore, it is better to defer specifications for repair until enough time has elapsed to be certain. Usually several days is an adequate period to wait. It is not recommended to assume the floor covering will come loose merely because it has been exposed to water, or water is lying on the surface.

Wood floors under tiling or linoleum may expand and become corrugated if water reaches them, as in flooding, and the ridges will eventually show through the floor covering.

Linoleum or tile floors that are damaged in part, may have to be completely replaced if they cannot be satisfactorily matched.

Chapter 21.

Anatomy of Unit Costs

ANALYZING UNIT COSTS

The practice of using unit costs is becoming more prevalent, and because much of the material in this book treats with the unit cost system, it is important to know how to analyze unit costs before attempting to make an audit of a building loss estimate.

A unit cost has been defined as the combined cost of the material and labor needed to install a unit of material. A pure unit cost should not include such items as overhead, profit, waste, or demolition and removal of debris. If any of these charges are embodied in the unit cost, it will be necessary to deduct them and reduce the unit to a *net* figure; that is, one which contains only the cost of the material and labor. Generally it will be found that the inclusion of such additional items is limited to profit or overhead, or both. It is always good practice, however, to inquire of the person who has submitted the unit cost being checked, whether or not it includes anything other than labor or material.

When *overhead* and *profit* are not added at the end of an estimate, they are undoubtedly included in the unit costs. If only profit is shown at the end of the estimate, then the *overhead* must be in the unit costs.

The first step, therefore, in analyzing a unit cost is to reduce it to a *net* basis. Any percentage allowance for either overhead or profit that is in-

485

cluded in the unit cost should be deducted. This is done by dividing the
unit cost by 100 percent plus the percentage of overhead and profit. If, for
example, a unit cost of $.12 includes 20 percent overhead and profit, the
net unit cost is obtained by dividing $.12 by 120 percent.

$$\$.12 \div 1.20 = \$.10$$

The *net* unit cost of $.10 is now presumed to contain only the material
and labor cost and it may readily be compared with other *net* unit costs for
the same work. Table 21-1 shows the figure by which a unit cost should be
divided to obtain the *net* unit cost when the percentage of overhead and
profit is known.

Table 21-1
Amount by Which a Unit Cost Should Be Divided to Reduce it to a
Net Basis

Percentage of Overhead and/or Profit	Amount by Which Unit Cost Is to Be Divided
5	1.05
10	1.10
15	1.15
20	1.20
25	1.25
30	1.30

The net unit cost is made up of labor and material and, since the cost
of the material is *constant,* any difference in net unit costs that are being
compared must be in the cost of labor. This presupposes that the kind,
quality, and price of the material have been established. The hourly rate of
wages for labor in a particular locality is usually standard. Therefore, the
real difference in two unit costs is to be found, in most instances, in the
number of hours allowed for labor. In other words the difference is in the
rate of production, the number of square feet a painter can paint in an
hour, the number of bricks a mason can lay in a day, or the number of
hours it takes a roofer to lay a square of shingles. The unit costs developed
in previous chapters for the specific building trades are based on average
rates of production under average conditions. Any variation from those
rates of production is subject to some explanation. That there may be
variations is not questioned, but such variations should be justified by some
unusual circumstances. The purpose of analyzing unit costs is to try to
disclose what those unusual conditions are and to what degree they will
affect the rate of production.

The arithmetical process for determining the rate of production that
has been used is to *deduct the unit price of material from the unit cost and
divide the remainder (the labor cost) by the hourly wage rate.*

Illustrative Example

Assume that a unit cost of $400.00 per thousand FBM ($.40 per FBM) has been used for framing 2" x 10" rafters on a flat roof. The price of the rafters is $150.00 per thousand, nails are $.25 per pound, and carpenter's wages are $6.00 per hour. The formula for computing a unit cost for this type of framing (Chapter 9) is:

Table				
	1,000 FBM lumber	@ $	= $	
9-14	Lbs nails per 1,000	@	=	
9-15	Hours carpenter labor	@	=	_____
	Unit cost per 1,000 FBM		= $	

The number of hours of labor allowed in a unit cost of $400.00 per 1,000 FBM may be determined from this formula.

Table				
	Unit cost for 1,000 FBM		= $400.00	
	Deduct material cost			
	1,000 FBM lumber	= $150.00		
9-14	8 Lbs nails @ $.25	= 2.00	= 152.00	
	Labor @ 6.00 per hr.		= $248.00	

Number of hours per 1,000 FBM = $\dfrac{\$248.00}{\$\ 6.00}$ = 41⅓ hrs.

Table 9-15 (Chapter 9) shows 24 hours of carpenter labor for installing 1,000 FBM of 2" x 10" rafters on a flat roof. An allowance of 41 1/3 hours used in a unit cost of $400.00 appears excessive and would bear checking carefully against actual job conditions.

When the labor factor is based on area covered, as it is in painting, plastering, wallboard, flooring, sheathing, and roof boards, the rate allowed in the unit cost may be readily tested against either the actual area involved, or against an arbitrary area such as 100 square feet or 100 square yards.

Illustrative Example
(Actual Area Involved)

Assume that a unit cost of $.50 per board foot ($500 per 1,000 FBM) has been used for laying 1" x 6" matched subflooring in a room 8' - 4" x 12'. If the boards are $150 per 1,000 FBM, the nails are $.20 per lb., and carpenter's wages are $6.00 per hour, the rate of production (number of board feet per hour) would be analyzed in the following manner.

Table			
	1,000 FBM matched boards	@ $ = $	
9-14	Lbs nails per 1,000 FBM	@ =	
9-15	Hours carpenter labor	@ =	_____
	Unit cost per 1,000 FBM = $		

The labor allowed in a unit cost of $500 per 1,000 FBM may be determined from this formula.

Table

Unit cost for 1,000 FBM			= $500.00
Deduct material cost			
1,000 FBM 1″ x 6″ matched boards	=	$150.00	
9-14 30 lbs 8d com. nails @ $.20	=	6.00	= 156.00
Labor (@ $6.00 per hour)		$344.00	= $344.00
Number of hours per 1,000 FBM	=	$\frac{\$344.00}{\$6.00}$	= 57.33 hrs.

The number of board feet of matched 1″ x 6″ subfloor required in a room 8′ - 4″ x 12′ is:

Table

8′ - 4″ x 12′		= 100 sq ft
9-3 Waste 24%		= 24 sq ft
		124 FBM

At the rate of 57.33 hours per 1,000 FBM it would take a carpenter 7.2 hours to lay the subflooring in this room.

$$\frac{124}{1000} \times 57.33 = 7.2 \text{ hours}$$

Table 9-15 (Chapter 9) shows a rate of 15 hours per 1,000 FBM for laying this type of subflooring. For 124 FBM it would require 1.86 hours.

$$\frac{124}{1000} \times 15 = 1.86 \text{ hours}$$

The question confronting the person making the analysis is whether a man can lay the subflooring in this particular room in approximately 2 hours or whether it will take 7.2 hours.

Such a rate is extremely unrealistic unless the working conditions are such that the boards must be scribed around numerous objects and openings in the floor, and that the job conditions vary considerably from normal straight work. It should be emphasized again that the rates of labor production shown in the various tables in this book are for the average normal working conditions and contemplate that all debris has been removed and the application or installation of the new material is to be done under circumstances comparable to new construction.

In some situations it may be more desirable to test a rate of production used in a unit cost by applying it to a specific area rather than to the entire job, or to the area involved.

Illustrative Example

Assume that a unit cost of $.30 per square foot ($30 per square) of asphalt strip shingles is to be analyzed. Assume also that the price of the shingles is $8.00 per square, nails are $.20 per pound, and a roofer's wages are $6.00 per hour.

The formula for obtaining a unit cost of asphalt strip shingles (Chapter 13) is:

1 square of shingles	@ $	= $
2 lbs roofing nails	@	=
Hours labor	@	= _____
	Unit cost per square	= $

The number of hours of labor allowed in the unit cost of $30 per square would be determined as follows:

Unit cost for one square $30.00
Deduct material cost
1 square asphalt shingles = $8.00
2 lbs roofing nails@ $.20   =   .40 =   8.40
                                  Labor (@$6.00 per hour) = $21.60

$$\text{Number of hrs per square} = \frac{\$21.60}{\$\,3.50} = 3.6 \text{ hrs}$$

If $30.00 per square is a unit cost applied to an entire side or whole roof, the rate of 3.6 hours per square would appear excessive, although for patching in one or two squares it may be entirely in order. By visualizing an area 10' x 10' (1 square) it may be judged whether a roofer would actually require 3.6 hours, the better part of 1/2 a day, to lay one square of shingles on the particular roof in question.

ANALYZING UNIT COSTS BY SIMPLE FORMULA

The number of hours of labor that have been used to develop a unit cost can be determined very quickly from the formula used to make it up. This is done by expressing the formula in terms of "rate of production." For example, for most materials, the unit cost is arrived at by the formula:

Where: $u = m + wR$

u = unit cost
m = cost of materials
w = wage rate
R = number of hours of labor
        (rate of application)

This may be expressed in terms of "R"

$$R = \frac{u - m}{w}$$

When u, m, and w are known, R can be determined.

This provides a simple method of computing the number of hours of labor a contractor or builder has contemplated in his unit costs, and permits the person checking the estimate to judge the reasonableness of the labor charge.

ANALYZING UNIT COSTS OF CONCRETE FORMS

Illustrative Example

Assume that a unit cost of $1.50 per square foot is submitted for foundation forms. The average lumber price is $120.00 per 1,000 FBM, and carpenter's wages are $6.00 per hour. From Table 7-1 (Chapter 7) we find that about 250 FBM of lumber per 100 square foot are needed in foundation forms.

Where:

    u = Unit cost of forms per 100 sq ft
    m = FBM per 100 sq ft forms (Table 7-1, Chapter 7)
    w = Carpenter wages per hour
    R = Hours of labor to build, erect and strip 100 sq ft forms

Substituting values in $R = \dfrac{u - m}{w}$

$$R = \frac{\$150 - \dfrac{250}{1,000} \times \$120}{\$6.00}$$

R = 20 hours per 100 sq ft forms

Table 7-1 (Chapter 7) shows an average of 10 hours per 100 square feet forms is required under normal conditions to build, erect and strip. The use of 20 hours per 100 square feet in the present illustration would appear to need some explanation from the estimator.

The unit costs of other type forms may be analyzed by using the same formula and referring to Table 7-1 for the number of FBM for various types.

ANALYZING UNIT COSTS OF CONCRETE

The first step in checking the hours of labor in a unit cost per cubic yard of concrete is to obtain the price of the materials in a cubic yard. Table 7-2 (Chapter 7) shows the quantities of cement and aggregate needed for a cubic yard of concrete of 1:2 1/4:3 mix.

        6 1/4 Sacks cement
        .52 Cu yds sand
        .70 Cu yds stone

Illustrative Example

Assuming cement costs $1.40 per sack, sand is $3.75 per cubic yard and stone is $3.50 per cubic yard, the total cost of materials for a cubic yard of concrete is $13.15 (when ready-mix concrete is used the local price per cubic yard should be obtained).

Assume that a unit cost of $40.00 per cubic yard has been submitted for machine-mixed, hand-placed concrete in foundation forms. Also that the materials cost $13.15 as shown above, and labor wages are $5.00 per hour. Where:

u = Unit cost of concrete per cubic yard in place
m = Cost of 1 cubic yard of concrete materials
w = Laborer's wages per hour
R = Hours required to mix and place 1 cubic yard

Substituting values in $R = \dfrac{u - m}{w}$

$$R = \frac{\$40 - \$13.15}{\$5.00}$$

R = 5.4 hours labor per cubic yard

Table 7-4 (Chapter 7) shows approximately 1 hour labor for mixing a cubic yard of concrete by machine, and 4 hours labor placing it by hand in 12 inch foundation walls. This total of 5.0 hours labor per cubic yard compares well with 5.4 hours labor allowed in a unit cost of $40.00 per yard. Whether the difference of .4 hours per yard is significant will depend on the job conditions and would be subject to the judgment of the person checking the unit cost.

If ready-mix concrete is used, the labor to mix it is eliminated in making the comparison, and only the hours of labor for placing it are considered.

Unit costs for other types of concrete work can be analyzed by using the same formula and comparing with the hours of labor shown in Table 7-4 (Chapter 7).

ANALYZING UNIT COSTS OF CONCRETE BLOCK

When analyzing a unit cost of laying up concrete block, it is necessary to compute the cost of the materials for 1 cubic yard of mortar. Chapter 8 shows the following materials for estimating a cubic yard of mortar:

10 Sacks cement
1 Cubic yard sand
2 (50-lb) sacks hydrated lime

Illustrative Example

Assuming that cement is $1.40 per sack, sand is $3.50 per cubic yard and hydrated lime is $.85 per sack, the cost of the materials for 1 cubic yard of mortar is $19.20.

Assume that a unit cost of $90.00 per 100 (8″ x 8″ x 16″) concrete block is to be analyzed for the number of hours of labor allowed to lay them. Also assume that the price of the block delivered is $.30 each, mortar is $19.20 per yard, and wages are $6.00 per hour for a mason and $4.00 per hour for a mason's helper.

Where:

u = Unit cost for 100 (8″ x 8″ x 16″) concrete blocks

m = Price of 100 block plus .15 cubic yards of mortar (Table 8-2, Chapter 8)

w = The sum of hourly wages of a mason and a mason's helper.

R = Hours of labor for a mason and his helper working together to lay 100 block (based on equal number of hours for both mason and helper)

Substituting values in $R = \dfrac{u - m}{w}$

$$R = \frac{\$90 - (\$30 + .15 \times \$19.20)}{\$10.00}$$

R = 5.7 hours labor per 100 block for a mason and
      5.7 hours for his helper

Table 8-3 (Chapter 8) shows 5 hours each for a mason and a helper for this type of work. The justification of 5.7 hours each should be judged by the person making the analysis. Job conditions in the particular instance may or may not warrant the increase in rate.

When the unit cost of 8" x 8" x 16" concrete block is expressed in square feet rather than per block, the unit cost should be divided by 1.125 to convert the figure to a unit cost per block. After the unit cost per block is obtained, the cost per 100 is obtained by multiplying by 100.

ANALYZING UNIT COSTS OF BRICK MASONRY

Before analyzing a unit cost of brick masonry, the cost of the materials for 1 cubic yard of mortar is obtained as shown in analyzing concrete block, page 491. Using the same assumed prices, the cost of mortar materials is $19.20 per cubic yard.

Illustrative Example

Assume that a unit cost of $150.00 per 1,000 common brick in a 12-inch wall has been submitted. Also assume that common brick is $45.00 per 1,000, mason's wages are $6.00 per hour, and a mason helper's wages are $4.00 per hour.

Where:

u = Unit cost per 1,000 common brick

m = Price of 1,000 brick plus cost of .60 cubic yards of mortar (Table 8-2, Chapter 8)

w = Sum of the hourly wages of a mason and a helper working together

R = Hours of labor for a mason and a helper working together, to lay 1,000 brick

Substituting values in $R = \dfrac{u - m}{w}$

$$R = \frac{\$200 - (\$45 + .60 \times \$19.20)}{\$10.00}$$

R = 14.35 hours labor per 1,000 brick for a mason
and 14.35 hours for his helper.

Table 8-4 (Chapter 8) shows that normally 10 hours each for a mason and helper is required to lay 1,000 brick in a 12-inch wall with average number of openings. The hours in the unit analyzed, show approximately 44 percent more time allowed. An increase of this amount over the average of 10 hours each should be accounted for by a reasonable explanation.

In analyzing other kinds of brickwork, care must be exercised in the use of Table 8-4 (Chapter 8).

ANALYZING UNIT COSTS OF HOLLOW TILE BLOCK

The same procedure for analyzing concrete block unit costs is used to check unit costs submitted for hollow tile block.

The cost of the materials for a cubic yard of mortar is first obtained. The cost of the mortar for 100 hollow tile block (Table 8-2, Chapter 8), plus the price of 100 block is subtracted from the unit cost per 100 block. This is divided by the sum of a mason's and helper's hourly wage rates.

ANALYZING UNIT COSTS OF FRAMING

Unit costs for framing, when expressed in terms of cents per FBM, will have to be converted to dollars per 1,000 FBM before they can be analyzed for labor allowance. This is done by multiplying the cost per FBM by 1,000 as:

$.30 per FBM = $300 per 1,000 FBM

Illustrative Examples

Assume that a unit cost of $.30 per FBM for framing exterior studding is submitted to determine the hours of labor allowed. Also assume that studding costs $130.00 per 1,000 FBM, nails are $.20 per pound, and carpenter's wages are $6.00 per hour.

Where:

u = Unit cost per 1,000 FBM

m = Price of framing lumber per 1,000 FBM, plus price of required nails Table 9-14 (Chapter 9)

w = Hourly wage rate of carpenters

R = Hours of labor to cut and frame 1,000 FBM.

Substituting values in $R = \dfrac{u - m}{w}$

$$R = \frac{\$300 - (\$130 + \$.20 \times 10)}{\$6.00}$$

R = 27.7 hours per 1,000 FBM

Under average conditions Table 9-15 (Chapter 9) shows that a carpenter can cut and frame exterior wall studding at a rate of 25 hours per 1,000 FBM. Whether or not 27.7 hours per 1,000 FBM is excessive for the work contemplated in the unit cost is a matter for the person making the analysis to check, giving proper consideration to all conditions surrounding the specific job.

Other framing unit costs may be readily analyzed for labor allowances in the same manner. Table 9-14 (Chapter 9) shows the quantity of nails for each kind of framing. When making a comparison of the number of hours of labor per 1,000 FBM that have been allowed in a unit, care should be taken to use the appropriate kind of framing shown in Table 9-15 (Chapter 9). Tearing out and carting of debris is presumed taken care of under a separate item.

ANALYZING UNIT COSTS OF CARPENTRY SURFACE MATERIALS

Sheathing

Subflooring

Roof boards

Finish Flooring

Solid Paneling

Exterior Wood Siding

Analyzing the labor allowed in unit costs of all surface *lumber* whether interior or exterior, is done in the same manner. The unit cost per 1,000 FBM is used and when submitted in cents per FBM it should be converted by multiplying by 1,000. The quantity of nails per 1,000 FBM is shown in Table 9-14 (Chapter 9).

Illustrative Example

Assume that a unit cost of $.24 per FBM is submitted for horizontally applied 1″ x 8″ matched sheathing, and it is to be checked for labor allowance.

Assume also that sheathing is $120.00 per 1,000 FBM, nails are $.20 per pound and carpenter wages are $6.00 per hour.

Where:

u = Unit cost per 1,000 FBM

m = Material cost per 1,000 FBM, plus cost of nails per 1,000 FBM

w = Carpenter's wages per hour

R = Hours of labor to apply 1,000 FBM

Substituting values in $R = \dfrac{u - m}{w}$

$$R = \frac{\$240 - (\$120 + 25 \times \$.20)}{\$6.00}$$

R = 19.1 hours per 1,000 FBM

Table 9-15 (Chapter 9) shows 16 hours labor per 1,000 FBM for this type sheathing. The labor allowed in the unit cost of $.24 per FBM appears excessive unless conditions are unusual. At a rate of 19.1 hours per 1,000 FBM it would take a carpenter 2.35 hours to cover an area of 10' x 10'.

Area 10' x 10'	= 100 sq ft
Waste 18%	= 18 sq ft
	118 FBM

$$\frac{118}{1,000} \times 19.1 \quad = 2.35 \text{ hours}$$

When analyzing unit costs for other kinds of surface lumber, care should be taken to obtain the proper number of pounds of nails from Table 9-14, Chapter 9.

ANALYZING UNIT COSTS OF CARPENTRY TRIM ITEMS

Interior or exterior items of trim such as base board, picture molding, water table or cornice members that are expressed in terms of unit cost per lineal foot may readily be checked for hours of labor by expressing the formula u = m + wR in terms of R.

$$R = \frac{u - m}{w}$$

Illustrative Example

Assume that a unit price for a base molding composed of 1" x 6" base and ¾" one quarter round floor molding is submitted at $1.00 per lineal foot or $100.00 per 100 lineal feet. Assume also that 1" x 6" pine is $11.00 per 100 lineal feet, ¾" one quarter round is $2.00 per 100 lineal feet, and carpenter's wages are $6.00 per hour. (Nails are omitted).

Where:
u = Unit price per 100 lin ft
m = Price of 100 lin ft of material
w = Carpenter's wages per hour
R = Hours per 100 lin ft

Substituting in $R = \frac{u - m}{w}$

$R = \frac{\$100 - (\$11 + \$2)}{\$6.00}$

R = 14.5 hours per 100 lin ft

Table 10-6 (Chapter 10) shows 2-member softwood base board can be installed at an average rate of 7 hours per 100 lineal feet. If the unit cost is $1.00 for soft pine, a labor allowance of 14.5 hours is extremely high. The person checking the unit cost must decide whether the type of wood and working conditions warrant such an allowance for the particular job under consideration.

ANALYZING UNIT COSTS FOR WALL BOARDS AND PLYWOOD PANELING

Unit costs for wallboards and plywood paneling are expressed in square feet. Unit costs may be analyzed for labor allowance from

$R = \frac{u - m}{w}$

Illustrative Example

Assume a unit cost of $.30 per square foot is submitted for ½-inch gypsum board, and that the price of the material is $.08 per square foot and labor is $6.00 per hour.

Where:
u = Unit cost per 100 sq ft
m = Price of material per 100 sq ft
w = Wages per hour
R = Hours labor to apply 100 sq ft

Substituting in $R = \frac{u - m}{w}$

$R = \frac{\$30 - \$8}{\$6.00}$

R = 3.67 hours per 100 sq ft

Page 48 (Estimating labor) shows two hours of labor required to apply 100 square feet wallboard under average conditions. Whether 3.67 hours per 100 square feet is justified for the particular job is a matter of judgment.

ANALYZING UNIT COSTS OF WINDOWS, DOORS, CABINETS

Such items of carpentry as stairs, doors, windows, and cabinets are quickly analyzed for labor allowance by deducting from the cost installed, the price of the material.

Illustrative Example

An ordinary 1 3/8″ fir two-panel door 2′ 10″ x 6′ 8″ is submitted at a unit cost of $75 complete with frame, door, trim and hardware.

The material for the door is priced at $22.50 and carpenter's wages are $5.00 per hour.

Where:

u = Unit cost of item involved

m = Price of material

w = Carpenter's wages

R = Hours labor to install

Substituting in $R = \dfrac{u - m}{w}$

$$R = \frac{\$75 - \$22.50}{\$5.00}$$

R = 10.5 hours to install

Table 10-5 (Chapter 10) shows that an average time to fit, hang and install an interior door complete with frame and so forth is 4 hours. The hours of labor developed from a unit cost of $75 is 10.5 hours, almost three times as long. Some explanation is indicated.

ANALYZING UNIT COSTS FOR ROLL ROOFING AND SHINGLES

Illustrative Example

Assume that a unit cost for asphalt strip shingles is submitted at $.27 per square foot or $27.00 per square. Also that shingles are $8.00 per square, nails are $.20 per pound and wages are $6.00 per hour. Table 13-5 (Chapter 13) shows 2 pounds of nails per square.

Where:

u = Unit cost of roofing per square

m = Price of materials

w = Wages per hour

R = Hours to lay or apply one square

Substituting in $R = \dfrac{u - m}{w}$

$$R = \frac{\$27 - (\$8 + .40)}{\$6.00}$$

$R = 3.1$ hours per square

This labor allowance could appear to be excessive inasmuch as Table 13-5 (Chapter 13) shows 2 hours per square under ordinary conditions for plain roofs and 3 hours for difficult roofs.

If the shingles are 235-lb, a square contains 80 shingles. At 3.1 hours per square it would require 2.3 minutes to apply one shingle. This would not seem reasonable.

$$\frac{3.1 \times 60}{80} = 2.3$$

Unit costs of other types of shingles and also roll roofing may be similarly analyzed by inserting material costs and wages per hour in the same formula. Comparisons should be made between hours per square developed and those in Table 13-5 (Chapter 13).

ANALYZING UNIT COSTS FOR PAINTING

Following the same principle of deducting the cost of material from a unit cost and dividing the remainder (labor cost) by the hourly wage rate, the hourly rate of application of painting material may be obtained. The formula for interior and exterior painting (Chapter 14) when using a gallon of paint to develop the unit cost *per coat* is:

Table

1 gallon paint	@ =	\$
(14-1) Coverage per gal		

$$\dfrac{\text{(14-1) Coverage per gal}}{\text{(14-3) Square feet per hour}} \times \text{wages per hour} \qquad @ = \underline{\qquad}$$

Total cost per coat = \$

Cost per sq ft $\quad = \quad \dfrac{\$}{\text{Coverage per gal}} = \$$

(or) $\qquad u = \dfrac{m + \dfrac{cw}{R}}{c}$

Where:

u = Unit cost per square foot *per coat*
m = Cost of 1 gallon paint material
c = Square foot coverage of paint per gallon
w = Painter's wages per hour
R = Number of square feet a painter can paint per hour

Painting interior smooth walls with *two coats* of flat white has been estimated at a unit cost per square foot of $.20 on a gross area basis (openings not deducted). If paint is $6.00 per gallon and painter's wages are $6.00 per hour, the number of square feet per hour contemplated in the unit can be readily obtained. Table 14-1 (Chapter 14) shows coverage for this type of painting of 450 square feet per gallon. If $.20 is the unit cost per square foot for two coats, it would be $.10 for one coat.

Expressed in terms of R:

$$R = \frac{cw}{cu - m}$$

$$R = \frac{450 \times \$6.00}{(450 \times \$.10) - \$6.00}$$

$$R = \frac{2700}{39}$$

$$R = 70 \text{ sq ft per hour}$$

Table 14-3 (Chapter 14) shows a rate of 150 square feet for this kind of painting under average conditions. The rate of 70 square feet seems low and needs explanation.

ANALYZING LUMP SUM CARPENTRY LABOR FIGURES

Some builders and contractors prepare estimates which show a complete detail of materials by item and cost; labor for installing the material is then shown in a single lump sum amount. Carpentry, more than any other trade, is treated in this manner. The basis for estimating the number of hours of labor varies. There are those who estimate that the labor cost of installing carpentry is approximately equal to the cost of the materials. Others use different ratios of the cost of materials to the cost of the labor. A 60 percent labor and a 40 percent material cost is a ratio often encountered. It is not unusual to find that the cost of carpentry labor has been determined by approximating the number of days it would require a number of men to install all of the materials. There are also estimators who break down the labor on their work sheets for each item and operation, but when putting their estimate into the final form, labor cost is entered as a single lump sum.

When the labor cost is shown in a single sum of money for numerous items of carpentry, an analysis may be made of the labor cost for each individual item. The total may then be compared with the total labor cost in the estimate. Tables in Chapter 9 show the approximate hours of labor to install or apply various materials. These may be used as a guide where average or normal working conditions prevail. The hourly labor rates shown may be adjusted for situations where working conditions are other than average. Good judgment is required in each case.

Illustrative Example

An estimate contains an item which specifies laying a 1″ x 3″ oak floor in a living room 18′ x 22″. It is to be analyzed to obtain the number of hours of carpenter labor required.

The quantity of flooring required is:

Table

10-2 18′ x 22′	= 396 sq ft
10-6 Waste 33⅓%	= 132 sq ft
	528 FBM

Assume under average conditions a carpenter can lay 1″ x 3″ oak flooring at the rate of 35 hours per 1,000 FBM. The time to lay 528 FBM would be:

$$\frac{528}{1,000} \times 35 = 18.48 \text{ hours}$$

By analyzing each item in this manner the hours of carpenter labor may be developed for the entire estimate. The labor cost can then be computed by multiplying the total number of hours by the prevailing hourly wages for a carpenter.

The following analysis may be made of the labor in this estimate. The hourly rates of production are taken directly from the tables as indicated. No consideration has been given to any unusual working conditions which may be involved in the particular job for which this estimate has been made. Variations in hourly rates shown in the tables are to be made by the person making the analysis who, it is presumed, has seen the damaged property and is familiar with all of the problems that may be encountered in effecting repairs. Whether or not the labor charge of $600 in the estimate being analyzed is adequate or excessive is a matter to be judged by the person who is doing the checking.

Illustrative Example

Estimate to Repair Fire Damage to Dwelling

Real wall studs	18-2″ x 4″ x 12′	144 FBM			
New plate	1-4″ x 4″ x 16′	22 FBM			
New Sill	1-4″ x 6″ x 16′	32 FBM			
Gable rafters	6-2″ x 6″ x 12′	72 FBM			
Floor beams	3-2″ x 8″ x 12′	48 FBM			
		318 FBM	@$135.00 = $	42.93
Replace 1″ x 6″ wall sheathing		200 FBM	@ 125.00 =	25.00
Replace 1″ x 6″ roof sheathing		360 FBM	@ 125.00 =	45.00
Rear Extension				
1″ x 3″ N.C. Flooring		160 FBM	@ 200.00 =	32.00
Novelty siding (1 x 6)		400 FBM	@ 300.00 =	120.00
1-2′10″ x 6′ 8″ x 1 3/8″ fir door complete			@ 28.00 =	28.00
6 D.H. 2′ 8″ x 5′ 2″ windows complete			@ 30.00 =	180.00

Kitchen

1" x 8" 3-member baseboard	100 lin ft	@	.26 =	26.00
1" x 3" N.C. pine floor	180 FBM	@	210.00 =	37.80
			$	536.73
Carpenter Labor			=	600.00
			Total =	$1,136.73

Labor Analysis of Illustrative Example

Item	Quantity	Table No.	Labor Rate	Hours Hours
Rear wall studs	144 FBM	9-15	25/m	3.60
New plate	22 FBM	9-15	20/m	.44
New sill	32 FBM	9-15	20/m	.64
Rafter (2" x 6")	72 FBM	9-15	30/m	2.16
Floor beams (2" x 8")	48 FBM	9-15	22/m	1.06
Wall sheathing (1" x 6")	200 FBM	9-15	16/m	3.20
Roof sheathing (1" x 6")	360 FBM	9-15	20/m	7.20
1" x 3" N.C. flooring	160 FBM	10-3	25/m	4.00
Novelty siding (1" x 6")	400 FBM	12-2	30/m	12.00
1-2' 10" x 6' 8" x 1 3/8" door	1	10-5	4 ea	4.00
6 D.H. 2' 8" x 5' 2" windows	6	10-4	4 ea	24.00
1" x 8" - 3 member baseboard	100 lin ft	10-6	8/c	8.00
1" x 3" N.C. pine floor	180 FBM	10-3	25/m	4.50
			Total	74.80

Carpenter Labor 74.8 hours @$6.00 $448.00

Chapter 22.

Adjusting Building Losses

When an estimate is made of the cost to repair, rebuild or replace a damaged building or other structure covered under an insurance policy, the usual purpose is to substantiate a claim of the policyholder. Often the mere submitting of a properly prepared estimate will suffice, where the damage is not serious, and the matter will be closed with payment to the insured. When one estimate is prepared, on a serious loss, for the insurer and also one for the insured, they generally will have to be reconciled because of differences in specifications and/or amounts of individual items. The process of resolving such differences is referred to as *adjusting the loss*.

The purpose of this chapter is to acquaint the parties involved in adjusting losses with the principal areas of information and knowledge needed to perform effectively, and the importance of pursuing and expanding such knowledge.

INSURANCE POLICIES

Policies of insurance describe the property covered, the perils insured against, and stipulate various conditions under which there are exclusions, extensions or limitations.

Property Covered

Descriptions pertaining to "Building Coverage" vary, some being very broad as in certain mercantile and manufacturing forms which state "All real and personal property." Others, like the Homeowners Policy, HO-2, Ed. 9-70-SW, are a little more explicit, as seen.

COVERAGE A - DWELLING
This policy covers the described dwelling building, including additions in contact therewith, occupied principally as a private residence.
This coverage also includes:
1. if the property of the insured and when not otherwise covered, building equipment, fixtures and outdoor equipment all pertaining to the service of the premises and while located thereon or temporarily elsewhere; and
2. materials and supplies located on the premises or adjacent thereto, intended for use in construction, alteration or repair of such dwelling.
COVERAGE B - APPURTENANT STRUCTURES
This policy covers structures (other than the described dwelling building, including additions in contact therewith) appertaining to the premises and located thereon.
This coverage excludes:
1. structures used in whole or in part for business purposes; or
2. structures rented or leased in whole or in part or held for such rental or lease (except structures used exclusively for private garage purposes) to other than a tenant of the described dwelling.

While the HO-2 is more detailed than the "All real and personal property" wording, it is still very broad and is known as the "Broad Form."

The description of building coverage under Form No. 18-(Uniform Standard) is exceptionally detailed and names just about everything in the building.

BUILDING COVERAGE. When the insurance under this policy covers a building, such insurance shall cover on the building, including foundations (except as excluded below), machinery used for the service of the building only, plumbing, electric wiring, electric sound, communication, stationary heating, lighting, ventilating, refrigerating, air-conditioning and vacuum cleaning apparatus and fixtures, boilers, all only while contained therein; ovens, kilns, furnaces, retorts, forges, cupolas and driers of brick construction or brick encased, resting on independent foundations built from ground, all only while contained therein; awnings, signs and metal smokestacks (except as provided below), screens, storm doors and windows if the property of the owner of the building and belonging to said building, while attached thereto or stored therein or in other buildings on the premises; signs (except as provided below) if the property of the owner of the building in the open within one hundred (100) feet thereof; also all permanent fixtures, stationary scales and elevators belonging to and constituting a permanent part of said building.

Perils Insured Against

There are two ways that insurers handle *perils insured against* or *perils covered* as the term is commonly used. One is the *all-risk-of-loss* policy which insures against *all* perils with certain exceptions or exclusions. The *specified-peril* policy names the specific perils insured against. Here again the HO-2 is an ideal example of the *specified-peril* policy. Each peril is named and described. Embodied in each (fire, lightning and explosion excepted) are provisions pertaining to conditions, exclusions and exceptions.

No attempt will be made to discuss in detail all of the individual perils and their relationship to the making of building estimates except to point out that the provisions and conditions set forth in many policies are critical in preparing a proper estimate. A person with little experience in estimating damage to buildings and who has had little or no exposure to policies covering property, is apt to innocently include *all* the repairs and replacements deemed necessary to completely restore a building. He may include such things as awnings and fences when they are specifically excluded under the perils of *windstorm, weight of ice and snow,* and *collapse.* His estimate may include serious water damage to the interior of a building (with no damage to walls or roof) when such damage is excluded under the *peril of windstorm* unless the building shall first sustain a damage to the roof or walls by the direct force of the wind. He may include work necessary to comply with local building codes or ordinances such as complete rewiring when only a portion of the wiring has been injured by fire or water. The code may so stipulate but the policy may have an exclusion in it.

See page 516 for further discussion of Sources of Error in Estimates.

EXCLUSIONS

As outlined above, there are specific exclusions relating to perils which are generally placed in an insurance policy to relieve the underwriters from losses which are not contemplated in the rate structure. An example is the exclusion of wave-wash under the peril of windstorm. Also, the insurer must protect himself from catastrophic losses such as earthquake.

Policies also *exclude* certain property. Notable is the exclusion of the cost of excavations, footing and foundations below ground or the lowest basement floor where there is a basement. Similarly, underground pipes, ducts, wiring and drains are excluded. The HO policies exclude these items in the calculation of value to determine eligibility for full repair and replacement cost. Any loss to these items, however, *is* covered.

EXTENSIONS

An example of an extension of coverage is under the Special Multi-Peril Policies which cover, up to a specified limit, on new additions, buildings and structures constructed on the described premises and intended for the same occupancy.

Special Clauses

There are several special clauses in policy forms which should be examined to determine coverage.

Alterations and Repair Clause permits the insured to alter, repair, and erect additions to buildings covered under the policy and covers such in the event of loss.

Removal of Debris Clauses vary in language. Most of these clauses cover the expense of removal from the premises of the debris *of the property covered*. Debris of other buildings, trees etc., blown onto the insured's premises in a windstorm are generally not covered.

Electrical Apparatus Clause excludes loss resulting from electrical injury to electrical appliances, devices, fixtures or wiring caused by electrical currents artifically generated unless fire ensues—then only for the ensuing damage. The wording "artifically generated" means man-made. *Lightning* damage (generated in nature) does not apply in the exclusion.

NEW YORK STANDARD FIRE POLICY

Because of the importance of some of the provisions of the New York Standard Fire Policy as related to estimating building losses a brief discussion follows.

This policy, referred to as the *Fire Contract,* consists of an insuring agreement and 165 lines of provisions and stipulations. They form ground rules of most policies covering buildings.

The 1943 New York Standard Fire Policy has been established by law in 32 states and adopted with minor changes by 15 others.

Minnesota, Massachusetts and Texas have statutory policies similar in most respects to the 1943 New York form.

The majority of policy forms that cover buildings are either attached to a standard fire policy or have incorporated into their own *policy jackets* the essential conditions and exclusions of the standard fire policy. Those conditions and exclusions which are most important when estimating and adjusting building losses are discussed in the following order.

1. Loss must be direct
2. Insurer's limitations of liability are:
   amount of insurance,
   insured's interest,
   cost to repair or replace, and
   actual cash value
3. Loss not caused by increase of hazard
4. No increased cost of repairs by reason of local ordinances
5. Vacancy or unoccupancy clause
6. Replacement cost coverage
7. Coinsurance
8. Requirements of insured in case of loss
9. Company's option to repair or replace
10. Appraisal in event of dispute

1. Loss Must Be Direct

The policy covers direct loss by the peril insured against. In other words the peril must be the immediate or proximate cause, rather than the consequential cause. When fire, lightning or other perils set in motion an unbroken chain of events resulting in damage at the other end of the chain, the loss is direct.

Fire, for example, weakens the walls of a building and the firemen trying to pull the weakened wall into the structure cause it to topple outward onto an adjacent building. The damage to that building would be considered a direct loss by fire. Yet, if a few months after the fire, workmen in demolishing the weakened wall, caused it to topple onto the adjacent building there is grave doubt it would be a direct loss by fire.

A fire in an electric panel board causes a drop in voltage, resulting in damage to motors in other remote parts of the building. The loss is a direct loss by fire.

Friendly Fires vs. Hostile Fires

Sometimes serious damage is done to a building's decorations by what has been ironically called a *friendly fire*. Smoky fireplaces, stationary heating devices, oil lamps and stoves may operate defectively. There is no actual fire outside the device but smoke damage is extensive. This is not a direct loss by fire and except when specifically named or under all-risk-of-loss forms it would not be covered.

A friendly fire is said to be something burning in a place it was intended to be burning. Fire in a stove, a fireplace, a furnace, trash burner or incinerator is said to be friendly as long as it remains in its proper place of confinement.

Burning soot in a chimney or a roast on fire in an oven can hardly be considered friendly. Damage by smoke from these fires is covered under the policy.

Smoke from a hostile fire burning next door or even blocks away which results in damage to the interior or exterior decorations of an insured's building is a direct loss by fire. Such damage, however, caused by smoke from a commercial incinerator would not be covered as the fire would not be considered hostile.

2. Insurer's Limitations of Liability

Insurance policies contain several limits of liability. The *amount of insurance* is an obvious limit beyond which the insured cannot collect.

Three other limits are stated in standard fire policies: the insured's *insurable interest* in the property; the cost to *repair or replace* the property with like kind and quality; and the *actual cash value* of the property.

Insurable interest. The sole and unconditional owner of a building has an *insurable interest* in the property and may collect any loss up to the amount of insurance or any other specific limits. A lessee who under his lease is required to carry insurance and restore the property in event of loss has an insurable interest to that extent. A mortgagee has an insurable interest to the extent of the outstanding indebtedness under the mortgage. But an owner of an undivided portion of a building is limited to *that fraction* of its value. In other words he can collect any loss up to value of his interest, not more.

Repair or replace. The cost to repair or replace the damaged portion of the building fixes another limit under the policy. This would be the full amount of the estimate made.

Actual cash value. The third limit stated in standard fire policies is the *actual cash value.* This term is broadly construed to mean its value in money to the insured. While court decisions have been understandably slanted to give the insured the best possible interpretation, general practice has been to treat *actual cash value* as the cost to replace the building less reasonable depreciation (loss of value) for wear and tear and for obsolescence.

Buildings deteriorate physically from wear and tear; from exposure to the elements; and from injury by insects, (termites) or bacteria and fungus (rot). If such conditions are not repaired through normal and good maintenance, physical depreciation could be severe at the time of loss. On the other hand, some parts of a building such as roofs, exterior and interior decorations, and to a degree plumbing and heating, deteriorate with age and are subject to depreciation at the time of loss—part of their useful life has been used up. If a roof has a normal expected life of 15 years and is destroyed by fire or wind at the end of 5 years the depreciation (deduction for betterment) would be one third. The same principle applies to decorations. These two particular items in a building, though otherwise well cared for, are subject to depreciation based on life-expectancy. An exception to this general rule is explained under the heading "Replacement Cost Coverage."

Another form of depreciation, *obsolescence,* applies more to the whole building rather than its parts in adjusting a loss. Obsolescence reveals itself in buildings which are architecturally passé or out of date and generally no longer serve their original purpose. A fine old theatre used for storage or for light manufacturing; mansion type dwellings with ornate woodwork, ceilings and numerous useless areas; and buildings condemned as unsafe for occupancy are examples of obsolescence. Changes in neighborhoods and in zoning laws can result in existing buildings showing obsolesence.

Obsolescence does not necessarily mean a building has little or no value. It does, however, affect its *cash value.*

3. Loss Not Caused by an Increase of Hazard

An insurer accepts a risk with known exposure to hazards existing or contemplated. If a loss occurs at a time when the hazard or hazards are increased *within the knowledge and control* of the insured, the insurer will not be liable for such loss. Such conditions can border on technicalities which may be repugnant to a court of law. They also can be serious increases of hazard on which the insurer may wish to deny coverage. The increase must be material and not insignificant. A change of occupancy from a bakery to a manufacturer of fireworks would be material. A change of occupancy from a bakery to an auto repair shop may not be. Any change in occupancy or hazard should be routinely reported to the insurer.

4. No Increased Cost of Repair by Reason of Local Ordinances

Many cities and towns in their building codes or local ordinances have special requirements when a building is damaged in excess of a specified percentage by fire (some include other perils such as windstorm). In some cases, when electrical wiring that does not comply with the latest code, is damaged partially it must be replaced completely to comply with the code. In other situations, if a building is damaged in excess of the specified percentage of its value, the building must be demolished and rebuilt. Sometimes chimneys, fireplaces, and heating devices must be torn out and replaced to meet requirements; fire-proofing and sprinklers are often required after a loss.

Any increased cost of repairs over and above what was there at the time of loss is not covered under the policy.

5. Vacancy or Unoccupancy Clause

The standard fire policies and many other policy forms limit or extend the period the premises can be vacant or unoccupied. Some forms tie vacancy or unoccupancy into specified perils as in the HO forms as it is considered an increase of hazard. Other forms grant full coverage during vacancy—unoccupancy thereby negating the standard fire policy conditions.

6. Replacement - Cost Coverage

This coverage, either in the form itself or by endorsement, supersedes the *actual cash value* limitation in that it grants full replacement without deduction for depreciation provided certain provisions and conditions in the form are complied with. Those provisions and conditions are as follows in most forms or endorsements.

a. The insured will maintain insurance equal to a specified percentage of the full *replacement-cost* of the building (in some forms the actual cash value).

b. Liability is limited to replacement-cost of the property or any part thereof identical with the property on the same premises and intended for the same use.

c. Or, liability is limited to the amount actually and necessarily spent to repair or replace the property.

d. Actual repair or replacement usually must be completed before payment of the excess over the depreciated value is paid.

e. Insured may claim the actual cash value loss and within a specified time make additional claim under the replacement-cost endorsement or clause.

There are some variations in the wording and provisions and conditions of various replacement-cost coverages. They should be carefully scrutinized. Some forms exclude certain property from the full replacement-cost coverage. The Homeowners Policies exclude carpeting, cloth awnings, domestic appliances and outdoor equipment. Other forms exclude such things as floor coverings, air conditioners, and outdoor equipment.

The provisions respecting coinsurance clauses as related to replacement-cost coverage should be examined carefully.

The provisions of these replacement-cost coverages spell out how the replacement-cost of the building, for eligibility, is to be computed, and also how the liability under the form is determined.

7. Coinsurance or Contribution Clauses

Coinsurance or contribution clauses in a policy limit the amount the insurer would be called upon to pay if adequate insurance were carried. Another way of saying this is that the insurer is protected from paying maximum losses while getting minimum premium. A coinsurance clause in a policy, for which a lower rate is charged, requires the insured to maintain an amount of insurance equal to a stipulated percentage of the actual cash value. If he neglects or fails to do this he becomes a *coinsurer* of any loss. The usual percentages are 80, 90 and 100 percent, the reduction in rate increasing with the higher percentages. A simple illustration would be a situation where a building has a cash value of $50,000 with insurance of $30,000. There is a loss of $8,000. The policy has an 80 percent coinsurance clause.

$$\text{Insurer pays} \quad \frac{\$30,000}{80\% \times \$50,000} \quad \times \$8,000 \quad = \$6,000$$

Insured is a coinsurer for the difference $\quad = \underline{\quad 2,000\quad}$
$\$8,000$

8. Requirements of the Insured in Case of Loss

Insurance policies stipulate what an insured must do in case of loss. The standard fire policy sets forth these requirements in some detail. Unless the time is extended by the insurer in writing, the insured must file a *proof of loss*[1], signed and sworn to stating his knowledge and belief as to

a. the time and origin of the loss
b. the insured's interest and the interest of all others on the property
c. the actual cash value and amount of loss
d. all encumbrances

---

[1] See page 535

e. all other insurance, valid or not covering the property
f. any changes in title, use, location, possession or exposure since issuance of the policy
g. by whom and for what purpose any building(s) described were occupied at the time of loss
h. whether or not it stood on leased ground

Insured is further required to substantiate any claim on buildings by furnishing verified plans and specifications. Also, insured, as often as may be reasonably required, must exhibit to any person designated by the insurer, all that remains of the property described in the policy. The latter requirement is not in the standard policies of Massachusetts or Minnesota. Plans and specifications are not required under the standard fire policies of those states nor under the standard fire policy of Texas.

9. Company's Option to Repair or Rebuild

While the insurer has this option it is very seldom exercised unless the insured is wholly agreeable, in fact prefers it. Nevertheless, it is essential to keep the option in mind.

10. Appraisal in the Event of Dispute

Either the insured or the insurer may demand an appraisal to settle disagreement over value and/or loss under policy provisions for such appraisal. The standard policy sets forth the provisions which should be read carefully. Basically, the arrangement is for the insured and insurer each to select a competent and disinterested appraiser notifying the other appraiser within 20 days of the demand. The appraisers then select an umpire, also competent and disinterested. If the appraisers fail to agree on an umpire for 15 days, either the insurer or the insured can request a judge of a court of record in the state in which the loss occurred to select an umpire.

The appraisers then appraise the loss and the value. They submit their differences only to the umpire. An award in writing by any two of the three shall determine the cash value and loss.

CHECK COVERAGE

The importance of checking the insured's original policies cannot be over-emphasized. The foregoing discussion of the differences in the conditions and provisions of policies and forms makes it apparent that such listing and checking of coverage should precede any attempt to prepare an estimate or to enter into a discussion of the adjustment. This is standard operating procedure with most experienced and careful adjusters. Nothing can be taken for granted.

INVESTIGATION AND SURVEY [2]

The adjusters first step in investigating a reported loss is to contact the insured to find out what happened, what property is involved, the insured's interest in the property, what steps have been taken to protect the property against further loss, what insurance is in force, encumbrances on the property and in general to determine the nature and extent of damage.

If everything is in order, the ground rules for procedure can be agreed upon to set the stage for adjusting the loss.

Should there be any question as to coverage, origin of the loss, or similar matters that may raise doubt concerning liability, the adjuster should report to his principal before proceeding further unless he has the authority and knowledge to make on-the-spot decisions.

The following subjects should be reviewed under Fieldwork and Notes, Chapter 3, inasmuch as they have an important bearing in the investigation and on any adjustment in which they are involved.

1. Damage That Existed Prior to the Occurrence of the Loss
2. Damage from a Previous Loss not Repaired
3. Violations of Local Building Codes or Ordinances
4. Repairs Made Prior to Inspection
5. Constructive Total Losses
6. Estimating Losses Caused by Mixed Perils

METHOD OF ADJUSTMENT

Circumstances in each case will govern the adjuster's decision on procedure. There are several methods available:

1. Where damage is minor, agree with insured on the work to be done and the cost to do it; then let the insured do the work, if agreeable.
2. Have insured submit two or more estimates from reliable contractors or subcontractors; check the details, and reach an agreement.
3. Adjuster to make his own estimate and discuss with insured. On run-of-the-mill losses agreement is frequently reached.
4. Adjuster to make his own estimate and discuss with insured's contractor or subcontractor to reach an agreement.

---

[2] Reed-Thomas, *Adjustment of Property Losses,* Mc-Graw-Hill, Inc., 1969.

5. Adjuster to engage a contractor to prepare an estimate under his guidance. Confront insured with the estimate, and a proposal that *insured* hire the contractor to do the work.
6. Have a reliable builder, painter, etc., do the work with insured's written approval on a cost-plus basis not to exceed a specific amount.
7. Have insured obtain an estimate, and adjuster to have a contractor or subcontractor submit an estimate. A meeting is then held between contractors with adjuster present to iron out differences.

PREPARATION FOR DISCUSSING THE CLAIM

On losses of any magnitude it is advisable to analyze estimates submitted, to substantiate a claim, before attending a meeting to discuss details with the insured or the insured's representative. It will save considerable time and when done in advance better concentration on details without distractions will produce a better result. Also, the essential points for questioning or discussion can be listed.

It is not unusual to be able to attend a meeting and state that the estimate submitted is in order with the exception of a couple of specific items that need better explanation. At other times several hours of discussion will be required to reach a common ground on which the adjustment can be concluded.

A building estimate that is submitted as a basis of claim under a property damage policy is subject to the same analysis and audit as any other type of claim made against an insurance company. The carrier is entitled to know what repairs are contemplated, the kind and quantity of materials, the price charged for the materials, the cost of the labor to install them and the profit and overhead that is included in the estimate. The company stands in no different position in this respect than a prospective buyer. They cannot be dealt with as strangers by submitting lump sum figures or estimates that are lacking in detail. The person who has sustained the loss is entitled to the same detailed explanation in the estimate so that he may know exactly what the repairs to his property will consist of, and what they will cost. There are still many contractors and builders, however, who are of the old school which held that the items, materials, and labor in an estimate are personal and private information not to be disclosed to anyone. An estimate which states a single sum of money or a series of lump sums without supporting details contributes very little to the insured or the insurer as a means of adjusting a loss.

As a result of many years of experience in adjusting building losses, those who prepare estimates of damage for either the policyholder, or the insurance company, or both have come to appreciate the advantages of a completely detailed estimate. While there may be differences of opinion about what work is necessary to restore a property, or what is a proper charge for the materials and labor, the detailed estimate does provide an intelligent basis for reconciling those differences. Without it, there is nothing to discuss except the adequacy of the lump sum amounts. This leads to trading tactics rather than to a proper adjustment that is equitable to all interested parties.

It is, therefore, essential that, before an audit or an analysis can be made, the estimate must be detailed. All sizes of rooms or areas should be indicated so that quantities may be checked. The room-by-room listing of items recommended in Chapter 2, "Fundamental Considerations," should be followed where applicable, with all the quantities shown in number of squares, square yards, square feet, and so forth. *Unit costs,* wherever used, should be indicated because it will save the person analyzing the figures the time required to compute them by dividing the cost by the quantity. Subcontractors' estimates, submitted in support of the general estimate, also should be itemized regarding the work to be done and the cost of each item. Electric, painting, plumbing, or heating sub-bids that show a single sum for several hundred or several thousand dollars are of little value in reconciling them against other estimates or sub-bids. A loss may be only partly adjusted if half of the figures are analyzed while the other half are negotiated on the lump sum estimates of subcontractors. In many of the larger cities it has become standard procedure for subcontractors to itemize their bids. They have come to realize that any discussion of their figures requires the details of the repair work contemplated and also the cost of each operation expressed in labor and materials.

PURPOSE OF ANALYZING ESTIMATES

Verification of the details of a building loss estimate is as important in adjusting a claim for damage to real property as is verification of the cost of repairs or replacement to household goods when adjusting a personal property claim. The insured and the insurer are both entitled to know how much plaster is to be repaired or replaced, how many brick or board feet of lumber are required, or how many square feet of painting is contemplated in restoring a building. Both have the same interest in the adequacy of the quantities as well as in the propriety of the charges for the labor and material. The only way of knowing that the quantity, the quality, and cost of the materials are sufficient to repair or replace the property is to subject such details to competent scrutiny and analysis.

No loss estimate may be considered inviolate if for no other reasons than the possibility of human error in judgment, in measurement, in knowledge, or in powers of observation. As has already been pointed out, contractors who compete in estimating new construction, using the same plans and specifications, may differ as much as 30 percent in their total figures. In estimating the cost of repairing building losses, each contractor writes his own specifications and takes his own measurements, thereby increasing the possibility of errors or differences of opinion. Since estimating by definition consists in approximating, none is beyond questioning, and the ultimate test of accuracy must in reality await the completion of the work.

The most common errors that are found in building loss estimates generally may be classified under 23 headings. All loss estimates should be examined to determine whether or not they contain any of these errors, bearing in mind the following important consideration: *All estimates contain items that are excessive and also items that are lower than the probable actual amount of money needed to restore the property. It would not be proper, therefore, to reduce all items which appear high without at the same time increasing those items which appear low.*

Because of the variation in adjusting practices throughout the country, the analyzing of building loss estimates may be required when:

1. One or more estimates submitted on behalf of the *insured* are to be compared with each other and checked against the loss.
2. A single estimate that has been submitted on behalf of the *insurer* is to be checked against the loss.
3. An estimate submitted on behalf of the *insurer* is to be compared with another that has been submitted on behalf of the *insured*.
4. A single estimate has been submitted on behalf of both *insured* and *insurer*.
5. A set of details and/or bills have been submitted after repairs are completed.

SOURCES OF ERROR IN ESTIMATES

Very few building estimates are made that do not contain errors of one kind or another—either in judgment or in the mechanics of preparing the figures. When a builder submits a competitive construction bid on a specific job, if he is not awarded the contract, he may never know what mistakes, if any, were made in his estimate. If he is awarded the contract on the basis of his figures, mistakes that were made in estimating will very likely be disclosed before the work is completed, provided, of course, that he keeps careful cost records and checks them against his original details.

Estimates of building losses usually are not on a sealed bid basis. Their main purpose is to provide a means for determining the amount of loss which an insured has sustained, so that he may be indemnified adequately under his policy of insurance. Once the dollar amount of loss and damage has been agreed to, the insurer is not concerned with the method of restoring the property, nor in the precise disposition of the money that has been paid. That is a matter between the policyholder and whatever builder he eventually chooses to employ to do the work.

A building loss estimate that is presented by an insured is, therefore, a claim under his policy. It becomes an instrument subject to examination for errors or mistakes made by the person who prepared it.

Experience has demonstrated that most mistakes fall under one of twenty-three classifications, which may be subdivided into three major headings:

1. Taking field notes.
2. Specifications of work to be done.
3. Computing the cost of the work.

It is seldom possible to check an estimate accurately without visiting the scene of the loss in order to verify measurements and specifications. Two contractors who have made estimates can compare their details but, in the event of a disagreement, they will generally have to reinspect the damaged property to reconcile differences. It is recommended that the checking of estimates involving anything more than minor repairs be done at the loss. Careful and orderly notes should then be made of all additions, deletions, and adjustments in the original figures so that an intelligent explanation may be given at any time as to the exact basis of the revisions. An insured may have an exaggerated estimate given to him by a contractor and, if it is later substantially altered in amount, he is thoroughly justified in requesting a detailed accounting of the changes. Whenever a single estimate is being checked out by an adjuster, his detailed notations of changes in quantities, costs, and so forth, should be recorded in a manner that can be clearly understood by anyone who might have occasion to review them.

The following list and the specific explanation of each source of error in estimates should be studied carefully so that they will automatically come to mind when an estimate is being analyzed or prepared.

Errors Made in Taking Field Notes

Wrong measurements and dimensions.
Inclusion of items that do not exist.
Inclusion of damage from a previous unrepaired loss.
Inclusion of property which is not covered under the policy.
Inclusion of previously existing structural defects.
Inclusion of damage not caused by an insured peril.
Overlooked items of repair or replacement.

Errors Made in Specifications

Including improvements to the property.

Replacing items that can be repaired.

Including work to comply with local ordinances, or building code requirements.

Not replacing with like kind and quality of material.

Allowances for unseen damage (contingencies).

Over emphasizing details.

Duplications by subcontractors.

Duplications of repairs covered under contents insurance.

Errors Made in Computing the Cost of the Work

Improper consideration given to the *class* of workmanship involved.

Inefficient methods and equipment used by builder.

Using improper wage rates.

Materials not priced correctly.

Inadequate or excessive labor allowances.

Inadequate or excessive overhead charged.

Inadequate or excessive profit charged.

Arithmetical errors in computations.

Wrong Measurements and Dimensions

An error that has been made in taking measurements, or in recording them, will result in a corresponding error in the cost of the items based on those measurements. If a room in a dwelling is actually 12' x 16' x 8' (high), but it is shown in the estimate as 13' x 17' x 9', the difference in the gross wall and ceiling areas will be 121 square feet.

	(A) *Room 13' x 17' x 9'*		
Wall area	2(13' + 17') x 9'	= 540 sq ft	
Ceiling area	13' x 17'	= 221 sq ft	
			761 sq ft
	(B) *Room 12' x 16' x 8'*		
Wall area	2(12' + 16') x 8'	= 448 sq ft	
Ceiling area	12' x 16'	= 192 sq ft	
			640 sq ft
		Difference	121 sq ft

While only one foot has been added to the length, width and height, it results in a gross wall and ceiling area of 19 percent more than actually exists. If such a room were to be plastered, painted, and a new 1" x 3" oak floor put down, the difference in the cost would be substantial.

Illustrative Example

Assume that lath and plaster is $.40 per sq ft ($3.60 per sq yd), painting 2 coats is $.10 per sq ft, and that the cost of laying 1" x 3" oak floor is $.50 per sq ft

Plastering		Area	Cost @ $.40 sq ft
Room (A)	13' x 17' x 9'	761 sq ft	$304.40
Room (B)	12' x 16' x 8'	640 sq ft	256.00
	Difference	121 sq ft	$ 48.40
New Oak Floor		Area + Waste	Cost @ $.50 sq ft
Room (A)	13' x 17' x 9'	295 FBM	$147.50
Room (B)	12' x 16' x 8'	256 FBM	128.00
	Difference	39 FBM	$ 19.50
Painting		Area	Cost @ $.10 sq ft
Room (A)	13' x 17' x 9'	761 sq ft	$ 76.10
Room (B)	12' x 16' x 8'	640 sq ft	64.00
	Difference	121 sq ft	$ 12.10

Summary of Differences

Plastering	$48.40
New oak floor	19.50
Painting	12.10
Total	$80.00

When overhead and profit of 20 percent is added, the difference is further increased to $96.00. It is readily seen that if a number of rooms are involved, and measurements are improperly taken or recorded, significant differences will develop even though identical unit costs are used.

Most errors in measuring lengths, widths, and heights, are the result of not carefully reading a tape or a rule. They also occur when attempts are made to pace off distances, outright guesses are made, or an inaccurate sketch of the layout is prepared.

Dimensions of the *sizes* of materials usually make a difference in their cost which is directly proportional to the error in measurement. If a 6" concrete slab floor is recorded as being 8" thick, the increase in the cost will be 33 1/3 per cent. When 2" x 8" floor joists are assumed to be 2" x 10", the cost will be increased 25 percent, because a lineal foot of 2" x 10" has 1 2/3 board feet, while a lineal foot of 2" x 8" has 1 1/3 board feet. The difference is 1/3 FBM which is 25 percent of 1 1/3 FBM. The accurate measurements of framing lumber, heavy mill timbers, and the thickness of boards, flooring, insulation, glass, metal flashing, linoelum, roofing materials, gypsum sheet rock and lath, doors, and windows are essential to determine the proper prices to be used in the estimate.

Inclusion of Items That Do Not Exist

An estimate may include one or more items that did not exist at the time of the loss, but which the person who made the estimate assumed were part of the building. This often happens when a structure is totally destroyed by fire and the estimator reconstructs it on paper in accordance with what he considers good or standard building practices. Such items as bridging, subflooring, sheathing, collar beams and fire stops may appear in an estimate but may not have been in the destroyed building. While in some instances assumptions must be made, many times questioning of various persons who were acquainted with the building, perhaps the original builder, and also painstaking examination of the debris will disclose the true conditions. It is not unusual for owners or tenants to remove doors between rooms, or excess radiators, and place them in the garage or basement for storage. In a seriously damaged building an estimator may automatically include such items in the cost of repair and replacement.

Inclusion of Damage from a Previous, Unrepaired Loss

An insured who has suffered a building loss does not always make full repairs, particularly if the damage is not structural. If it can be concealed satisfactorily by covering up or, in the case of charred timbers, by scraping and whitewashing or painting, he may be content to spend no more than is necessary. The same condition may be found in the case of damage to a roof by hailstorm or wind. Following an adjustment, the insured may decide to patch the roof, or do nothing at all. A second loss may later involve the same roof.

Discovery of a previous unrepaired loss is made either on direct inspection, through questioning the insured, from the personal knowledge of the one making the inspection, or from some outside source. When such a condition is suspected, anyone having information about the previous loss should be carefully questioned.

These persons may be the:

Insured, owner, or tenant, at time of previous loss.
Agent or broker for the insured.
Adjuster who handled the previous loss.
Contractor who estimated previous loss for the insured or insurer.
Person who made repairs if any were made.
Neighbors.
Local fire department.

Whenever possible the estimates, adjuster's report, and other papers in connection with the adjustment of the previous loss, should be secured and reviewed to determine what items of repair were included and the cost of those repairs. The purpose of investigating these things is to avoid making payment a second time for repairs of the old damage.

Any evidence of a previous loss that can be disclosed by inspection will depend on the physical conditions. After a fire, charred wood, for example, maintains about the same appearance over a period of years. The old char, however, will collect dust and cobwebs and can be distinguished from newly charred wood. Where old char has been scraped and whitewashed or painted, the change in the appearance is readily noticeable. If, in patching charred joists, rafters or studs, new wood has been nailed to the damaged timbers, the difference in appearance, even though exposed to smoke and heat, is easily recognized. The most common places where fire damage is repaired on a patch basis will be found to involve floor joists and roof rafters which have been slightly charred. As a rule considerable savings can be made by scraping and whitewashing, or covering up with wall board.

When examined, a roof surface that has been damaged by hail and not repaired can usually be detected by the pitting. In areas of the Middle West states where hailstorms are not infrequent, it is good policy to carefully check roofs for a previous unrepaired loss.

Roofs that have had wind damage of a substantial nature are generally replaced by the owners. However, sometimes a full allowance may have been made for an entire roof which has had rather severe scattered damage over its surface. If it has been patched up rather than replaced, the person making an inspection following a second loss may be completely unaware that the old roof has been paid for, unless his attention is attracted to numerous new or off-color shingles that have replaced those that were damaged. (See also page 90, Chapter 3.)

Inclusion of Property Which Is Not Covered Under the Policy

Many of the builders and contractors who make estimates of a loss are not familiar with all of the conditions of the policy relating to the property covered. They may inadvertently include the repair or replacement of items that are not covered, or which are specifically excluded. They may also *omit* items that are considered part of the building but appear on the surface to be contents. Wall-to-wall carpeting, for example, under certain conditions is a building item and should be included in the estimate.

The inclusion of items that are not covered by a policy is more common in estimates of damage to buildings occupied for mercantile and manufacturing purposes. Tenants' property, or improvements and betterments covered by the tenants' insurance, is frequently confused with the property of the building owner. Duct work for collectors, ventilating and air-conditioning equipment, power wiring, shelving or trade fixtures belonging to tenants, are often found in an estimate of damage to the building. A careful analysis of items is required to eliminate those that are not covered by the policy.

Inclusion of Previously Existing Structural Defects

When a builder or contractor makes an inspection of property damaged by an insured peril, his primary interest is to estimate the cost of putting the building back in a sound condition. Many buildings, particularly old ones, may have structural defects that are in no way related to the subject loss, and were in existence before it occurred. Since the person making the estimate prepares his own specifications for restoring the property, he may unintentionally include the cost of repairing the previously existing defects. Such items may include sagging floors or roof timbers, cracked masonry walls or plaster, defective roofs and flashing, broken window glass, or eroded mortar joints in brickwork. Only a careful check of the items at the place of loss will disclose the inclusion of these and other similar items. (See also page 89, Chapter 3.)

Damage Not Caused by a Covered Peril

Estimates will sometimes specify the repairing or replacing of items which on close inspection will be found caused by an uninsured peril. A very common damage that occurs during windstorms accompanied by rain, and one which is not covered under policies that insure the specific peril of windstorm, is that caused by rain driven in around leaky windows, under doors, and through leaky or defective roofs, flashings and skylight. High water, wave wash, backing up of sewers, and seepage of ground water through basement walls are excluded under most forms. Electrical breakdown involving motors, generators and panel boards are also excluded under most policies of insurance. Damage from a friendly fire (smoke from a bonfire or an incinerator) is a type of loss that is not covered in most cases. It is therefore important for the person making an estimate to be familiar with the coverage as well as exclusions in the form involved in order to relate the damage sustained with a peril insured against.

Overlooked Items

Persons who make estimates of building losses may overlook items as a result of haste in taking off the details, or not making a thorough inspection. In some instances the ownership of property is taken for granted as to whether it belongs to the building owner or tenant. Failure to inspect possible damage to out buildings or yard fixtures which may also be covered under the policy, leads to the omission of damaged items.

Failure to Allow for Salvage

Occasionally certain building materials have a salvage value, mainly as scrap. Iron, steel, copper cable, or roofing in quantity may have a salvage value which should be credited to the loss estimate. Much will depend on the labor necessary to get it out of the premises, and on the location and prevailing condition of a market to dispose of it.

In some sections of the country common brick has a good salvage value where labor for cleaning is not high, and where there exists a ready market. The market is best in a region where there is a demand for second-hand brick for veneer on new homes.

Inclusion of Improvements to the Property

The occasion of a serious loss to part or all of a building frequently presents an opportunity for the insured to make changes in the construction or layout of the building. These changes may be minor such as cutting a door into a partition where none existed prior to the loss. They may, however, represent major changes such as extending the building, closing in an open porch, putting up a fire wall, laying a concrete floor in a secondary structure that previously had an earth floor, installing cupboards where there were none, or even putting in another bathroom.

Many times an insured is unaware that these changes will cost more than it would to restore the property as it was prior to the loss. Often the builder or contractor inadvertently follows "the insured's" instructions when preparing his estimate of the damage. As a result he may include a number of changes or improvements. Generally an explanation that the loss under the insurance policy is limited to putting the property back in the same condition as it was prior to the loss will result in the items of improvement being deleted from the estimate.

The task of the person analyzing the estimate is not so much that of identifying structural changes or improvements, as it is properly excluding them, because the labor and materials may be intermingled with other general repairs. In many cases the best approach is to refigure the cost of repairing the damage strictly on replacement.

Replacing Items That Can Be Repaired

If a structural part of a building can be repaired satisfactorily, the estimate should be made on that basis rather than figuring complete replacement. When a stile or a panel of an expensive door has been damaged, and it can be satisfactorily restored by replacing the stile or panel, the cost should be limited to repairs. Broken sash or mullions in a window do not necessarily mean that a complete new window is required. Finished floors that have been chopped can usually be repaired by cutting out the damaged section, inserting new boards and refinishing the entire floor. Many times framing timbers that are broken or burned can be spliced, or new timbers installed alongside those that have been damaged.

Spalled or broken masonry units can often be cut out and new ones set in place. Most builders experienced in working on damaged buildings have numerous and unique methods of making repairs. The shoring of upper floors, partitions, and fixtures, while new joists, sills, or plates are installed below, seems hardly possible to one who has never encountered the problem; yet it is commonplace to those who have done such work.

Building equipment such as elevators, exhaust systems, ventilating and air-conditioning units, plumbing and heating fixtures that are damaged should be carefully examined by an expert or a factory representative when the items are seriously affected, to determine whether they are subject to repair or must be completely replaced.

Wallpaper, wall paint, murals, asphalt, plastic or vinyl tile, ceramic tile, masonry and other surfaces that have been soiled by smoke or other stains should be test-cleaned to determine if washing with solutions will restore them. The test-cleaning frequently can be done by the person making the inspection. In doubtful cases experts in cleaning have to be called in to experiment.

Including Work to Comply with Local Ordinances

Because most policies exclude any increased cost of repairs or reconstruction made necessary by local building ordinances it is important to eliminate such items from loss estimates. This means that the specifications of repairs should be confined to work necessary to put the property back with materials like those that have been damaged or destroyed. If there is a combustible enclosure around a heating boiler, and the building ordinance requires that it be replaced with one which is fire-resistive, the difference in cost must be borne by the insured. Many older properties are wired with knob and tube or 14 gauge electrical cable whereas the ordinance stipulates that, if it is damaged by fire, the affected portion or in some situations the entire building must be rewired with 12 gauge cable. The estimate of repair should be confined to repairing or replacing that portion which has been damaged, and the cost of the material should be based on the same kind of wiring that existed at the time of loss. The same procedure applies to distribution boxes, service entrance wiring and fixtures.

Problems sometimes arise when a fire involves an unlined chimney, or one which is built contrary to the local ordinance. The estimate should not include the increased cost of complying with the requirements of the ordinance. Sometimes parapet or fire walls are specified by the building code if a structure is partially damaged or entirely destroyed. In congested business areas, or in public buildings, automatic sprinklers, stairway and elevator enclosures are a prerequisite to restoring the property. All of these increased costs are excluded under the policy conditions unless provided for by endorsement. (See also page 91, Chapter 3.)

Not Replacing with Like Kind and Quality of Material

The fire policy limits recovery to repairing or replacing with like kind and quality of material. While the quality of materials in a building are not always readily determined, even by experienced persons, every effort should be made to estimate repair costs on the basis of the same kind and quality of materials that existed at the time of the loss. A careful examination of wallpaper, framing lumber, wood trim, plaster thickness, wood floors, glass thickness, type of paint, grade of shingles, number of plies in a built-up roof, and grades of other materials is necessary before a proper price can be determined. It is not unusual for an estimator to take samples of shingles, wallpaper and other items in order to have the quality confirmed by a local supplier.

Allowance for Unseen Damage—Contingencies

Builders who are unfamiliar with repairing structures that have been damaged, particularly by fire, frequently include a lump sum of money for "contingencies." Being inexperienced, they are fearful of encountering damage which is unseen, or which might be overlooked when they made up their estimate. While there are instances when it is impossible to accurately determine the work necessary to restore a property, the occasions are rare. When uncertainties present themselves, it is generally obvious to all concerned. It may be a dwelling that has been twisted from its foundation by a hurricane or tornado, or spot patching the shingles on a large expensive type of roof. Where an item for contingencies appears in an estimate, the person who is analyzing the details should call for a reasonable explanation.

Overemphasis of Details

Illustrative Example

A room 12' x 16' x 8' is to be painted two coats. On a gross area basis the net cost is estimated to be $.107 per square foot. This is based on painter's wages of $6.00 per hour and the cost of paint at $8.00 per gallon. The area of the room is 640 square feet. The total cost would be 640 x $.107 or $68.48. A second painter estimates the cost in the following manner:

Material

4 gal paint	@ $8.00 = $32.00	
1 Box detergent	.50	
patching plaster	1.00	
1 lb putty	.50	
	$ 34.00	

Labor

Remove furniture and pictures	3.50
Prepare walls	7.00
Putty holes in trim	3.50
Patch cracks in plaster	3.50
Paint wall and ceiling	30.00
Paint 2 windows and 2 doors @ $3.50	14.00
Paint baseboard	3.50
Remove and reset radiator	1.50
Paint radiator	3.50
Clean up	7.00
	$ 77.00
	$111.00

While accuracy in estimating depends largely on a careful breakdown of material and labor for each item, over-detailing of items can result in a high figure. The reason is that too much detail produces overlapping and also slightly excessive amounts which are cumulative.

This illustration is exaggerated to emphasize what over-detailing is, and its effect on the over-all cost. It is not unusual for estimates to contain several small items to build up the cost of painting an ordinary room.

Each item in itself, though figured on a liberal basis, may not appear excessive unless carefully analyzed. Yet the effect is a cumulative error which in the aggregate makes a substantial difference. Over-detailing is readily apparent to an experienced estimator, but can be overlooked by the less experienced person unless he understands what to look for and learns to recognize the characteristics.

Duplications by Subcontractors

Among the numerous reasons why a subcontractor's estimate should be in detail, one of the most important is to detect and eliminate duplications of work and materials which are also included in the estimate of the general contractor or other subcontractors. In most instances an electrician, plumber, plasterer, painter, or mason, when called in to prepare an estimate as a subcontractor will make up his specifications without the guidance or supervision of the general contractor. He takes off the loss as he sees it, including all of the repairs which he feels fall within his particular trade. The result is that there is frequently overlapping with other trades. A plumber may take in duct work, not knowing that a sheet metal worker has been asked to estimate that item. Painters may include glazing, an electrician may include elevator or air-conditioning wiring and motors unaware that they are being figured by someone else. A floorlayer may include sanding, which is being taken care of by a floor sander. Subcontractors will often include the tearing out of damaged materials pertaining to their trade, if they are not told that it has been included by the general contractor in his item of "wrecking and debris" for the entire job.

Most subcontractors add overhead and profit to their estimate. A general contractor is entitled to his normal profit on top of the sub-contractor's bid, but seldom is he also entitled to add all of his regular overhead.

For these reasons it is always advisable, when analyzing or checking estimates, to review the specifications on which subcontractors base their figures.

Duplications of Items Covered Under Contents Policies

There are a number of items in a building that normally might be considered solely part of the structure but which under certain conditions, may also be wholly or partly covered under the contents insurance. The provisions in a lease between a landlord and tenant will often offer a solution for the proper disposition of such items. In the absence of a lease, ownership may be established by common law. The Guiding Principles, designed to take care of overlapping coverages, may be the basis for determining whether a portion or all of the cost of the particular item falls upon the building or the contents insurance.

There are items in mercantile and manufacturing buildings that have been installed by a tenant and pertain to the occupancy: these may be improvements and betterments of a permanent nature; they may be removable trade fixtures or equipment that can be dissembled and taken out when the tenant vacates the premises. While it would not be feasible to list all of the items that are frequently covered by a tenant's insurance, the following are those most generally encountered:

Improvements and Betterments
Painting and decorating.
Floor coverings (tile, carpeting, etc.).
Partitions.
Store fronts.
Paneling.

Improvements and Betterments (Continued)
> Air-conditioning systems.
> Elevators and escalators.
> Electrical light wiring.
> False ceilings.
> Plumbing (sinks, toilets, etc.).
> Heating plants.

Trade Fixtures
> Shelving.
> Stock bins.
> Display booths.
> Electrical light fixtures.
> Awnings and signs.

Equipment
> Power wiring.
> Blowers, ducts, etc.
> Air-conditioning units.
> Refrigeration equipment.
> Ranges and stoves.
> Heating equipment for processing.
> Plumbing equipment for processing.

An estimator who is unfamiliar with policy coverages, may include these and similar items in the building estimate. At the same time they may appear in an inventory or in an estimate of damage to the contents. There is no way to eliminate overlapping except by careful comparison of the building and contents claims. When the occupant is also the building owner, very little difficulty is encountered. Improvements and betterments are automatically eliminated as they are part of the building and normally the owner and mortgagee are the only interested parties.

A separation of fixtures and equipment from the building estimate can be accomplished best by conference. When a tenant occupies the premises, an effort should be made to get together all of the interested parties, including those who represent the contents insurance carrier, in order to make comparisons and through questioning and examining of leases reach an understanding.

Improper Consideration Given to Class of Workmanship

A contractor or builder, accustomed to erecting or altering expensive dwellings where a high grade of workmanship is a requirement, may unwittingly apply the same standards to lower grade or secondary structures when estimating the cost of repairing them after a loss. There are some contractors who are unable to estimate on cheap buildings because they are not set up to handle anything but work of the highest quality.

Their workmen are usually skilled craftsmen who take pride in each job they do. They carefully plan and lay their work out with deliberation. Their rate of production may be relatively slow, but the finished product is of high quality. Workmanship of this kind is more costly than that turned out by the average mechanic.

Conversely, the contractor or builder who normally works on cheaper buildings is handicapped when he comes to estimate damage to an expensive residence where only the finest quality of workmanship is acceptable.

Failure to give proper consideration to the quality of workmanship that a building warrants leads to overestimating or underestimating repair and replacement costs. Many times differences in estimates are impossible to reconcile until each estimator has made it clear, either in his specifications or through inquiry, precisely the grade of workmanship contemplated in his figures. Since there is usually no great variation in the cost of materials, the difference in the quality of workmanship will be disclosed in the cost of labor.

Inefficient Methods and Equipment Used by a Builder

The repairing of buildings that have been damaged is a specialized branch of construction. Experienced builders can visualize clearly what has to be done and how they will proceed with the work. They quickly recognize operations that lend themselves to short cuts and they note, while making their estimate, places where time-saving tools and equipment can be put to use. They have available portable table-saws, electrical hand-saws, drills and mortising machines, aluminum ladders, scaffold brackets, concrete and plaster mixing machines, and other labor saving equipment. A contractor who lacks modern equipment will, as a rule, have a higher labor charge for operation that can be performed more rapidly by use of labor-saving devices.

Using Improper Wage Rates for the Locality

The hourly wages in the building trades vary throughout the country. The union scale for a particular trade can be ascertained through local union offices, supply yards, and builders. There should be no difficulty in determining the prevailing wage rate for any trade in a local area except in cases where non-union labor exists. In that type of area the rates for a trade may vary in accordance with the skill and ability of the individual. The wage rate for a first grade man however is usually uniform. Whenever a person is estimating or checking an estimate in a territory with which he is unfamiliar, he should inquire of reliable sources the scale of wages for the different trades involved in the job.

Materials Not Priced Correctly

Most errors in the pricing of materials are the result of a wrong description or dimension of the material, or surmising what the price is instead of verifying it at a local source of supply. It is important at all times to be certain that the material is properly described as to kind, quality, size and dimension, and that the correct local price is used.

Inadequate or Excessive Labor Allowances

The most common source of error and disagreement in building loss estimates is in the number of hours of labor allowed for doing the work. The subject has been discussed at considerable length throughout this book. The reader is urged to review those sections dealing with labor under Chapter 2 and Chapter 3, and also those chapters that deal with specific building trades.

Moderate differences of opinion are inevitable when judgment and individual experience play such an important part. However, rates of production that are wholly inadequate for the work to be performed, or which are exaggerated to the point of excess, should be questioned. As demonstrated under the analysis of labor, it is not difficult to show up labor rates that are unrealistic by relating them to a specific area to be covered, or to a specific quantity of material to be installed. When serious differences arise regarding a proper labor allowance, it is advisable to consult with a competent specialist in the particular trade, rather than to resort to a distasteful compromise which could readily establish an undesirable precedent.

Inadequate or Excessive Overhead Charged

The overhead of a contractor includes the various general expenses which are not allocated to a particular operation. (See Chapter 2.) Most of the contractors show at the end of their estimate an overhead charge of 5 or 10 percent of the total labor and material cost. When an item of overhead does not appear as such, it is probably included in the unit costs or is buried in the various items throughout the estimate. If it is not shown anywhere in the details, the person who made up the estimate should be questioned concerning how it has been treated. On small repair jobs it is not unusual for a painter, roofer, or plasterer to work for wages plus a small profit, in which event he may not add anything for overhead. Normally when the charge for overhead exceeds 10 percent of the combined labor and material cost, an analysis should be made to determine whether or not it is justified, and also that the charges included are in fact true overhead items. There are overhead charges which may range as high as 30 percent or more of the job cost, but these are usually confined to certain types of subcontractors for specialized equipment such as elevators, refrigeration systems, certain power installations and boilers. The actual parts and labor costs may be nominal, but additional charges are made for an engineer's travel time and expense from the factory, plant overhead, and so forth.

It is not usual for a general contractor to charge his full overhead on top of a subcontractor's estimate. It is customary for him to charge his profit. In certain instances that require careful supervision of a subcontractor, the general contractor will add a supervisory charge in addition to his profit.

Inadequate or Excessive Profit Charged

Much of what has been said regarding inadequate or excessive overhead also applies to profit. When it is not shown as an item at the end of the estimate, it is included, in all probability, somewhere in the main body of the estimate. It may be in the unit costs or in the various items or, if labor is in a lump sum, it may be found there. Only by analysis and by questioning the person who prepared the details can it be disclosed. An estimate which shows a profit and overhead of 10 percent, or in some cases 15 percent, will be found to contain, as a rule, an additional allowance buried in the details of the estimate.

Throughout the country an acceptable profit for the run-of-the-mill work is 10 percent of the combined material and labor cost. There are variations to this. (See Chapter 2.)

Arithmetical Errors in Computations

A building loss estimate is an arithmetical computation of the material and labor required to repair or replace a damaged structure. It contains numerous operations involving simple addition, subtraction, multiplication, and division. Many of the computations are done in the head, many on scratch paper, and frequently they are done on a calculating machine. The possibility of human error is always present. Even when an adding machine is used, items may be omitted, or added twice. Subtotals are not always cleared in the machine before starting a new series of additions. A machine that records on a tape is always best because the figures on the tape can be doublechecked against the ones in the estimate. It is considered good practice to attach the tape to the original estimate sheet from which the figures were taken. In this way verification can be made by anyone several days or weeks later should the occasion arise.

One should not assume that the quantity computations, the extensions, or column additions are correct. There are many occasions when a single estimate is being checked, or two estimates are being compared, and they appear to be entirely in order as to detail sizes, unit costs, and so forth. When the arithmetic is checked, however, they are found higher or lower than indicated in the total.

When making an estimate, *all* arithmetic computations should be doublechecked and, when practical, it should be done by another person. When checking estimates, the operation cannot be considered complete until all arithmetic has been carefully checked.

THE MECHANICS OF CHECKING ESTIMATES

Anyone who undertakes to check an estimate made by someone else should consider himself to be actually *making* the estimate while using the other person's details as a guide. Nothing in the estimate that is being checked should be taken at face value without verifying areas of repair needed, measurements, sizes, kind and quality of materials, and all other factors upon which the final cost is based.

Checking a Single Estimate

The checking of an estimate should always be done at the place of loss unless the person doing it is completely familiar with every aspect of damage. While it is not always practical to do so, the ideal procedure is to check an estimate in the company of the person who made it, so that questionable items can be inspected and discussed on the spot.

The process of checking requires considerable patience and time. Each item should be reviewed individually. Room sizes, measurements of area, and the kind of materials should be verified. Particular attention should be given to the specifications set forth in the estimate. The checker should be certain that the work specified to be done is adequate to effect restoration, and also that it is not excessive. Because all estimates are generally high on some items and low on others, it is important not to confine the adjustment of the estimate solely to those that are high. Proper consideration must be given to any items which in the opinion of the checker are low.

The estimate should be examined in light of the twenty-three sources of error outlined in this chapter. Many of the errors discussed may be automatically eliminated because of the particular type of loss or coverage. Many others, however, are usually inherent in all estimates.

Checking Two or More Estimates by Comparison

When one or more estimates have been submitted by an insured, and also one or more have been prepared for the insurer, it is good procedure to make a general comparison of the estimates prior to any formal discussion. This will often disclose where the differences are, their nature, and their extent. While differences in estimates may be scattered throughout, it is not uncommon to find major variations in one or two items or trades.

The first step in making the comparison is to prepare a recapitulation of each estimate by trade: that is, by rough and finished carpentry, masonry, plastering, decorating, roofing, sheetmetal, plumbing, heating, electric, wrecking and debris, and any others that are involved. This procedure is explained in Chapter 4. Considerable care should be exercised in checking off each item when it is taken out of the estimate for summarizing under a particular heading. The total of the recapitulation should always be identical to the total of the detailed estimate. If it is not the same, the fault may be in the addition of the items in the original estimate, or it may be the result of omitting or duplicating items in the summary.

It is, of course, essential that any comparison be made on the same basis. If the overhead and profit are in the unit costs or in each individual item of one estimate, it will be convenient for the purpose of comparison to add it into each item of the other estimate, or to add it to the recapped total for each trade. If the wrecking and debris costs are included in each item in one estimate, they should be deleted or they will have to be included in the individual items of the other estimate to make them comparable.

After each estimate has been recapitulated, the differences in the totals for the trades are examined. In most cases one estimate will be higher in some trades and lower in others. Frequently these variations are not great and one will compensate for the other. In other words a number of items may show differences in several trades, but the total of these items in each recapitulation may be close enough to require no further analysis.

When the major differences are found in one, two, or more trades it reduces the work by restricting further analysis to those items. This analysis is best accomplished by reviewing the details of the trades carefully, checking out quantities and costs, and using the twenty-three sources of error previously outlined as a guide.

If the difference in the totals of the estimates is substantial and most, or all of the items in one recapitulation are higher than those in another, the job of determining the reasons can be very time consuming unless one of the estimators has intentionally kept his figures high, or low, for negotiating purposes.

Checking Subcontractor's Estimates

For reasons that are not clear, it is very common to have a completely detailed estimate of everything to be done except the work of a sub-contractor. Estimates of subcontractors are submitted separately or embodied in the general estimate as a lump sum amount. They may appear as "Electrical Repairs," "Plastering," "Painting," "Plumbing and Heating" or some other trade classification.

These subcontract estimates may be sizable items, but in a lump sum form they cannot be checked beyond comparing them with the cost submitted by another subcontractor. This comparison, while looked upon by some as a satisfactory method of checking, is not without pitfalls inasmuch as both could be in error. When there is a considerable difference between the estimates of two subcontractors, they should be called upon to review their details jointly to reconcile differences.

A far more desirable procedure is to explain to subcontractors before they make up their figures that a complete breakdown is required showing materials, labor, overhead and profit. There is nothing more confidential about a subcontractor's estimate than one made by the general contractor. The probable reason for lump sum sub-estimates is the failure to explain to the electrician, plumber, or whoever may be involved, the real purpose of requiring details. Subcontractors' estimates are subject to all of the twenty-three "Sources of Error" outlined and explained in this chapter. When these estimates are in detail, they can be better analyzed.

VALUE FOR INSURANCE REQUIREMENTS

Insurance underwriters are interested in the value of a building covered for three reasons.

1. To assure that they are receiving adequate premium for the exposure.
2. If the policy contains a coinsurance or standard average clause, it is important to know the value of the building, as these clauses impose a limit on the insurer's liability.
3. If there is full replacement-cost coverage, the replacement cost of the building must be established in accord with the policy provisions to determine eligibility for the coverage and also the amount the insured is entitled to under *that* coverage.

It is one of the functions of the adjuster to establish the value under the policy and report it to the insurer. There is also an item to be completed in the Proof of Loss, which the insured executes, stating the *Actual Cash Value* of the property at the time of loss.

Methods for Determining Value

The several "Kinds of Estimates" discussed in Chapter 1, offer a choice for establishing the replacement cost of buildings. *Square foot,* and *cubic foot* systems are those most commonly used as they are fairly accurate when used by persons of experience. They are also methods for quick calculations.

The *detailed* method is seldom used except where the building is a total loss or nearly so and the loss figures themselves establish much of the value. The remaining portion of the structure is then detailed and added to the values reflected in the damaged section.

PROOF OF LOSS AND FINAL PAPERS

When agreement has been reached as to value and loss, and the amount collectible by the insured giving consideration to any applicable deductible clauses or coinsurance or other limiting clauses, a proof of loss is prepared by the adjuster or the insured's representative.

Two proof of loss forms are shown in the following pages. The first on page 536 is the form recommended by the American Insurance Association and printed by the General Adjustment Bureau Inc. It is a widely used form and is printed on one side only.

The second proof of loss form, on page 537, designed by the Kemper Insurance Companies for their use, provides for all the information required to be furnished by the insured, but in addition is a combination proof of loss form and "Subrogation Agreement." It also has at the bottom of the backside of the proof of loss a "Satisfaction Agreement." This eliminates printing and keeping on hand subrogation and satisfaction agreement forms for use when needed.

The Subrogation Agreement, item No. 10 in the form, transfers to the insurer any claim which the insured has to recover against person or persons who may be primarily responsible for causing the loss—to the extent, of course, of the insurer's payment.

The Satisfaction Agreement is signed by the insured whenever repairs to the property have been made to the insured's satisfaction by a contractor, subcontractor, or other repair firm, engaged directly by the insured with approval of the insurer.

It will also be noted that this proof of loss form provides for insertion of full replacement cost of both loss and value when such figures are considered in the adjustment of the loss. (See Replacement Cost Coverage—page 538).

Provision is also made to insert the amount of any deductible.

Form 22-1

POLICY NO.

G. A. B. FILE NO.

AMOUNT OF POLICY AT TIME OF LOSS

SWORN STATEMENT

COMPANY CLAIM NO.

$_____
DATE ISSUED

IN

AGENT

PROOF OF LOSS

DATE EXPIRES

AGENCY AT

To the _____

of _____
At time of loss, by the above indicated policy of insurance, you insured—

_____

_____

against loss by_____to the property described according to the
terms and conditions of said policy and of all forms, endorsements, transfers and assignments attached thereto.

TIME AND ORIGIN   A_____loss occurred about the hour of_____o'clock_____M.,
on the_____day of_____19_____, the cause and origin of the said loss were:_____

_____

OCCUPANCY   The building described, or containing the property described, was occupied at the time of the loss as follows, and for
no other purpose whatever:_____

TITLE AND INTEREST   At the time of the loss, the interest of your insured in the property described therein was_____
_____ . No other person or persons had any interest therein or
incumbrance thereon, except:_____

_____

CHANGES   Since the said policy was issued, there has been no assignment thereof, or change of interest, use, occupancy, possession, location or exposure of the property described, except _____

_____

_____

TOTAL INSURANCE   THE TOTAL AMOUNT OF INSURANCE upon the property described by this policy was, at the time of the loss,
$_____, as more particularly specified in the apportionment attached, besides which there
was no policy or other contract of insurance, written or oral, valid or invalid.

VALUE   THE ACTUAL CASH VALUE of said property at the time of the loss was . . . . . $_____

LOSS   THE WHOLE LOSS AND DAMAGE was . . . . $_____

AMOUNT CLAIMED   THE AMOUNT CLAIMED under the above numbered policy is . . . . $_____

STATEMENTS OF INSURED   The said loss did not originate by any act, design or procurement on the part of your insured, or this affiant; nothing
has been done by or with the privity or consent of your insured or this affiant, to violate the conditions of the policy,
or render it void; no articles are mentioned herein or in annexed schedules but such as were destroyed or damaged at
the time of said loss; no property saved has in any manner been concealed, and no attempt to deceive the said company,
as to the extent of said loss, has in any manner been made. Any other information that may be required will be furnished and considered a part of this proof.

The furnishing of this blank or the preparation of proofs by a representative of the above insurance company is not a waiver of any
of its rights.

State of_____   _____

County of_____   _____
                                                                                                 Insured

Subscribed and sworn to before me this_____day of_____, 19_____

Form recommended by the   _____
American Insurance Association   Notary Public
765 (6/69)

Form 22-2

SWORN STATEMENT IN PROOF OF LOSS
AND SUBROGATION AGREEMENT

$_____
AMOUNT OF POLICY AT TIME OF LOSS

_____
DATE ISSUED

_____
DATE EXPIRES

POLICY NUMBER
_____

AGENCY AT
_____

AGENT
_____

To the_____(Insurer)
By the above policy you insured_____
against loss by_____to the property described under Schedule "A."

1. **Time and Origin:** A_____loss occurred about the hour of_____o'clock____M.,
on the_____day of:_____19____. The cause and origin of the said loss were:_____
_____

2. **Occupancy:** The building described, or containing the property described, was occupied at the time of the loss as follows, and for no other purpose whatever:_____

3. **Title and Interest:** At the time of the loss the interest of your insured in the property described therein was _____No other person or persons had any interest therein or incumbrance thereon, except:_____

4. **Changes:** Since the said policy was issued there has been no assignment thereof, or change of interest, use, occupancy, possession, location or exposure of the property described, except:_____

5. **Total Insurance:** The total amount of insurance upon the property described by this policy was, at the time of the loss, $_____, as more particularly specified in the apportionment attached under Schedule "C," besides which there was no policy or other contract of insurance, written or oral, valid or invalid.

(FULL REPLACEMENT COST FIGURES TO BE INSERTED ONLY WHEN CONSIDERED IN THE ADJUSTMENT)	FULL REPLACEMENT COST	ACTUAL CASH VALUE
	$	$
6. The value of said property at the time of loss was.....................	$	$
7. The whole loss and damage was...	$	$
		$

8. The amount (less Ded. of $_____) claimed under this policy is $_____

9. The said loss did not originate by any act, design or procurement on the part of your insured, or this affiant; nothing has been done by or with the privity or consent of your insured or this affiant, to violate the conditions of the policy, or render it void; no articles are mentioned herein or in annexed schedules but such as were destroyed or damaged at the time of said loss; no property saved has in any manner been concealed, and no attempt to deceive the said company, as to the extent of said loss, has in any manner been made. Any other information that may be required will be furnished and considered a part of this proof.

10. The insured hereby assigns, transfers, and sets over to the Insurer any and all claims or causes of action of whatsoever kind and nature which the Insured now has, or may hereafter have, to recover against any person or persons as the result of said occurrence and loss above described, ro the extent of the payment above made; the Insured agrees that the Insurer may enforce the same in such manner as shall be necessary or appropriate for the use and benefit of the Insurer, either in its own name or in the name of the Insured; that the Insured will furnish such papers, information, or evidence as shall be within the Insured's possession or control for the purpose of enforcing such claim, demand, or cause of action. The Insured covenants that no release or settlement of any such claim, demand, or cause of action has been made.

11. The statements and agreements on the reverse side hereof or attached hereto are made a part of this instrument.

12. The furnishing of this blank or the preparation of proofs by a representative of the above insurance company is not a waiver of any of its rights.

State of_____   _____

County of_____   _____Insured

Subscribed and sworn to before me this_____day of_____19____

_____Notary Public

(OVER)

Form 22-2 (cont.)

SCHEDULE "A"—POLICY FORM

Policy Form No._____Dated_____

Item 1. $_____on_____

Item 2. $_____on_____

Item 3. $_____on_____

Item 4. $_____on_____

Situated_____

Coinsurance, Average, Distribution, or Deductible Clauses, if any_____

Loss, if any, payable to_____

SCHEDULE "B"
STATEMENT OF ACTUAL CASH VALUE AND LOSS AND DAMAGE

		ACTUAL CASH VALUE	LOSS AND DAMAGE
Totals:			

SCHEDULE "C" — APPORTIONMENT

POLICY NO.	EXPIRES	NAME OF COMPANY	ITEM NO.____		ITEM NO.____	
			INSURES	PAYS	INSURES	PAYS
Totals:						

_____Adjuster

SATISFACTION AGREEMENT

The insured hereby acknowledges repair or replacement of loss and damage to the Insured's entire satisfaction and agrees that the payment of the sum of_____Dollars, ($_____),
by the Insurer to_____, the person or firm making the repair or replacement, shall constitute a full performance of the obligation of the Insurer under its said policy.

Dated_____19____   _____

THE INSURED

EK 115 6-73 15M   PRINTED IN U.S.A.

The proof of loss is signed by the insured and sworn to before a Notary if the amount claimed requires it. On small to moderate losses most companies require only a signature. Each insurer's attitude in this respect should be verified. A statement of how the loss was determined is either written or typed on the back of the proof of loss, or on a separate sheet or sheets attached to it. In support of the adjusted amount the adjuster encloses all estimates.

If more than one insurer is involved, an apportionment schedule is either completed on the back of the proof or embodied in the statement of loss.

Glossary of Building Terms

Adobe. An aluminous earth from which unfired bricks are made, especially in the western part of the United States; an unfired brick dried in the sun; a house or other structure built of such materials or clay.

Aggregate. In mixing concrete, the stone or gravel used as a part of the mix is commonly called the *coarse* aggregate, while the sand is called the *fine* aggregate. Most aggregates contain varying degrees of moisture. To prevent an excessive amount of water from finding its way into the concrete the amount held by the aggregate must be determined, and this amount subtracted from that specified for the batch.

Air-dried lumber. Lumber that has been stacked on end or piled in yards or sheds for any length of time. For the United States as a whole, the minimum moisture content of thoroughly air-dried lumber is 12 to 15 percent and the average is somewhat higher.

Air slaking. In masonry, the process of exposing quicklime to the air, as a result of which it will gradually absorb moisture and break down into a powder.

Alabaster. A fine-grained gypsum, usually white but sometimes delicately tinted, often carved for mantle ornaments or other decorative construction features.

Alkyd resins. Have generally replaced oils. They are treated vegetable oils in a form of glyptal resin, used principally for lacquer, paints, varnishes, and metal finishes.

Alligatoring. Coarse checking pattern characterized by a slipping of the new coating over the old coating to the extent that the old coating can be seen through the fissures.

Alphaduct. In electricity, a flexible nonmetal conduit.

Alternating current. An electric current which reverses its direction of flow at regular intervals.

American bond. A method of bonding brick in a wall whereby every fifth, sixth, or seventh course consists of headers; other courses being stretchers. This type of bond is used extensively because it is quickly laid.

Ampere. The practical unit which indicates the rate of flow of electricity through a circuit.

Anchor bolts. Bolts to secure steel beams or columns or wooden sills to concrete or masonry floors, walls or piers.

Annealed wire. A soft pliable wire used extensively in the building trade for tie wires, especially for wiring concrete forms.

Anodize. An electrolytic oxide protective coating on aluminum.

Apron. A plain or molded piece of finish below the stool of a window, put on to cover the rough edge of the plaster. Also the pitched portion of the garage floor in front of the door to carry off the water.

Arcade. A range of arches, supported either on columns or on piers and detached or attached to the wall.

Arch-buttress. Sometimes called a flying buttress; an arch springing from a buttress.

Arch. An arrangement of building materials in the form of a curve which preserves its given form when resisting pressure and enables it, supported by piers or beams, to carry weights.

Architrave. The lowest of the principal divisions of an entablature, resting immediately on the columns or pilasters. In modern use it is the molding above and on both sides of the door.

Armored cable. Rubber-insulated wires which are wrapped with a flexible steel covering; often called BX.

Armored concrete. Concrete which has been strengthened by reinforcing with steel rods or steel plates.

Arris fillet. A triangular piece of wood for raising the slates or tiles of a roof near the eaves to throw off rain water; a sort of canting strip.

Asbestos. A variety of mineral fiber occurring in long and delicate fibers or fibrous masses. It is a poor conductor of heat and can withstand high temperatures. Used for insulating wires exposed to a high temperature.

Also mixed with portland cement under pressure for asbestos-cement shingles.

Ashlar. The facing of thin slabs of stone or terra cotta, which covers the rough brick and structural steel in the exterior walls of a building.

Asphalt. A bituminous material found in native form or as a residue from evaporated petroleum. It is insoluble in water but is soluble in gasoline and melts when heated. Used widely in building for waterproofing roof coverings of many types, exterior wall coverings, flooring tile, and the like.

Astragal. A molding attached to one of a pair of swinging doors against which the other door strikes.

Attic. A low story above the main entablature of a building with walls vertical to ceiling; in some building laws defined as any one story, in whole or part, which is situated in the roof.

Attic ventilators. In home building, usually openings in gables or ventilators in the roof; also, mechanical devices to force ventilation by the use of power-driven fans. (Also see Louver.)

Awning. A term sometimes used for marquise; any covering intended as a screen from the sun or protection from the rain.

Awning type window. A type of window in which each light opens outward on its own hinges, which are placed at its upper edge. Such windows are often used as ventilators in connection with fixed picture windows.

Backfill. The replacement of excavated earth into a pit or trench or against a structure.

Balcony. A projection from the wall of a building, supported by columns, consoles, or cantilevers and usually covered at its extremity by a balustrade.

Balloon framing. A type of framing with studs extending from sill to roof in a 1-1/2 or 2 story building.

Balusters. Small vertical spindles or members forming the main part of a railing for stairway or balcony, fastened between a bottom and top rail.

Balustrade. A row of balusters with the rails, generally used for stairs, porches, balconies, etc.

Banister. The balustrade of a staircase; a corruption of the word baluster.

Bargeboard. The decorative board covering the projecting portion of a gable roof; the same as a verge board; during the late part of the nineteenth century, bargeboards frequently were extremely ornate.

Base or baseboard. A board placed against the wall around a room next to the floor. May consist of a single piece, two or three pieces, base shoe, baseboard and base molding.

Base molding. Molding used to trim the upper edge of interior baseboard.

Batten. Narrow strips of wood or metal used to cover joints, or as decorative vertical members over plywood or wide boards.

Batter board. One of a pair of horizontal boards nailed to posts set at the corners of an excavation, used to indicate the desired level, also as a fastening for stretched strings to indicate outlines of foundation walls.

Bay. The wall space between two columns; the whole space between column centers.

Bay window. Any window projecting outward from the wall of a building and commencing from the ground. If they are supported on projecting corbels, they are called Oriel windows.

Beam. A piece of timber, steel, or other material placed across an opening or from post to post to support a load.

Bearing. The portion of a beam, or truss, etc., that rests on the supports.

Bearing partition. A partition that supports any vertical load in addition to its own weight.

Bearing wall. A wall that supports any vertical load in addition to its own weight.

Bed. The horizontal surfaces on which the stones or bricks are laid in a wall.

Bedding. A filling of mortar, putty, or other substance in order to secure a firm bearing.

Belt course—in masonry. A course of stones or brick projecting from the face of a brick or stone wall.

Belt course—in carpentry. A horizontal board across or around a building, usually made of a flat member and a molding.

Bevel board (pitch board). A board used in framing a roof or stairway to lay out bevels.

Bevel siding. Page 289.

Bidet. A bathroom basin-like fixture to be straddled to wash the perineum following use of the toilet.

Blank flue. If the space on one side of a fireplace is not needed for a flue, a chamber is build in and closed off at the top in order to conserve material and labor and to balance weight.

Blinds (shutters). Light wood sections in the form of doors to close over windows to shut out light, give protection, or add temporary insulation. Commonly used now for ornamental purposes in which they are fastened rigidly to the building.

Blind-nailing. Nailing in such a way that the nailheads are not visible on the face of the work.

Blind stop. A rectangular molding, usually 3/4 to 1-3/8 inches or more, used in the assembly of a window frame.

Block bridging (solid bridging). In building construction, blocks or short pieces of wood nailed between joists or between studding to serve as bridging braces. In case of fire, such blocks also help to check the fire and prevent its spreading to other parts of the structure; also called fire blocks or fire stops.

Bloom. An efflorescence which sometimes appears on masonry walls, especially on a brick wall. Also, a defect on a varnished surface usually caused by a damp atmosphere.

Blue stain. A bluish or grayish discoloration of the sapwood caused by the growth of certain moldlike fungi on the surface and in the interior of the piece, made possible by the same conditions that favor the growth of other fungi.

Blushing. In painting, a condition in which a bloom or gray cloudy film appears on a newly finished surface, on hot humid days, usually caused by the condensation of moisture or by the too rapid evaporation of the solvents.

Board. Lumber less than two inches thick.

Board and batten. A type of siding composed of wide boards and narrow battens. The boards (generally twelve inches wide) are nailed to the sheathing so that there is one-half inch space between them. The battens (generally three inches wide) are nailed over the open spaces between the boards.

Board foot. Page 202.

Board measure. Page 202.

Boarding in. The process of nailing boards on the outside studding of a house. Also applying boards or plywood over all openings in a building to protect it from elements and intruders.

Boiled linseed oil. Linseed oil in which enough lead, manganese, or cobalt salts have been incorporated to make the oil harden more rapidly when spread in thin coatings.

Boiler. A vessel in which water is heated and circulated either as steam or hot water.

Bond—in masonry. The connection between bricks, stones, or other materials formed by laying them upon one another in carrying up the work, so as to form a single mass of wall.

Bow window. A window in the bow of a building such as a bay window, especially a bay with a curved ground plan.

Box column. A type of built-up hollow column used in porch construction; it is usually square in form.

Box cornice. Page 268.

Box gutter. A gutter build into a roof, consisting of a horizontal trough of wood construction lined with galvanized iron, tin, or copper to make it watertight; sometimes called concealed gutter.

Box joist (box sill). A header nailed on the ends of joists and resting on a wall plate or sill. Used in frame building construction.

Brace. Pieces fitted and firmly fastened to two or more other pieces, at any angle, in order to strengthen the angle thus created.

Bracket. A projecting support for a shelf or other structure.

Break. Any projection from the general surface of wall.

Breezeway. A covered passage between house and garage.

Brick veneer. A facing of brick laid against frame, cement block, or other dissimilar wall construction.

Bridging. Page 220.

British thermal unit. The quantity of heat required to raise the temperature of one pound of pure water one degree Fahrenheit at or near the temperature of maximum density of water 39 degrees Fahrenheit. Abbreviation B.t.u.

Broken joints. In building construction, joints are arranged so they will not fall in a straight line. Broken joints tend to all strength and stiffness to a structure, as in masonry, roof shingles, siding, etc.

Brown coat. Page 400.

Buck. Often used in reference to rough frame opening members. Door bucks used in reference to a door frame.

Building paper. Cheap, rosin-sized or asphalt saturated paper used to insulate a building before the siding or roofing is put on; sometimes placed between double floors.

Built-up roof. A roofing composed of three to five layers of rag felt or jute saturated with coal tar, pitch, or asphalt. The top may be finished with crushed slag or gravel. Generally used on flat or low-pitched roofs.

Built-up timber. A timber made of several pieces fastened together and forming one of larger dimension.

Butt joint. The junction where the ends of two timbers or other members meet in a square-cut joint. Where the ends of two pieces of any material butt together.

Butterfly roof. A roof constructed so as to appear as two shed roofs connected at the lower edges to form a "V."

Buttering. In masonry, the process of spreading mortar on the edges of a brick before laying it.

Buttress. Masonry projecting from a wall to gain additional strength against the thrust of a roof or vault.

Buttress, flying. A detached buttress or pier of masonry at some distance from a wall and connected to the wall by an arch or part of an arch.

BX cable-(in electrical). Page 449.

Cabinet. A shop- or job-built unit for kitchens, bathrooms, etc. Cabinets usually include combinations of drawers, doors, and the like.

Caisson. A sunken panel or coffer in ceilings, vaults, and domes; the term is also used for concrete cylindrical foundations or tubular piers filled with concrete.

Camber. The convexity sometimes placed in beams or trusses, to prevent them becoming concave under their loads.

Canopy. An ornamental covering over a niche; a name sometimes given to a marquise.

Cant strip. A wedge or triangular-shaped piece of lumber used at gable ends under shingles or at the junction of the house and a flat deck under the roofing. At the junction of a flat roof deck and a parapet wall or the junction between the roof deck and a chimney.

Canted wall. A wall built at an angle to the face of another wall.

Cantilever. A projecting beam or truss, sometimes called console or bracket.

Cantilever joists. Short joists used to support a projecting balcony or cornice where the overhang is parallel to the second-story joists. Cantilever joists are also used to support a bay window which has no supporting foundation.

Canting strip. A projecting molding near the bottom of a wall to direct rain water away from the foundation wall; in frame buildings, the same as a water table.

Cap. The upper member of a column, pilaster, door cornice, molding, and the like.

Capital. The upper part of a column, pilaster, or pier, usually ornamented with moldings or foliage or combined.

Carport. An open sided roof shelter for automobiles frequently attached to one side of the house.

Cased. A building term used to describe a structural member or part covered with a different material, usually of a better quality.

Cased opening. Any opening finished with jambs and trim, but without doors.

Casein. The dried curd of a cow's milk; the principal ingredient in cheese. It is used extensively as an adhesive in making glue; also used in paint.

Casement. A window, the sash of which is hinged to the vertical sides of the frame into which it is fitted.

Casement frames and sash. Frames of wood or metal enclosing part or all of the sash, which may be opened by means of hinges affixed to the vertical edges.

Casing. The trimming around a door or window opening, either outside or inside, or the finished lumber around a post or beam, etc.

Caulking. Various types of elastic compounds introduced into cracks and openings around doors, windows etc., to make waterproof and airtight.

Causeway. A raised or paved way.

Cavity wall. A hollow wall, usually consisting of two brick walls erected a few inches apart and joined together with ties of metal or brick. Such walls increase thermal resistance and prevent rain from driving through the inner face; also called hollow wall.

Ceiling. The upper horizontal or curved surface of a room or hall, opposite the floor, which conceals or ornaments the construction of the floor above or that of the roof.

Cement—in building. A material for binding other material or articles together, usually plastic at the time of application but hardens when in place; any substance which causes bodies to adhere to one another such as portland cement, stucco, and natural cements; also, mortar or plaster of Paris.

Cement—Keene's. The whitest finish plaster obtainable that produces a wall of extreme durability. Because of its density, it excels for wainscoting plaster for bathrooms and kitchens and is also used extensively for the finish coat in auditoriums, public buildings, and other places where walls will be subjected to unusually hard wear or abuse.

Center-hung sash. A sash hung on its centers so that it swings on a horizontal axis.

Center stringer. The center horse in a flight of stars. The same as carriage or roughstring.

Ceramic tile. Page 479.

Chair rail. A wooden molding around the wall of a room at chairback height, to afford protection against damage when chairs are pushed back or rubbed against the wall.

Chamfer. To bevel the edge of anything originally right-angled; to round off or bevel an edge to fit on or into a connecting piece.

Checking. Fissures that appear with age in many exterior paint coatings, at first superficial, but which in time may penetrate entirely through the coating.

Checkrails. Meeting rails sufficiently thicker than a window to fill the opening between the top and bottom sash made by the parting stop in the frame. They are usually beveled.

Chimney cricket. (See cricket).

Chimney hood. A covering for a chimney to make it more ornamental; also, to prevent rain water from entering the flues. The design of the hood may be circular in form or it may be flat.

Chimney throat. That part of a chimney directly above the fireplace where the walls of the flue are brought close together as a means of increasing the draft.

Chord. In building construction, the bottom beam of any truss; in a Howe truss, where two parallel beams are placed one above the other, both beams are called chords.

Circuit breaker. Page 442.

Circuits. Page 442.

Cistern. An artificial reservoir or tank, often underground, for the storing of rain water collected from a roof.

Clamp. A mechanical device used to hold two or more pieces together.

Clapboards. (Similar to bevel siding.)

Class. Class A, Class B, etc., designed by the building codes for a building, denotes the character of its construction, under certain limiting conditions of height or area for each, to fulfill the minimum requirements relative to strength fireproofing and beauty.

Cleat. A strip of wood or metal fastened across a door or other object to give it additional strength; a strip of wood or other material nailed to a wall usually for the purpose of supporting some object or article fastened to it. In electricity, a piece of insulating material used to fasten wires to flat surfaces.

Clout nail. In building, a nail with a large flat head, used principally for fastening sheet metal and nailing on gutters.

Coat. A thickness or covering of paint, plaster, or other work done at one time.

Coffer. A sunken panel (See Caisson.)

Coffer dam. A frame surrounding an excavation; or a frame placed in water, the water inside the frame being pumped out to build masonry piers.

Collar beam. A beam connecting pairs of opposite rafters above the attic floor.

Column. In architecture, a perpendicular supporting member, circular or rectangular in section, usually consisting of a base, shaft, and capital. In engineering, a structural compression member, usually vertical, supporting loads acting on or near and in the direction of its longitudinal axis.

Comb. In carpentry, the ridge of a roof; a comb board. In masonry, a tool used to give a finish to the face of stone, a drag. In house painting, an instrument used for graining surfaces.

Combination frame. A combination of the principal features of the full and balloon frames.

Concrete. A mass composed of cement, sand, and broken stone or gravel making an artificial stone.

Concrete block. Page 173

Condensation. In a building, beads or drops of water, and frequently frost in extremely cold weather, that accumulate on the inside of the exterior covering of a building when warm moisture-laden air from the interior reaches a point where the temperature no longer permits the air to sustain the moisture it holds.

Conductors. Pipes for conducting water from a roof to the ground or to a receptacle or drain; downspout. Electrical, see page 442.

Conduit, (electrical). Page 448.

Conifer. A tree with needle-like foliage. A member of the soft wood family.

Continuous header. The top plate is replaced by 2" x 6" turned on edge and running around the entire house. This header is strong enough to act as a lintel over all wall openings, eliminating some cutting and fitting of stud lengths and separate headers over openings. This is especially important because of the emphasis on one-story, open planning houses.

Coped joint. See Scribing.

Coping. The highest and covering course of masonry in a wall.

Corbel. A short piece of wood or stone projecting from the face of a wall to form a support for a timber, or other weight; a bracketlike support; a stepping out of courses in a wall to form a ledge; any supporting projection of wood or stone on the face of a wall.

Corner bead. A strip of formed galvanized iron, sometimes combined with a strip of metal lath, placed on corners before plastering to reinforce them. Also, a strip of wood finish three-quarters round or angular placed over a plastered corner for protection.

Corner boards. Page 268.

Corner braces. Diagonal braces let into studs to reinforce corners of frame structures.

Cornerite. Metal-mesh lath cut into strips and bent to a right angle. Used in interior corners of walls and ceilings on lath to prevent cracks in plastering.

Cornice. A decorative element made up of molded members usually placed at or near the top of an exterior or interior wall.

Cornice return. That portion of the cornice that returns on the gable end of a house.

Corridor. The passageway which gives communication between the various parts of a building.

Counterflashing. A flashing usually used on chimneys at the roofline to cover shingle flashing and to prevent moisture entry. They also allow for expansion and contraction without danger of breaking the flashing.

Countersink. The cavity for the reception of a plate or the head of a screw or bolt, so that it will not project beyond the face of the work.

Course. A continued layer of bricks, stones, terra cotta, slate, shingles, etc.

Court. An uncovered area in front, behind, or in the center of a building.

Cove. The curved, concave portion of a cornice or ceiling as distinguished from the square parts or corners. Cove-molding, cove-ceiling, etc., refer to the concave portion of the molding, ceiling, etc.

Cove molding. A three-sided molding with concave face used wherever interior angles are to be covered.

Crawl space. A shallow space below the living quarters of a house. It is generally not excavated or paved and is often enclosed for appearance by a skirting or facing material.

Crawling. In painting, a defect appearing during the process of applying paint in which the film breaks, separates, or raises, as a result of applying the paint over a slick or glassy surface. Such defects may also be due to surface tension caused by heavy coatings or the use of an elastic film over a surface which is hard and brittle.

Cricket. In architecture, a small false roof, or the elevation of part of a roof as a means of throwing off water from behind an obstacle. As a watershed built behind a chimney or other roof projection.

Cripple. In building construction, any part of a frame which is cut less than full size, as a cripple studding over a door or window opening.

Cripple jack rafter. A jack rafter that is cut in between a hip and valley rafter. A cripple jack rafter touches neither the ridge not the plate, but extends from a valley rafter to a hip rafter.

Crosslap. A joint where two pieces of timber cross each other. This type of joint is formed by cutting away half the thickness of each piece at the place of joining so that one piece will fit into the other and both pieces will lie on the same plane.

Crossover. In plumbing, a U-shaped fitting having the ends turned outward. It is used for passing the flow of one pipe past another when the two pipes lie in the same plane.

Crown molding. A molding used on cornice or wherever a large angle is to be covered.

Cup. In lumber, a distortion of a board in which there is a deviation flatwise from a straight line across the width of a board.

Cupola. A small structure surmounting a roof.

Curb. The dividing line between the sidewalk and the street, the edge of an opening in a floor.

Curb roof. The mansard roof taking its name from the architect who designed it. This type of roof has a double slope on each side, with the lower slope almost vertical. Frequently the lower slope contains dormer windows, which make possible the addition of another story to the house.

Curb stringer. In stair building, the three-member assembly sometimes used in an open stair, consisting of one faced stringer surmounted with a shoe rail to receive balusters and one housed stringer.

Curtain wall. A non-bearing wall built between columns.

Cypress. The wood of an evergreen tree which grows in the southern part of the United States. It is one of the most durable of woods, somewhat resembling the cedar, and is used for both inside and outside work, in the building trades.

Dado. A rectangular groove in a board or plank. In interior decoration, a special type of wall treatment.

Dado joint. A joint formed by the intersection of two boards, usually at right angles, the end of one of which is notched into the side of the other for a distance of half the latter's thickness.

Damageability. Susceptibility to damage—the quality of being damageable.

Damper. An adjustable plate in the flue to regulate draft.

Deadman. A stake, post or anchor driven into the ground to serve as a brace against which an angled brace may be positioned.

Deadening. Construction intended to prevent the passage of sound.

Decay. Disintegration of wood or other substance through the action of fungi.

Deck paint. Paint with a high degree of resistance to mechanical wear, designed for use on such surfaces as porch floors.

Decking. Boards, plywood etc., applied to roofs. Roof sheathing.

Density. The mass of substance in a unit volume. When expressed in the metric system, it is numerically equal to the specific gravity of the same substance.

Dimension lumber. Lumber as it comes from the saws, two inches thick and from four to twelve inches wide; also, lumber cut to standard sizes or to sizes ordered. (See also scantlings and planks.)

Direct current. Page 442.

Direct nailing. To nail perpendicular to the initial surface or to the junction of the pieces joined. Also termed face nailing.

Dome. The projection on the top of buildings in the form of an inverted cup.

Door butt. Door hinge.

Doorjamb—interior. The surrounding case into which and out of which a door closes and opens. It consists of two upright pieces, called jambs, and a head, fitted together and rabbeted.

Door sill. Member at bottom of door opening that serves as a threshold.

Dormer. An internal recess, the framing of which projects from a sloping roof, to form a vertical wall suitable for windows.

Door stop. A devise attached to door, floor or wall to prevent door from opening too far.

Double-hung window. One with two sash that move up and down in tracks.

Douglas fir. A tree also known as Oregon pine. A strong wood of light weight, used extensively for flooring and for framing timbers.

Dovetailing. The method of joining boards or other material together; the joints, tooth or dovetail shaped, fitting into each other.

Dowel. A pin used in the joining of two pieces of material for the purpose of holding them in place.

Downspout. A pipe of metal or plastic for carrying rainwater from roof gutters.

Drainage. Relates to system of pitched roofs or floors, gutters, downspouts, conduits, etc., for the purpose of shedding water.

Dressed lumber. Lumber that has been machined and surfaced at the mill.

Drier (paint). Usually oil-soluble soaps of such metals as lead, manganese, or cobalt, which, in small proportions, hasten the oxidation and hardening (drying) of the drying oils in paints.

Drip cap. A molding placed on the exterior top side of a door or window frame to cause water to drip beyond the outside of the frame.

Drop siding. Page 290.

Dry-wall interior construction. Page 280.

Drywell. Hole in the ground lined with stone used to catch waste or rainwater.

Dutch door. A door divided horizontally, so the lower section can be closed and fastened while the upper part remains open; commonly used on barns.

Dutchman. An odd piece inserted to fill an opening or to cover a defect in woodwork.

Eaves. The margin or lower part of a roof projecting over the wall.

Eave trough. Gutter.

Eaves board. A strip of wedge-shaped wood, used at the eaves of a roof to back up the first course of shingles, tiles, or slates.

Elevation. The drawing of the external walls of a building; the vertical projection of any member or structure; the distance from datum to a given height.

Expansion joint. Material used to separate blocks or units of concrete to prevent cracking due to expansion as a result of temperature changes. Also used on concrete slabs.

Facia or fascia. A flat board, band, or face, used sometimes by itself but usually in combination with moldings, often located at the outer face of the cornice. Generally, the board of the cornice to which the gutter is attached or adjoins.

Filler wall. A term in some municipal building laws for a non-bearing wall between columns and supported at each floor; a partition.

Filler (wood). A heavily pigmented preparation used for filling and leveling off the pores in open-pored woods.

Fir. A name commonly applied to a variety of trees yielding a soft wood which is used extensively for structural lumber, especially for interior finish and also for framing purposes.

Fire door. In building construction, a door made of fire-resisting material which is slow burning or difficult to ignite.

Fire-resistive. In the absence of a specific ruling by the authority having jurisdiction, applies to materials for construction not combustible in the temperatures of ordinary fires and that will withstand such fires without serious impairment of their usefulness for at least one hour.

Fire stop. A solid, tight closure of a concealed space, placed to prevent the spread of fire and smoke through such a space. Usually consists of 2" x 4" cross blocking between studs.

Flakeboard. A high quality particleboard. (See Particleboard.)

Flagstone (flagging or flags). Flat stones, from one to four inches thick, used for rustic walks, steps, floors, and the like.

Flashing. Piece of lead, tin, or sheet metal, either copper or galvanized iron, used around dormers, chimneys, or any rising projection such as window heads, cornices, and angles between different members or any place where there is danger of leakage from rain water or snow.

Flat paint. An interior paint that contains a high proportion of pigment, and dries to a flat or lusterless finish.

Floating. The equal spreading of plaster or cement on a surface by means of a board, called a float.

Flue. The space or passage in a chimney through which smoke, gas, or fumes ascend. Each passage is called a flue, which together with any others and the surrounding masonry, make up the chimney.

Flue lining. Fire clay or terra-cotta pipe, round or square, usually made in all of the ordinary flue sizes and in two-foot lengths, used for the inner lining of chimneys with the brick or masonry work around the outside. Flue lining should run from the concrete footing to the top of the chimney cap.

Flush. Adjacent surfaces even, or in same plane (with reference to two structural pieces).

Flush door. A door with two flat surfaces—no panels—sometimes it has a hollow core.

Folding door. One whose panels fold on each other when opened.

Footing. The spreading course or courses at the base or bottom of a foundation wall, pier, or column. An enlargement at the bottom of a wall to distribute the weight of the super-structure over a greater area and thus prevent settling.

Footing courses. The bottom and heaviest courses of a piece of masonry.

Formica. A brand name for laminated plastic material used extensively in the building trades for wall covering, as a veneer for plywood panels, for kitchen cabinet tops.

Foundation. The supporting portion of a structure below the first-floor construction, or below grade, including the footings.

Frame. A term used extensively in building, with a qualifying noun as window frame, steel frame; when used in specifications as a verb, it relates to the skeleton-like connections of various members in a building.

Frame construction. A type of construction in which the structural parts are of wood or dependent upon a wood frame for support. In codes, if brick or other combustible material is applied to the exterior walls, the classification of this type of construction is usually unchanged.

Framing. The rough timber structure of a building, including interior and exterior walls, floor, roof, and ceilings.

Framing—balloon. A system of framing a building in which all vertical structural elements of the bearing walls and partitions consist of single pieces extending from the top of the soleplate to the roofplate and to which all floor joists are fastened.

Framing—platform. A system of framing a building in which floor joists of each story rest on the top plates of the story below or on the foundation sill for the first story, and the bearing walls and partitions rest on the subfloor of each story.

Frieze. Any sculptured or ornamental band in a building. Also the horizontal member of a cornice set vertically against the wall.

Frostline. The depth of frost penetration in soil. This depth varies in different parts of the country. Footings should be placed below this depth to prevent movement.

Full pitch. In roof framing, a term applied to a roof with a pitch having a rise equal to the width of the span of the roof.

Fungi—wood. Microscopic plants that live in damp wood and cause mold, stain, and decay.

Fungicide. A chemical that is poisonous to fungi.

Furring. Strips of wood or metal applied to a wall or other surfaces to even it, to form an air space, or to give the wall an appearance of greater thickness.

Fuse. Page 442.

Fuse box. Page 442.

Gable. That portion of a wall contained between the slopes of a double-sloped roof.

Gable roof. Page 314.

Gage (gauge). A tool used by carpenters to strike a line parallel to the edge of the board. A device for measuring.

Galvanizing. A coating of zinc applied to iron or steel to prevent rusting.

Gambrel. A symmetrical roof with two different pitches or slopes on each side.

Gingerbread work. A gaudy type of ornamentation in architecture, especially in the trim of a house.

Girder. A large timber or steel member, either single or built up, used to support floors or walls over an opening.

Girt (ribband). The horizontal member of the walls of a full or combination frame house which supports the floor joists or is flush with the top of the joists.

Glazier. A workman whose business is that of cutting panes of glass to size and fitting them in position in frames for doors or windows.

Gloss (paint or enamel). A paint or enamel that contains a relatively low proportion of pigment and dries to a sheen or luster.

Grade. The term is used to denote the established street and sidewalk planes or surfaces; the natural surface of the ground or finished surface of the ground where it is cut away or added to; the elevation above the datum.

Grade. The designation of the quality of a manufactured piece of wood or of logs.

Grain. The direction, size, arrangement, appearance, or quality of the fibers in wood.

Grain—edge (vertical). Edge-grain lumber has been sawed parallel to the pith of the log and approximately at right angles to the growth rings; i.e., the rings form an angle of 45 degrees or more with the surface of the piece.

Grain—flat. Flat-grained lumber has been sawed parallel to the pith of the log and approximately tangent to the growth rings; i.e., the rings form an angle of less than 45 degrees with the surface of the piece.

Grain—quartersawn. Another term for edge grain.

Greenwood. A term used by woodworkers when referring to timbers which still contain the moisture or sap of the tree from which the wood was cut.

Grillage. A framework of heavy timbers or beams laid longitudinally and crossed by similar beams laid upon them, for sustaining walls to prevent irregular setting.

Groin. The intersection of two vaulting surfaces.

Groove. A long hollow channel cut by a tool into which a piece fits or in which it works.

Ground floor. The floor of a building on a level, or nearly so, with the ground.

Grounds. Strips(s) of wood the same thickness as lath and plaster which are attached to walls before the plastering is done. Used around windows, doors and other openings as a plaster stop and guide for plaster thickness.

Grout. Mortar made so thin by adding water that the mixture will run into joints or cavities of the mason-work and fill it up solid.

Gumwood (gum tree). A dark-colored wood used for interior work.

Gunite. A construction material composed of cement, sand, or crushed slag and water mixed together and forced through a cement gun by penumatic pressure. Sold under the trademark Gunite.

Gusset. A brace or angle bracket used to stiffen a corner or angular piece of work.

Gutter. The channel for carrying off rain water from the roofs.

Gypsum. A mineral, hydrous sulphate of calcium. In the pure state, gypsum is colorless. When part of the water is removed by a slow heating process, the product becomes what is known as plaster of Paris.

Gypsum blocks. A type of building material usually grayish white in color; because of its friable texture, it is used only in nonload-bearing partition walls.

Gypsum board. Page 282.

Half-pitch roof. A roof having a pitch which has a rise equal to one half the width of the span.

Hardboard. A hard rigid board made from refined wood fibers into sheets 4' x 8' and larger and thicknesses 1/16" to 3/4".

Header. In building, a brick or stone laid with the end toward the face of the wall; one or more pieces of lumber used generally around openings to support free ends of floor joists, studs, or rafters and transfer their load to other parallel joists, studs, or rafters.

Header joist. In carpentry, the large beam or timber into which the common joists are fitted when framing around openings for stairs, chimneys, or any openings in a floor or roof; placed so as to fit between two long beams and support the ends of short timbers.

Head room. The distance between the top of finished floor and the finished side of a beam or girder, or joists, in floor above.

Hearth. The floor of a fireplace, usually made of brick tile or stone.

Heartwood. The wood extending from the pith to the sapwood, the cells of which no longer participate in the life processes of the tree.

Heel of the rafter. The end or foot that rests on the wall plate.

Hemlock. An inexpensive wood resembling spruce in appearance and used for framing timber.

Herring-bone work. Bricks, tile, woodflooring or other materials laid in a pattern of rows of short slanted parallel lines in the direction of the slant alternating row by row.

Hickory. A hard, tough wood used for structural members which require bending.

Hip rafters. Rafters which form the hip of a roof as distinguished from the common rafters. A hip rafter extends diagonally from the corner of the plate to the ridge.

Hip roof. Page 319.

Horse. In building and woodworking, one of the slating supports of a set of steps to which the treads and risers of a stair are attached; also a string.

Housed joint. A joint made by cutting out a space in the end of a piece of wood to receive the tongue cut on another piece to which the first piece is to be attached; any fitted joint such as one made with a mortise and tenon.

Housed string. A stair string with horizontal and vertical grooves cut on the inside to receive the ends of the risers and treads. Wedges covered with glue often are used to hold the risers and treads in place in the grooves.

Hydrostatic. The branch of physics having to do with the pressure and equilibrium of water and other liquids.

I-beam. A steel beam with a cross section resembling the letter "I."

Inlaid linoleum. Page 481.

Insulation—building. Any material high in resistance to heat transmission that, when placed in the walls, ceilings, or floors of a structure, will reduce the rate of heat flow.

Jack rafter. A short rafter of which there are three kinds. (1) those between the plate and a hip rafter; (2) those between the hip and valley rafters; (3) those between the valley rafters and the ridge board. Jack rafters are used especially in hip roofs.

Jalousie windows, jalousie doors. Windows and doors with movable glass louvers adjustable to slope upward to admit light and air yet exclude rain or snow.

Joints:

Butt—Squared ends or ends and edges adjoining each other.

Dovetail—Joint made by cutting pins the shape of dovetails which fit between dovetails upon another piece.

Drawboard—A mortise-and-tenon joint with holes so bored that when a pin is driven through, the joint becomes tighter.

Fished—An end butt splice strengthened by pieces nailed on the sides.

Halved—A joint made by cutting half of the wood away from each piece so as to bring the sides flush.

Housed—A joint in which a piece is grooved to receive the piece which is to form the other part of the joint.

Glue—A joint held together with glue.

Lap—A joint of two pieces lapping over each other.

Mortised—A joint made by cutting a hole or mortise in one piece, and a tenon, or piece to fit the hole, upon the other.

Rub—A glue joint made by carefully fitting the edges together, spreading glue between them, and rubbing the pieces back and forth until the pieces are well rubbed together.

Scarfed—A timber spliced by cutting various shapes of shoulders, or jogs, which fit each other.

Joint cement. A powder that is usually mixed with water and used for joint treatment in gypsum-wallboard finish. Often called "spackle."

Joist. One of a series of parallel beams used to support floor and ceiling loads, and supported in turn by larger beams, girders, or bearing walls.

Joist hangers. A steel or iron stirrup used to support the ends of joists which are to be flush with the girder.

Junction box. Page 441.

Kerf. Cut made by a saw.

Kerfing. The process of cutting grooves or kerfs across a board so as to make it flexible for bending. Kerfs are cut down to about two-thirds of the thickness of the piece to be bent. An example is found in the bullnose of a stair riser which frequently is bent by the process of kerfing.

Key-stone. The stone placed in the center of the top of an arch.

Kiln. A large oven or heated chamber for the purpose of baking, drying, or hardening, as a *kiln* for drying lumber, a *kiln* for baking brick, a lime *kiln* for burning lime.

Kiln dried. A term applied to lumber which has been dried by artifically controlled heat and humidity to a satisfactory moisture content.

Kilowatt. Page 441.

Kilowatt hours. Page 441.

King rod. In roof framing, a steel or iron tie rod used in place of a king post.

King-post truss. A truss framed with one tie member in the center.

Knee. In building construction, a member placed diagonally between a post or wall and joist or truss to relieve the weight or secure rigidity.

Knee brace. In building construction, a member placed across the inside of an angle in a framework to add stiffness to the frame, especially at the angle between the roof and wall of the building.

Knee walls. Partitions of varying length used to support roof rafters when their span is so great that additional support is required to stiffen them.

Knot. That portion of a branch or limb that has become incorporated in the body of a piece of lumber.

Laced valley. In building, a valley formed in a tile roof by interlacing tile-and-a-half tiles across a valley board.

Lacing course. In masonry, a course of brickwork built into a stone wall for bonding and leveling purposes.

Lally column. A cylindrically shaped steel member, sometimes filled with concrete, used as a support for girders or other beams.

Laminate. In home construction, the building up with layers of wood, each layer being a lamination or ply; also, the construction of plywood.

Landing. A platform between flights of stairs or at the termination of a flight of stairs.

Lath—gypsum. Page 397.

Lath—metal. Page 389.

Lath—wood. Page 396.

Lattice. An assemblage of wood or metal strips, rods, or bars made by crossing them to form a network.

Lavatory. A basin or bowl as in a bathroom.

Leaching trenches. In plumbing, trenches which carry waste liquids from sewers. Such trenches may be constructed in gravelly or sandy soils which permit the liquids to pass into the surrounding soil by percolation; or the trenches may be dug in firm ground to the required depth, and then be filled with broken stones, gravel, and sand.

Leader. See downspout.

Lean-to. A small building whose rafters pitch or lean against another building or against a wall.

Ledgerboard. The support for the second-floor joists of a balloon-frame house, or for similar uses; ribband.

Ledger strip. A strip of lumber nailed along the bottom of the side of a girder on which joists rest.

Level. A term describing the position of a line or plane when parallel to the surface of still water; an instrument or tool used in testing for horizontal and vertical surfaces, and in determining differences of elevation.

Light. Space in a window sash for a single pane of glass. Also, a pane of glass.

Linear. Relating to a straight line—one dimension.

Linear measure. Page 143.

Linoleum. Page 481.

Lintel. A piece of wood, stone, or steel placed horizontally across the top of the door and window openings to support the walls immediately above the openings.

Live load. The moving load or variable weight to which a building is subjected, due to the weight of the people who occupy it; the furnishings and other movable objects as distinct from the dead load or weight of the structural members and other fixed loads; the weight of moving traffic over a bridge as opposed to the weight of the bridge itself. Live load does not include wind load or earthquake shock.

Load-bearing walls. Any wall which bears its own weight as well as other weight; same as supporting wall; also called bearing wall.

Lookout. A short wood bracket or cantilever to support an overhanging portion of a roof or the like, usually concealed from view.

Louver. An opening with a series of horizontal slats so arranged as to permit ventilation but to exclude rain, sunlight, or vision. See attic ventilators.

Lumber. Lumber is the product of the sawmill and planing mill not further manufactured other than by sawing, resawing, and passing lengthwise through a standard planing machine, crosscut to length and matched.

Lumber—dressed size. The dimensions of lumber after shrinking from the green dimension and after planing. Less than the nominal or rough size.

Lumber—matched. Lumber that is edge-dressed and shaped to make a close tongue-and-groove joint at the edges or ends when laid edge to edge or end to end.

Lumber—nominal size. As applied to timber or lumber, the rough-sawed commercial size by which it is known and sold in the market.

Lumber—patterned. Lumber that is shaped to a pattern or to a molded form in addition to being dressed, matched, or shiplapped, or any combination to these workings.

Lumber—plainsawed. Another term for flat-grained lumber.

Lumber—rough. Lumber as it comes from the saw.

Lumber—shipping-dry. Lumber that is partially dried to prevent stain and mold in transit, and reduce freight costs.

Lumber—structural. Lumber that is two or more inches thick and four or more inches wide, intended for use where working stresses are required. The grading of structural lumber is based on the strength of the piece and the use of the entire piece.

Lumber—surfaced. Lumber that is dressed by running it through a planer.

Lumber—vertical grained. Another term for edge-grained lumber.

Main rafter. A roof member extending at right angles from the plate to the ridge. Same as common rafter.

Male plug. The two and sometimes three-prong connector used to establish an electrical circuit fitting into a receptacle.

Mansard. Page 320.

Mantel. The shelf above a fireplace. Originally referred to the beam or lintel supporting the arch above the fireplace opening.

Maple. A tough, hard wood used in building construction for flooring and veneer.

Marine glue. In woodworking, an adhesive substance composed of crude rubber, pitch, and shellac; the proportions are: 1 part rubber, 2 parts shellac, and 3 parts pitch. Used where exposed to the weather.

Marquise (Marquee). The hood or canopy projecting over a carriage or other entrance to a building as a protection from the weather.

Masonite. A trade name for a building board used for insulation purposes.

Masonry. Stone, brick, concrete, hollow-tile, concrete-block, gypsum-block, or other similar building units or materials or a combination of the same, bonded together with mortar to form a wall, pier, buttress, or similar mass.

Mastic. A term applied to a thick adhesive, consisting of a mixture of bituminous preparations such as asphalt and some foreign matter, usually fine sand; used for bedding and pointing window frames, bedding wood-block flooring and structural glass, and for repairing flat roofs.

Meeting rail. The bottom rail of the upper sash and the top rail of the lower sash of a double-hung window. Sometimes called the check rail.

Mezzanine. From the Italian word meaning "middle," a low story between two regular floors.

Millwork. Generally, all buildings made of finished wood and manufactured in millwork plants and planing mills are included under the term "millwork." It includes such items as inside and outside doors, window and door frames, blinds, porchwork, mantels, panelwork, stairways, moldings, and interior trim. It does not include flooring, ceiling or siding.

Mineral wool. Insulation—see page 462.

Miter. Two pieces with ends usually cut at 45 degrees to form a miter joint.

Molding:

Base—The molding on the top of a baseboard.

Bed—A molding used to cover the point between the plancier and frieze; also used as a base mold upon heavy work and sometimes as a member of a cornice.

Lip—A molding with a lip which overlaps the piece against which the back of the molding rests.

Rake—The cornice upon the gable edge of a pitch roof, the members of which are made to fit those of the molding of the horizontal eaves.

Picture—A molding shaped to form a support for picture hooks often placed at some distance from the ceiling on the wall to form the lower edge of the frieze.

Moisture content of wood. Weight of water contained in the wood, usually expressed as a percentage of the weight of the oven-dry wood.

Mop board. See scrub board.

Mortise. A slot, a notch, a hole, or an opening cut into a board, plank, or timber, usually edgewise, to receive tenon (projecting piece) of another board, plank, or timber, shaped to fit, to form a joint.

Mullion. A slendar bar or pier forming a division between panels or units of windows, screens, or similar frames. Often confused with muntin.

Muntin. The vertical member between two panels of the same piece of panel work. The vertical sash-bars separating the different panes of glass.

Natural finish. A transparent finish, usually a drying oil, sealer, or varnish, applied on wood for the purpose of protection against soiling or weathering. Such a finish may not seriously alter the original color of the wood or obscure its grain pattern.

Neat cement. In masonry, a pure cement uncut by a sand admixture.

Neat plaster. A term applied to plaster made without sand.

Neat work. In masonry, the brickwork above the footings.

Newel (newel post). Any post to which a railing or balustrade is fastened.

Nonbearing wall. A wall supporting no load other than its own weight.

Nosing. The rounded edge of a board or step. Usually applied to the projecting edge of a stair tread.

Oak. A strong, hard, durable wood obtained from the oak tree. This wood is suitable for many purposes and is used in the building trades for flooring and trim.

Oakum. Hemp or untwisted rope used for caulking joints.

O.C.—(on center). The measurement of spacing for studs, rafters, joists, and the like in a building from center of one member to the center of the next member; i.e., 16" o.c. - 24" o.c.

Offsets. When the face of a wall grows higher and thinner, the jogs are called offsets.

O.G.—Ogee. A molding with a profile in the form of a letter "S" having the outline of a reversed curve.

Open-string stairs. Stairs which are so constructed that the ends of the risers and treads are visible from the side, as opposed to close-string stairs.

Open-web studs. Studs made of open-web steel which permit pass-through of plumbing pipes and vents, electrical conduit, without notching or cutting. Studs snap into place in base and ceiling runners.

Open wiring (knob and tube). Electric wires fastened to surfaces by the use of porcelain knobs; a circuit supported by insulators; wiring which is not concealed.

Orange peel. In painting, a term applied to a pebble effect in sprayed coats of paint or lacquer similar to the peel of an orange. This is caused by too much air pressure, holding the gun too close to the surface, spraying lacquer which is cooler than room temperature, or using a thinner which dries too quickly and prevents the proper flow of the solids.

Oriel. A recessed window that ordinarily projects beyond the exterior face of the wall, is in plan octagonal or hexagonal, is commonly corbelled or cantilevered out. One starting from its own foundation is usually called a bay window.

Outlookers. A cantilever to build the cornice on or to hang the cornice to.

Outriggers. A projecting beam used in connection with overhanging roofs. A support for rafters in cases where roofs extend two or more feet beyond the walls of a house.

Overhanging eaves. A type of roof in which the rafters and roofing extend two or more feet beyond the exterior face of a building. Often used for ornamental effect and to guard against snow and rain. The wide overhang of such eaves permits freer use of window walls. They cast a large shadow area, preventing direct sunlight from entering the room, while allowing sufficient natural illumination.

Overhead door. A door which may either be mounted on a sliding track or pivoted canopy frame which moves upward to an overhead position when opened.

Paint. A combination of pigments with suitable thinners or binders.

Paints—Binders:

Alkyd resins
Epoxies
  Esters
  Polyesters
Linseed oil
Phenolic alkyds
Phenolic resins
Water-thinned resins (Latex)

Paints—Pigments:
Carbonate white lead (phasing out due to laws and cost)
Titanium dioxide (replaced zinc oxide)
Lithopone
Zinc Sulfide

Panel. A large, thin board or sheet of lumber, plywood, or other material. A thin board with all its edges inserted in a groove of a surrounding frame of thick material. A portion of a flat surface recessed or sunk below the surrounding area, distinctly set off by molding or some other decorative device. Also, a section of floor, wall, ceiling, or roof, usually prefabricated and of large size, handled as a single unit in the operations of assembly and erection.

Panel box (service panel or panel board). A box in which electric switches and fuses for branch circuits are located.

Panel heating. A method of heating a building by using heating units or coils of pipes concealed in special panels, or built in the wall or ceiling plaster. Also called concealed heating.

Panel saw. A carpenter's handsaw with fine teeth, making it especially suitable for cutting thin wood. A term also applied to a type of cross-cut handsaw.

Parapet wall. That portion of any exterior wall, party wall, or fire wall which extends above the roof line; a wall which serves as a guard at the edge of a balcony or roof.

Particleboard. A rigid board made of wood chips and an adhesive (glue) formed under hot pressure into various thicknesses.

Parting stop or strip. A small wood piece used in the side and head jambs of double-hung windows to separate upper and lower sash.

Party wall. A wall built upon the dividing line between adjoining buildings for their common use.

Patio. An open area—usually paved—for outdoor living. It may be entirely or partially surrounded by parts of the house.

Pebble dash. In the building trade, a term used for finishing the exterior walls of a structure by dashing pebbles against the plaster or cement.

Penthouse. The roof houses on office and apartment buildings, covering stairways, elevator shafts, etc., irrespective of the shape of the roof; frequently used as living quarters.

Perimeter heating. A system of heating in which ducts radiate from a central plenum chamber and release warm air through registers located along the outer walls.

Perling. See Purlin.

Perlite. Page 462.

Pier. The part of a wall between windows and doors; any detached mass of masonry to support an arch, girder, or column; a heavy column used to support weight.

Pigment. See paints.

Pilaster. A flat, square, or rectangular column attached to the wall and projecting from it to reinforce or strengthen the wall it attaches with or is a part of.

Pile. Timber or concrete shafts sunk into soft ground upon which foundations are built.

Pin. A cylindrical piece of wood, steel, or any other material used to hold two or more members together by passing through a hole in each of them.

Pitch. Page 312.

Plain rail. Meeting rails which are of the same thickness as the balance of the window. It is the opposite of a check rail.

Plank. Material two or three inches thick and more than four inches wide such as joists. Often used for scaffolding.

Plaster. A mixture of lime, hair, and sand to cover lath work for interior walls and ceilings.

Plate. (1) A horizontal structural member placed on a wall or supported on posts, studs, or corbels to carry the trusses of a roof or to carry the rafters directly. (2) A shoe or base member as of a partition or other frame. (3) A small, relatively flat member placed on or in a wall to support girders, rafters, etc.

Plate glass. Page 389.

Plinth block. A small block slightly thicker and wider than the casing for interior trim of a door. It is placed at the bottom of the door trim against which the baseboard or mopboard is butted.

Plough. To cut a groove in the same direction as the grain of the wood.

Plumb. Exactly perpendicular; vertical.

Plumb cut. Any cut made in a vertical plane. The vertical cut at the top end of a rafter.

Ply. A term to denote the number of thicknesses or layers of roofing felt, veneer in plywood, or layers in built-up materials, in any finished piece of such materials.

Plywood. A piece of wood made of three or more layers of veneer joined with glue and usually laid with the grain of adjoining plies at right angles. Almost always an odd number of plies are used to provide balanced construction.

Pointing. A term used in masonry for finishing of joints in a brick or stone wall.

Polystyrene. A thermoplastic polymer, an excellent insulator in board and in foam.

Portal. The arch over a door.

Portland cement. Hydraulic cement made by heating a mixture of limestone and clay in a kiln and pulverizing the resulting clinker.

Post. Generally, any vertical piece whose function is to sustain a vertical load. In structural work, members composed of single angles, rolled shapes, or built-up sections and carrying stresses in compression, whether inclined or vertical are designed as posts or columns.

Prefabricated modular units. Units of construction which are prefabricated on a measurement variation base of 4" or its multiples, and can be fitted together on the job with a minimum of adjustments. Modular units include complete window walls, kitchen units complete with installations as well as masonry, wall panels, and most of the other components of a house. Units are usually designed in such a way that they will fit functionally into a variety of house sizes and plan types.

Preservative. Any substance that, for a reasonable length of time, will prevent the action of wood-destroying fungi, borers of various kinds, and similar destructive life when the wood has been properly coated or impregnated with it.

Prestressed concrete. Prestressing is the imposition of preliminary interval stresses in a structure before working loads are applied, in such a way as to lead to a move favorable state of stress when the loads come into action.

Primer. The first coat of paint in a paint job that requires two or more coats.

Pulley stile. The member of a window frame which contains the pulleys and between which the edges of the sash slide.

Purlins. A piece of timber laid horizontally to support or strengthen the common rafters of a roof; also layed on roof trusses.

Push plate. A metal plate fastened on the surface of a door to protect the door from wear and soiling, as a result of its being pushed by persons when opening it. In public buildings and business houses, the word "push" is usually a part of the face design of the push plate.

Putty. A type of cement usually made of whiting and boiled linseed oil, beaten or kneaded to the consistency of dough and used in sealing glass in sash, filling small holes and crevices in wood, and for similar purposes.

Quarry tile. In masonry, a name given to machine-made, unglazed tile. Also called promenade tile.

Quarter round. A molding that presents a profile of a quarter circle.

Quonset hut. A type of hut or shelter semicircular in shape, made of corrugated metal.

Rabbet. A rectangular longitudinal groove cut in the corner of a board or other piece of material.

Radiant heating. A method of heating usually consisting of coils or pipes placed in the floor, wall or ceiling.

Radiate. To emit heat or light radially as the spokes of a wheel.

Rafter. One of a series of structural members of a roof designed to support roof loads. The rafters of a flat roof are sometimes called roof joists.

Rag felt. In building construction, a type of heavy paper composed of rags impregnated with asphalt; used in the manufacture of waterproofing membranes, for making asphalt shingles, and other types of composition roofing such as asphalt roofing.

Rail. A horizontal bar or timber of wood or metal extending from one post or support to another as a guard or barrier in a fence, balustrade, staircase, etc. Also, the cross or horizontal members of the framework of a sash, door, blind, or any paneled assembly.

Rake. The trim members that run parallel to the roof slope and form the finish between wall and gable roof.

Raked joint. In brick masonry, a type of joint which has the mortar raked out to a specified depth while the mortar is still green.

Ramp. An inclined roadway.

Random work. A term used for stones fitted together at random without any attempt at laying them in courses.

Range work. Ashlar laid in horizontal courses; same as coursed ashlar.

Red oak. A coarse-grained wood, dark in color, used for interior trim in buildings.

Reflective insulation. *Sheet material* with one or both surfaces of comparatively low heat emissivity that, when used in building construction so that the surfaces face air spaces, reduces the radiation across the air space. This type of insulation also acts as a vapor barrier.

Reinforcing. Steel rods or metal fabric placed in concrete slabs, beams, or columns to increase their strength. (Page 160).

Resin-emulsion paint. Paint, the vehicle (liquid part) of which consists of resin or varnish dispersed in fine droplets in water.

Resins. One of the many viscous substances of plant origin such as copal, rosin, and amber used in lacquers, varnishes, synthetic plastics, adhesives, etc.

Return. The continuation of a molding, projection, etc., in an opposite direction.

Reveal. The vertical sides of an opening between the front of the wall and the frame; used also the designate the return of a pilaster or pier, between the face of the pilaster or pier and the main wall face.

Ribbon. A narrow board let into the studding to add support to joists.

Ridge. The horizontal line at the junction of the top edges of two sloping roof surfaces. The rafters at both slopes are nailed at the ridge.

Ridge board. The board placed on edge at the ridge of the roof to support the upper ends of the rafters.

Ridge course. The last or top course of slates or tiles on a roof, cut to length as required.

Ridge roll. A strip of sheet metal, composition roofing, tiling, or wood used to cover and finish a roof ridge.

Rigid conduit. Page 448.

Rise of roof. Page 312.

Rise and run. Page 315.

Riser. Each of the vertical boards closing the spaces between the treads of stairways.

Riser pipe. In building, a vertical pipe which rises from one floor level to another floor level for the purpose of conducting steam, water, or gas from one floor to another.

Rock wool. See mineral wool. Page 462.

Roll roofing. Page 343.

Roof sheathing (roof decking). See sheathing.

Roof tiles. See "collar beam."

Roof truss. Page 222.

Roofing. The material put on a roof to make it watertight.

Rotten stone. A decomposed, brittle limestone from which the calcium carbonate has been removed by the solvent action of water. Marketed in the form of a fine powder and used in the polishing of varnished surfaces.

Rough coat. The first coat of plastering applied to a surface.

Rough hardware. All of the concealed hardware in a house or other building such as bolts, nails, and spikes which are used in the construction of the building.

Roughing-in. In building, a term applied to doing the first or rough work as roughing-in plumbing; installing pipes for water, sewage and drainage up to the plumbing fixtures.

Rough opening. An unfinished window or door opening; any unfinished opening in a building.

Rowlock. In masonry, a term applied to a course of bricks laid on edge. Also, the end of a brick showing on the face of a brick wall in a vertical position.

Rubber-emulsion paint. Paint, the vehicle of which consists of rubber or synthetic rubber, dispersed in fine droplets in water.

Rubble work. Masonry of rough, undressed stones.

Run. Page 315.

Saddle board. The finish of the ridge of a pitch-roof house; sometimes called comb board.

Sanitas. A wall covering somewhat similar to lightweight oilcloth, manufactured and sold under the trade name of Sanitas.

Sash. The framework which contains and holds the glass of a window.

Sash balance. A device, usually operated with a spring, designed to counter-balance window sash. Use of sash balances eliminates the need for sash weights, pulleys, and sash cord.

Saturated felt. (See building paper).

Sawyer. One whose occupation is that of sawing wood or other material; sometimes used in a restricted sense meaning one who operates one of several saws.

Scabbing. Slang expression—see sistering. To reinforce a structural member.

Scaffold or staging. A temporary structure or platform enabling workmen to reach high places.

Scantling. Lumber with a cross section ranging from 2" x 4" to 4" x 4".

Scarf. The joint in timber construction to make two pieces appear as one. In steel, when two plates are lapped, one edge is thinned down to a feather edge or "scarfed" so that the two surfaces will be brought into the same plane.

Scotia. A hollow molding used as a part of a cornice and often under the nosing of a stair tread.

Scratch coat. The coat of plaster which is scratched to form a bond for the next coat.

Screed coat. In plastering, a coat of plaster laid level with the screeds.

Screeds. Narrow strips of plaster put on a wall as guides for the workmen. The strips usually are about 8 inches wide with a thickness of two coats of plaster, serving also as thickness guides when applying the remainder of the plastering; more often, a strip of wood to act as a guide for plaster of concrete work.

Scribing. Fitting woodwork to an irregular surface.

Scuttle hole. A framed opening with its cover through the roof or ceiling.

Sealer. A finishing material, either clear or pigmented, that is usually applied directly over uncoated wood for the purpose of sealing the surface.

Seasoning. Removing moisture from green wood.

Section. A drawing showing the kind, arrangement, and proportions of the various parts of a structure. It is assumed that the structure is cut by a plane, and the section is the view gained by looking in one direction.

Semigloss paint or enamel. A paint or enamel made with a slight insufficiency of nonvolatile vehicle so that its coating, when dry, has some luster but is not very glossy.

Septic tank. In plumbing, a tank in which sewage is kept to effect disintegration of organic matter by natural bacterial action which dissolves most of the solids into liquids and gases within about 24 hours.

Service drop (**electrical**). Page 443.

Service entrance. The place where the service wires are run into a building.

Service panel. (electrical). The main distribution box in a building from which the many circuits distribute electrical service to various outlets.

Shake. A handsplit shingle, usually edge-grained.

Sheathing. In construction work, a term usually applied to surface materials or boards nailed to studding or roofing rafters, as a foundation for the covering of the outer surface of the side walls or roof of a house.

Sheathing paper. See building paper.

Shellac. A transparent coating made by dissolving lac, a resinous secretion of the lac bug (a scale insect that thrives in tropical countries, especially India), in alcohol.

Shingles. Roof covering of asphalt, asbestos, wood, tile, slate, or other material cut to stock lengths, widths, and thicknesses.

Shiplap. In carpentry, a term applied to lumber that is edge dressed to make a close rabbeted or lapped joint.

Shoe mold. For interior finish, a molding strip nailed to the baseboard close to the floor; also called base shoe.

Shore. A piece of timber usually placed in an inclined position to support a building, roof or wall temporarily while it is being repaired or altered.

Siding. Page 289.

Siding (bevel). Page 289.

Sill. The lowest member of the frame of a structure, resting on the foundation and supporting the uprights of the frame. The member forming the lower side of an opening, as a door sill, window sill, etc.

Sill anchor. In building construction, a bolt embedded in a concrete or masonry foundation for the purpose of anchoring the sill to the foundation; sometimes called a plate anchor.

Sill cock. Water faucet on the outside of the house to which a hose may be connected.

Siphon trap. In plumbing, a trap fitted to water closets and sinks, having a double bend like an "S", the lower bend containing the water seal which prevents the reflux of foul odors and gases.

Sistering. A slang expression to reinforce a structural member, nail or affix a strengthening piece to a weakened piece.

Sizing. Working material to the desired size. A coating of glue, shellac, or other substance applied to a surface to prepare it for painting or other methods of finish.

Skirt. In architecture, the border or molded piece under a window stool commonly called apron; also, a baseboard. The same as skirting.

Skylight. A frame supporting glass sash, placed in a roof to light a passage or rooms below.

Slag. A vitreous mass as a by-product of smelting.

Slag strip. In roofing, a strip of wood or metal nailed around the edges of a graveled roof to give the edge a finish and to prevent the gravel from rolling off the roof. Also called gravel strip.

Slate shingles. Page 340.

Sleeper. Timber laid on the ground to receive joists; or pieces of wood imbedded in concrete to fasten finished floors to.

Smoke shelf. A ledge in the front lower part of a chimney that prevents cold air from flowing down the chimney into the fireplace.

Soffit. The underside of the members of a building such as staircases, cornices, beams, and arches, relatively minor in area as compared with ceilings.

Soil cover (ground cover). A light roll roofing used on the ground or crawl spaces to minimize moisture permeation of the area.

Soil pipe (soil stack). A vertical drain pipe conveying waste matter from a water closet to the drainage system of a building. Same as soil stack.

Sole plate. See Plate.

Span. The distance between structural supports such as walls, columns, piers, beams, girders, and trusses.

Spandrel. Those portions of the exterior walls, side or court walls, which lie between the piers and between the window spaces of the successive stories.

Specification. A description of the kind, quality and quantity of materials and workmanship that are to govern the fabrication and erection of a building or other construction.

Spread footing. A footing whose sides slope gradually outwards from the foundation to the base.

Splash block. A small masonry block laid with the top close to the ground surface to receive roof drainage and to carry it away.

Split shakes. In carpentry, a type of shingle split by hand.

Square. A unit of measure—100 square feet.

Staging. (scaffolding)

Stain. A form of oil paint, very thin in consistency, intended for coloring wood without forming a coating of significant thickness or gloss.

Stairs. A series of steps.

Stair landing. A platform between flights of stairs or at the termination of a flight of stairs.

Starter strip. In roofing, the first strip of composition roofing material applied to a roof.

Stepped footings. If a house is built on sloping ground, the footings cannot all be at the same depth; hence, they are stepped.

Stile. The upright piece in framing or paneling. One of a vertical member in a door or sash into which secondary members are fitted.

Stool (window). The flat, narrow shelf forming the top member of the interior trim at the bottom of a window. The base or support of a window as the shelflike piece inside and extending across the lower part of the opening; sill.

Storm sash or Storm window. An extra window usually placed on the outside of an existing window as additional protection against cold weather.

Stretcher. A brick or block of masonry laid lengthwise of a wall.

String, stringer. A timber or other support for cross members. In stairs, the support on which the stair treads rest; also, springboard.

Strut. In carpentry, any piece fixed between two other pieces to keep them apart, as a member which is designed to resist pressure or compressive stress endwise in a frame or structure.

Stucco. Most commonly refers to an outside plaster made with portland cement as its base.

Stud. One of a series of slender wood or metal structural members placed as supporting elements in walls and partitions. Usually 2" x 4" in standard dwelling construction. (Plural: studs or studding.)

Subfloor. Boards or sheet material laid on joists over which a finish floor is to be laid.

Sump. A hole or depression to collect water, usually in the basement floor. A pump on the sump automatically keeps it emptied.

Surface. To make plane and smooth.

Sweat joints. In plumbing, a type of joint made by the union of two pieces of copper pipe which are coated with solder containing tin. The pipes are pressed together and heat applied until the solder melts.

Tail beam. A relatively short beam or joist supported in a wall on one end and by a header on the other.

Templet. A form to lay out work; piece of timber or stone to distribute pressures over a larger area. When applied to steel, the templets are known as "bearing plates" or "slabs" as the case may be.

Termites. Insects that superficially resemble ants in size, general appearance, and habit of living in colonies; hence, they are frequently called "white ants." If unmolested, they eat out the woodwork, leaving a shell of sound wood to conceal their activities, and damage may proceed so far as to cause collapse of parts of a structure before discovery.

Termite shield. A shield, usually of noncorrodible metal, placed in or on a foundation wall or other mass of masonry or around pipes to prevent passage of termites.

Terra cotta. Baked clay of a fine quality.

Terrazzo. A type of Venetian marble mosaic in which portland cement is used as a matrix. Though used in buildings for centuries, terrazzo is a modern floor finish used also for bases, borders, and wainscoting, as well as on stair treads, partitions, and other wall surfaces.

Thermostat. An electric instrument that controls heat and air-conditioning temperature.

Threshold. A strip of wood or metal beveled on each edge and used above the finished floor under outside doors. On interior doors sometimes called a carpet strip.

Timbers. Lumber at least five or more inches in dimension.

Toenailing. To drive a nail at a slant with the initial surface in order to permit it to penetrate into a second member.

Tongue. A projection of a board or other material to be inserted in a groove.

To the weather. In building, a term applied to the projecting of shingles or siding beyond the course above; that part of a shingle, siding, etc., exposed to the elements.

Top plate. In building, the horizontal member nailed to the top of the partition studding.

Transom. The bar or horizontal construction which divides a window. More commonly applied to the sash over the door.

Tread. The horizontal board in a stairway on which the foot is placed.

Trim. The finish materials in a building such as moldings applied around openings (window trim, door trim) or at the floor and ceiling of rooms (baseboard, cornice, picture molding).

Trimmer. The beam or joist into which a header is framed.

Truss. Page 222.

Twist. A defect in lumber consisting of a form of warp resulting in distortion caused by the turning or winding of the edges of a board.

Undercoat. A coating applied prior to the finishing or top coats of a paint job. It may be the first of two or the second of three coats. In some usage of the word, it may become synonymous with priming coat.

Union joint. In plumbing, a pipe fitting in which two pipes are joined so they can be disconnected without disturbing the pipes themselves.

Valley. The internal angle formed by the junction of two sloping sides of a roof.

Valley rafter. The rafter immediately under the valley and to which the jack rafters or purlins connect.

Vapor barrier. Material used to retard the flow of vapor or moisture into walls and, thus, to prevent condensation within them. There are two types of vapor barriers, the membrane that comes in rolls and is applied as a unit in the wall or ceiling construction, and the paint type which is applied with a brush.

Varnish. A thickened preparation of drying oil or drying oil and resin suitable for spreading on surfaces to form continuous, transparent coating or for mixing with pigments to make enamels.

Veneer. Thin sheets of wood or other materials.

Vent. A pipe installed to provide a flow of air to or from a drainage system or to provide a circulation of air within such systems to protect trap seals from siphonage and back pressure.

Verge boards. The boards which serve as the eaves finish on the gable end of a building.

Vermiculite. A mineral closely related to mica with the faculty of expanding on heating to form lightweight material with insulation quality. Used as bulk insulation and also as aggregate in insulating and acoustical plaster and in insulating concrete floors.

Vertical grain. Same as edge grain.

Volatile thinner. A liquid that evaporates readily and is used to thin or reduce the consistency of finishes without altering the relative volumes of pigments and nonvolatile vehicles.

Wainscoting. Matched boarding or panel work covering the lower portion of a wall.

Wallboard. Woodpulp, gypsum, or other materials made into large, rigid sheets that may be fastened to the frame of a building to provide a surface finish.

Wall plates. Pieces of timber or steel which are placed on top of walls to form the support of the roof of a building.

Warp. To bend, or turn from a straight line; a piece of lumber, when improperly seasoned, may become curved, twisted or turned from a straight flat form; a defect or permanent distortion of a timber from its true form, caused usually by exposure to heat or moisture.

Water repellent. A liquid designed to penetrate into wood and to impart water repellency to the wood.

Water table. A ledge or offset on or above a foundation wall for the purpose of shedding water. Level of water below ground level.

Weatherboards. Boards shaped so as to be specially adaptable for overlapping at the joints to prevent rain or other moisture from passing through the wall; also called siding.

Weatherstrip. Narrow strips made of metal or other material so designed that when installed at doors or windows, they will retard the passage of air, water, moisture, or dust around the door or window sash.

Weep hole. In retaining walls and other similar structures, a small hole through which surplus water drains to the outside; hence, preventing damage to the wall by pressure of accumulated water back of or under the structure.

Western framing. In building construction, a method of framing where the studding of each story rests on a sort of sill, as opposed to the balloon type of framing; also called platform framing.

Wind. A term used to describe the surface of a board when twisted (winding) or when resting upon two diagonally opposite corners, if laid upon a perfectly flat surface.

Wing. An offset of the main building.

Wiped joint. In plumbing, a joint formed between two pieces of lead pipe. Molten lead is poured upon the joint until the two pieces of pipe are of the right temperature. The joint is then wiped up by hand with a moleskin or cloth pad while the solder is in a plastic condition.

Wire-cut brick. Brick formed by forcing plastic clay through a rectangular opening designed for the purpose and shaping the clay into bars. Before burning, wires pressed through the plastic mass cut the bars into uniform brick lengths.

Wire glass. In building construction, a type of window glass in which wire with a coarse mesh is embedded to prevent shattering of glass in case it is broken; also, to protect a building against intruders. Wire glass is used in windows of buildings where valuables are kept; also as a safety measure in case of fire in adjacent buildings.

Wire mesh reinforcing. Page 160.

Wood flour. Wood reduced to finely divided particles approximately those of cereal flours in size, appearance, and texture and passing a 40/100 mesh screen.

Wooden brick. Piece of seasoned wood made the size of a brick and laid where it is necessary to provide a nailing in masonry walls.

Zonolite conrete. A form of concrete which acts as insulation; used as parts of floor slabs for houses without basements.

Bibliography

BOOKS

Adjustment of Property Losses
> By Reed and Thomas
> McGraw-Hill Book Company, Inc.
> New York, N.Y.

Building Construction Costs Data 1974
> By Robert Snow Means Company, Inc.
> P.O. Box G, Duxbury, Mass.

Building Construction Estimating
> By Alonzo Wass
> Prentice-Hall, Inc.
> Englewood Cliffs, N.J.

Building Construction Simplified
> By Cooper and Badzinski
> McGraw-Hill Book Company, Inc.
> New York, N.Y.

Building Estimator's Reference Book 1974
> By Frank R. Walker
> Frank R. Walker, Publishers
> Chicago, Ill.

Cabinet Making and Millwork 1970
> By John L. Feirer
> Charles A. Bennett Company, Inc. Publishers
> Peoria, Ill.

Construction Estimates and Costs
    By H. E. Pulver
    McGraw-Hill Book Company, Inc.
    New York, N.Y.

Construction Industry Production Manual
    By Taylor Winslow
    Craftsman Book Company of America, Publishers
    Los Angeles, California

Electrical Construction Wiring
    By Walter N. Alerich
    American Technical Society, Publishers
    Chicago, Ill.

Estimating For the Building Trades
    By Joseph Steinberg and Martin Stempel
    American Technical Society, Publishers
    Chicago, Ill.

House Wiring Simplified
    By Floyd M. Mix
    The Goodheart-Willcox Company, Inc.
    Chicago, Ill.

How To Build and Buy Cabinets For the Modern Kitchen
    By Robert P. Stevenson
    Arco Publishing Company, Inc.
    New York, N.Y.

Materials of Construction
    By Professor Albert P. Mills
    John Wiley & Sons, Inc.
    New York, N.Y.

National Construction Estimator
    Craftsman Book Company
    Los Angeles, California.

Practical Estimating For Painters and Decorators
    By William P. Crannel
    Frederick J. Drake & Company, Publishers
    Wilmette, Ill.

Practical Electric Wiring
    By H.P. Richter
    McGraw-Hill Book Company, Inc.
    New York, N.Y.

Wood Frame House Construction
    By L.O. Anderson, Engineer
    U.S. Department of Agriculture
        Agricultural Handbook No. 73, July 1970

ASSOCIATIONS

American Plywood Association, 1119 A, Tacoma, Wa.
    Plywood Underlayment
    Plywood for Floors
    Plywood Siding

Asphalt Roofing Manufacturers Association, New York, N.Y.
    Manufacture, Selection and Application of Asphalt Roofing and Siding Products, 1974

American Insurance Association
    Blasting Claims—A Guide For Adjusters

Brick Institute of America, 1750 Meadow Road, McLean, Va.
    Technical Notes Nos. 10 and 10B

Painting and Decorating Contractors of America
    7223 Lee Highway, Falls Church, Va.
    Estimating Guide, 1974

Red Cedar Shingle and Handsplit Shake Bureau, Seattle, Wa.
    Certi-Split Manual of Handsplit Red Cedar Shingles

United States Gypsum, Dept. 122, 101 S. Wacker Drive, Chicago, Ill.
    Drywall Construction Handbook

Western Wood Products Association, Portland, Ore.
    Products Use Manual

Western Red Cedar Lumber Association, Portland, Ore.
    Bevel Siding

GUIDE TO LABOR, QUANTITIES, AND TABLES

Index